TRANSBOUNDARY HARM IN INTERNATIONAL LAW

Many harms flow across the ever-more porous sovereign borders of a globalizing world. These harms expose weaknesses in the international legal regime built on sovereignty of nation states. Using the *Trail Smelter* arbitration, one of the most cited cases in international environmental law, this book explores the changing nature of state responses to transboundary harm. Taking a critical approach, the book examines the arbitration's influence on international law generally and international environmental law specifically. In particular, the book explores whether there are lessons from *Trail Smelter* that are useful for resolving transboundary challenges currently confronting the international community. The book collects the commentary of a distinguished set of international law scholars who consider the history of the *Trail Smelter* arbitration, its significance for international environmental law, its broader relationship to international law, and its resonance in fields beyond the environment.

Rebecca M. Bratspies holds a B.A. in Biology from Wesleyan University and graduated with honors from the University of Pennsylvania Law School, where she was elected to the Order of the Coif and awarded the Green Prize for Excellence in Torts. She was named a Luce Foundation Scholar and Seconded to Taiwan's Ministry of the Environment. Her scholarly research focuses on environmental regulatory regimes; she is particularly interested in the international dimensions of environmental regulation and the role of nonstate actors. She currently holds an associate professorship of law at CUNY School of Law where she teaches environmental, property, and administrative law. While on the faculty at the University of Idaho College of Law, she cofounded, with Russell Miller, the Annual Idaho International Law Symposium. The inaugural symposium gave rise to this book.

Russell A. Miller has degrees from Washington State University (B.A.); Duke University (J.D./M.A.); and Johann Wolfgang Goethe University, Frankfurt, Germany (LL.M.). He was the recipient of a 1999 Robert Bosch Foundation Fellowship. He is a frequent Visiting Scholar at the Max Planck Institute for Comparative Public Law and Public International Law. Professor Miller is the cofounder and Co-Editor-in-Chief of the *German Law Journal* (http://www.germanlawjournal.com). He is also the coeditor of the *Annual of German & European Law* and the coauthor of the forthcoming third edition of *The Constitutional Jurisprudence of the Federal Republic of Germany*. With Rebecca Bratspies he cofounded the Annual Idaho International Law Symposium. The inaugural symposium gave rise to this book.

Transboundary Harm in International Law

Lessons from the *Trail Smelter* Arbitration

Edited by

Rebecca M. Bratspies
CUNY School of Law

Russell A. Miller
University of Idaho College of Law

CAMBRIDGE
UNIVERSITY PRESS

CAMBRIDGE UNIVERSITY PRESS
Cambridge, New York, Melbourne, Madrid, Cape Town, Singapore,
São Paulo, Delhi, Dubai, Tokyo

Cambridge University Press
32 Avenue of the Americas, New York, NY 10013-2473, USA

www.cambridge.org
Information on this title: www.cambridge.org/9780521126427

First published 2006
This digitally printed version 2009

A catalog record for this publication is available from the British Library

Library of Congress Cataloging in Publication data

Transboundary harm in international law : lessons from the Trail Smelter arbitration / edited by Rebecca Bratspies and Russell Miller.
p. cm.
Includes index.
ISBN-13: 978-0-521-85643-0 (hardback : alk. paper)
ISBN-10: 0-521-85643-4 (hardback : alk. paper)
1. Jurisdiction (International law) 2. State responsibility. 3. Transboundary pollution – Law and legislation. 4. Consolidated Mining and Smelting Company of Canada Ltd. 5. United States – Claims vs. Canada. I. Bratspies, Rebecca, 1965– II. Miller, Russell (Russell A.)
III. Title.
KZ6148.T73 2006
341.4–dc22 2005028952

ISBN 978-0-521-85643-0 Hardback
ISBN 978-0-521-12642-7 Paperback

For my uncle Dennis Replansky –
who loved the law

Rebecca M. Bratspies

* * *

To my parents, for giving me the gift of the breathtaking rivers of the American northwest. Who would have thought those rivers might one day flow out into the world like this?

It pleases me, loving rivers.
Loving them all the way back
to their source.

RAYMOND CARVER,
Where Water Comes Together with Other Water,
WHERE WATER COMES TOGETHER WITH OTHER WATER: POEMS 17 (1986).

Russell A. Miller

Contents

Contributors *page* xi
Acknowledgments xvii
Foreword by David D. Caron xix

Introduction 1
Rebecca M. Bratspies and Russell A. Miller

PART ONE. THE *TRAIL SMELTER* ARBITRATION – HISTORY, LEGACY, AND REVIVAL

History

1 "An Outcrop of Hell": History, Environment, and the Politics of the *Trail Smelter* Dispute 13
James R. Allum

2 The *Trail Smelter* Dispute [Abridged] 27
John E. Read

Roots and Legacy

3 Of Paradoxes, Precedents, and Progeny: The *Trail Smelter* Arbitration 65 Years Later 34
Stephen C. McCaffrey

4 Pollution by Analogy: The *Trial Smelter* Arbitration [Abridged] 46
Alfred P. Rubin

5 Has International Law Outgrown *Trail Smelter*? 56
Jaye Ellis

6 The Flawed *Trail Smelter* Procedure: The Wrong Tribunal, the Wrong Parties, and the Wrong Law 66
John H. Knox

7 Rereading *Trail Smelter* [Abridged] 79
Karin Mickelson

8 *Trail Smelter* and the International Law Commission's Work on State Responsibility for Internationally Wrongful Acts and State Liability 85
Mark A. Drumbl

9 Derivative versus Direct Liability as a Basis for State Liability for Transboundary Harms 99
Mark Anderson

Return to Trail

10 Transboundary Pollution, Unilateralism, and the Limits of Extraterritorial Jurisdiction: The Second *Trail Smelter* Dispute 109
Neil Craik

PART TWO. *TRAIL SMELTER* AND CONTEMPORARY TRANSBOUNDARY HARM – THE ENVIRONMENT

11 *Trail Smelter* in Contemporary International Environmental Law: Its Relevance in the Nuclear Energy Context 125
Günther Handl

12 Through the Looking Glass: Sustainable Development and Other Emerging Concepts of International Environmental Law in the *Gabčikovo-Nagymaros* Case and the *Trail Smelter* Arbitration 140
James F. Jacobson

13 *Trail Smelter's* (Semi) Precautionary Legacy 153
Rebecca M. Bratspies

14 Surprising Parallels between *Trail Smelter* and the Global Climate Change Regime 167
Russell A. Miller

15 Sovereignty's Continuing Importance: Traces of *Trail Smelter* in the International Law Governing Hazardous Waste Transport 181
Austen L. Parrish

16 The Legacy of *Trail Smelter* in the Field of Transboundary Air Pollution 195
Phoebe Okowa

17 The Impact of the *Trail Smelter* Arbitration on the Law of the Sea 209
Stuart B. Kaye

PART THREE. *TRAIL SMELTER* AND CONTEMPORARY TRANSBOUNDARY HARM – BEYOND THE ENVIRONMENT

18 *Trail Smelter* and Terrorism: International Mechanisms to Combat Transboundary Harm 225
Pierre-Marie Dupuy and Cristina Hoss

19 The Conundrum of Corporate Social Responsibility: Reflections on the Changing Nature of Firms and States 240
Peer Zumbansen

20 A Pyrrhic Victory: Applying the *Trail Smelter* Principle to State Creation of Refugees 254
Jennifer Peavey-Joanis

21 Transboundary Harm: Internet Torts 268
Holger P. Hestermeyer

22 International Drug Pollution? Reflections on *Trail Smelter* and Latin American Drug Trafficking 281
Judith Wise and Eric L. Jensen

23 Application of International Human Rights Conventions to Transboundary State Acts 295
Nicola Vennemann

Annex A. Convention Between the United States of America and the Dominion of Canada Relative to the Establishment of a Tribunal to Decide Questions of Indemnity and Future Regime Arising from the Operation of Smelter at Trail, British Columbia 309

Annex B. *Trail Smelter* Arbitral Tribunal Decision, April 16, 1938 314

Annex C. *Trail Smelter* Arbitral Tribunal March 11, 1941, Decision 326

Index 337

Contributors

James R. Allum is a Senior Consultant in the Chief Administrator's Office for the City of Winnipeg, where he is actively engaged in municipal environmental management issues. He holds a Ph.D. in Canadian and Environmental History from Queen's University in Kingston, Ontario, and he remains active in academics and the heritage community. He has taught in the Department of History at the University of Winnipeg since 1999 and was appointed Chair of the Manitoba Heritage Council by the Minister of Culture, Heritage and Tourism in 2002. His Ph.D. thesis (Queen's University, 1995) examined the environmental politics of the *Trail Smelter* dispute. He has published articles on environmental history in *Conservation and Environmentalism: An Encyclopedia* as well as reviewed books on environmental history and politics in the *Urban History Review* and the *Queen's Journal*. He also has a long history of community and political activism; he has twice been a candidate for the Parliament of Canada and was elected President of the Manitoba New Democratic Party in March 2005.

Mark Anderson is a Professor at the University of Idaho College of Law, where he teaches Business Associations, Antitrust, and Criminal Law. He has published in the fields of antitrust, business associations, and natural resource policy. He received his B.A. from Macalester College in 1973 and his J.D. from the University of Chicago in 1977.

Rebecca M. Bratspies is an Associate Professor of Law at the CUNY School of Law. She lectures and publishes on the topics of genetically modified organisms, environmental liability, and international fisheries. She holds a B.A. in Biology from Wesleyan University and graduated with honors from the University of Pennsylvania Law School. She served as a law clerk to the Hon. C. Arlen Beam on the United States Eighth Circuit. As a 1994/95 Henry Luce Foundation Scholar, she spent a year as a legal advisor to the Republic of China Environmental Protection Administration and Ministry of Justice. With Russell Miller, she created and continues to convene the Annual Idaho International Law Symposium, which was launched with the proceedings that led to this book.

Neil Craik is an assistant professor at the Faculty of Law, University of New Brunswick, where he teaches and researches in the fields of international environmental law, domestic (Canadian) environmental law, and municipal and planning law. He holds degrees from McGill University (BA (hons.)), Dalhousie University (LLB), and University of Edinburgh (LL.M.), and he is an SJD candidate at the University of Toronto. In addition to authoring journal articles on topics relating to environmental matters, he is the coauthor of *Canadian Municipal and Planning Law* (2nd ed. 2004) and he is currently completing a manuscript on international commitments to conduct environmental impact assessments. Before his academic appointment, he practiced environmental law and municipal law at Cassels Brock and Blackwell, LLP, in Toronto.

Mark A. Drumbl is Associate Professor and Ethan Allen Faculty Fellow at the School of Law, Washington & Lee University, where he teaches public international law, global environmental governance, and transitional justice. His publications in the area of international law have appeared in a wide variety of periodicals, including legal journals such as the *Michigan Law Review*, *Northwestern University Law Review*, *New York University Law Review*, and the *Criminal Law Forum*; in social science journals such as *Human Rights Quarterly* and *Third World Quarterly*; and in a number of edited volumes. He has political science degrees from McGill University (B.A., M.A.) and law degrees from the University of Toronto (J.D.) and Columbia University (LL.M., S.J.D.). He is currently working on a book that explores state responsibility in the context of international crimes. In fall 2005, he was Visiting Fellow at University College, Oxford.

Pierre-Marie Dupuy is Professor of Public International Law at the European University Institute on leave from the University Panthéon-Assas, Paris II. He has a Ph.D. in law from Paris II University (Docteur d'Etat en droit) and a graduate diploma from the Paris Institute of Political Studies (Paris). From 1990 to 2000 he was the Director of the Institute for International Advanced Studies of Paris (Institut des Hautes Etudes Internationales de Paris). He has been a Visiting Professor at the University of Michigan School of Law, Ludwig Maximilian (Munich) University, and Complutense (Madrid) University. He has extensive experience as an advocate and counsel before the International Court of Justice. His main publications are the *Manual of Public International Law* (6th ed.), and he serves on the Board of Editors of the *Revue Générale de Droit International Public*. He is one of the founders of the *European Journal of International Law*. He gave the General Course of Public International Law at the Hague Academy of International Law (L'unité de l'ordre juridique international) in the year 2000.

Jaye Ellis is an assistant professor in the Faculty of Law and School of Environment at McGill University where she teaches public international law, international environmental law, global environmental politics, and ethics and the environment. She has published on the precautionary principle, international fisheries law, rhetoric, and discourse ethics. She has conducted research at the Max Planck Institute for Comparative Public Law and Public International Law in Heidelberg, Germany, and she has been a regular contributor to the *Yearbook of International Environmental Law*. She received a D.C.L. from McGill University, an LL.M. from the University of British Columbia, a B.C.L. and LL.B. from McGill University, and a B.A. (hons.) from the University of Calgary.

Günther Handl is the Eberhard Deutsch Professor of Public International Law at Tulane University Law School. He holds law degrees from the University of Graz (Dr. iur.), Cambridge (LL.B.), and Yale (SJD). He is the founder and former Editor-in-Chief of the *Yearbook of International Environmental Law* and has published extensively in the field of public international law, international environmental law, and law of the sea. He is the recipient of a number of awards, including, in 1998, the *Prix Elisabeth Haub* for "exceptional achievements in the field of international environmental law." His article "Territorial Sovereignty and the Problem of Transnational Pollution" was awarded the 1976 ASIL Francis Deak Prize, which recognizes a younger author for meritorious scholarship published in *The American Journal of International Law*.

Holger P. Hestermeyer is a clerk with the Appellate Court in Hamburg, Germany. He is also pursuing a doctorate degree in law with the University of Hamburg (Germany). Before his clerkship he worked as a Research Assistant at the Max Planck Institute for Comparative Public Law and Public International Law, Heidelberg, Germany. He has published articles in the fields of public international law and technology law. He received his J.D. equivalent from Münster University Law School (Germany), an LL.M. from the University of California at Berkeley, and he is admitted as an attorney in New York. He is a former Fulbright Fellow and a German National Merit Foundation Fellow.

Cristina Hoss holds a graduate diploma and Ph.D. in law from the University of Paris II and a diploma of the Institute for International Advanced Studies (Institut des Hautes Etudes Internationales de Paris). From 2000 to 2004 she served as a Research Fellow at the Max Planck Institute for Comparative Public Law and Public International Law in Heidelberg, Germany. She is currently an Associate Legal Officer at the International Court of Justice in The Hague.

James F. Jacobson is an attorney with Sasser & Inglis, P.C., a litigation firm in Boise, Idaho. He previously clerked for the Honorable Judge Darla S. Williamson, Chief District Judge for the Fourth District Court in the State of Idaho. He holds a J.D. from the University of Idaho College of Law and a B.A. in English from Brigham Young University.

Eric L. Jensen is professor of sociology at the University of Idaho. He attained the Ph.D. in sociology at Washington State University in 1978. He has coedited two books with Jurg Gerber: *Drug War, American Style: The Internationalization of Failed Policies and Its Alternatives* (2001) and *The New War on Drugs: Symbolic Politics and Criminal Justice Policy* (1998). He was a Fulbright Lecturer/Research Scholar at the School of Law, University of Aarhus in Aarhus, Denmark, during the spring semester of 2002. During the Fulbright stay he studied the development of drug policies in Eastern and Central European nations following the fall of communism and compared contemporary drug policies and their cultural origins in Sweden and Denmark. He served as Senior Advisor on juvenile justice policy to The Danish Institute for Human Rights in Copenhagen during the spring of 2004.

Stuart B. Kaye is Dean and Professor of Law at the University of Wollongong, Australia. He holds degrees in Arts and Law from the University of Sydney and a doctorate in Law from Dalhousie University. He has an extensive research interest in the law of the sea and international law generally, and he has published a number of books including *Australia's Maritime Boundaries* (1995, 2001), *The Torres Strait* (1997), and *International Fisheries Management* (2001). He is a member of the Editorial Board of *Ocean Development and International Law* and the *Australian International Law Journal.*

John H. Knox is a Professor of Law at the Dickinson School of Law of the Pennsylvania State University. His recent publications include: "The Myth and Reality of Transboundary Environmental Impact Assessment," *American Journal of International Law* (2002), which received the ASIL Francis Deak Prize, awarded to a young author for meritorious scholarship published in *The American Journal of International Law*; *Greening NAFTA: The North American Commission for Environmental Protection* (co-edited with David Markell, 2003); and "The Judicial Resolution of Conflicts Between Trade and the Environment," *Harvard Environmental Law Review* (2004). He is working on a book on private rights in international environmental law.

Stephen C. McCaffrey is Distinguished Professor and Scholar at the University of the Pacific, McGeorge School of Law in Sacramento, California. Professor McCaffrey served as a member of the International Law Commission of the United Nations (ILC) from 1982 to 1991 and chaired the Commission's 1987 Session. He was the ILC's special rapporteur on the Law of the Non-Navigational Uses of International Watercourses from 1985 until 1991, when the Commission provisionally adopted a full set of draft articles on the topic. The ILC's draft articles formed the basis for the 1997 United Nations Convention on the same subject. He served as Counselor on International Law in the Office of Legal Adviser, U.S. Department of State, from 1984 to 1985. Among other activities, he currently serves as Legal Counsel to the Nile River Basin Negotiating Committee and Legal Adviser to the Palestinian Authority/PLO. His publications include *The Law of International Watercourses: Non-Navigational Uses* (2001) and *International*

Environmental Law & Policy (with Edith Brown Weiss, Daniel Magraw, and others, 1998).

Russell A. Miller is Associate Professor at the University of Idaho College of Law, where he teaches international law, international environmental law, and constitutional law. He is the cofounder and Co-Editor-in-Chief of the *German Law Journal* (http://www.germanlawjournal.com) and the *Annual of German & European Law*. He has published articles in the fields of international law (*Columbia Journal of Transnational Law*) and comparative constitutional law (*Washington & Lee Law Review*). He is a regular research visitor at the Max Planck Institute for Comparative Public Law and Public International Law, Heidelberg, Germany. He was a 1999/2000 Fellow of the Robert Bosch Foundation and a law clerk to the Hon. Robert H. Whaley (U.S. Dist. Court, E.D. Washington). He received his B.A. from Washington State University in 1991, his J.D./M.A. from Duke University in 1994, and his LL.M. from Johann Wolfgang Goethe University (Frankfurt-am-Main, Germany) in 2002. With Rebecca Bratspies, he created and continues to convene the Annual Idaho International Law Symposium, which was launched with the proceedings that led to this book.

Phoebe Okowa (LLB, BCL, D. Phil.) is Senior Lecturer in International Law at Queen Mary, University of London. She is the joint editor of *Foundations of Public International Law* and the author of *State Responsibility for Transboundary Air Pollution in International Law* (2000).

Austen L. Parrish is an associate professor at Southwestern University School of Law. He is also the current Director of Southwestern's Summer Law Program in Vancouver, B.C., Canada, where he teaches international environmental law at the University of British Columbia. He received his B.A. from the University of Washington in 1994 and his J.D. from Columbia University in 1997. Prior to entering academia, he was an attorney with O'Melveny & Myers, LLP. His most recent publications have focused on international and comparative law, as well as on issues of jurisdiction and conflict of laws.

Jennifer Peavey-Joanis is an Assistant Attorney General for the State of Alaska, Human Services Section. She holds a J.D. from the University of Idaho College of Law and a B.A. in International Relations and Spanish from Mount Holyoke College. Formerly she was seconded to the ICRC during the Kosovo conflict and worked in Venezuela for the American Red Cross, International Services. She served as the student liaison for the first Annual Idaho International Law Symposium, which was launched with the proceedings that led to this book.

Nicola Vennemann is currently a clerk with the Appellate Court in Cologne, Germany. She was a research assistant at the Max Planck Institute for Comparative Public Law and Public International Law in Heidelberg, Germany. Under the supervision of Prof. Dr. Rüdiger Wolfrum, she is pursuing a Doctorate in law at the

University of Heidelberg. Her research interests and areas of publication include international and European human rights law, European Community and Union law, and general public international law. She received her J.D. equivalent from Heidelberg University Law School and an LL.M. from both the University of Paris-Sorbonne and the University of Cologne. She is a German National Merit Foundation Fellow.

Judith Wise is an assistant professor at Willamette University College of Law. She has worked as a Corporate Associate in Mergers and Acquisitions at the law firm of Skadden, Arps, Slate, Meagher & Flom, LLP, in New York; has taught as a visiting assistant professor at Chapman University School of Law; and has served as a law clerk to the Honorable Judge (now Chief Judge) Jane A. Restani of the United States Court of International Trade. She received her B.A. with honors and graduated Phi Beta Kappa from the University of California at Berkeley, holds an M.A. in Sociology from the University of Chicago, and holds a J.D. from the University of Chicago.

Peer Zumbansen is the Canada Research Chair for Transnational and Comparative Corporate Governance at Osgoode Hall Law School of York University, Toronto, Canada. He holds law degrees from the University of Paris X Nanterre (Licence en Droit, 1991), from the University of Frankfurt (Legal State Exam Diploma, 1995; Dr. Iur., 1998; Habilitation [postdoctoral, full professor qualification)], 2004), and from Harvard Law School (LL.M., 1998). He is the founder and director of the Comparative Research in Law and Political Economy Research Network at Osgoode Hall Law School. He is also the coeditor of the *CLPE Research Paper Series*, cofounder and Co-Editor-in-Chief of the *German Law Journal* (http://www.germanlawjournal.com), coeditor of the *Annual of German & European Law* (Berghahn Books), and coeditor of *Kritische Justiz*. Recent publications include: "Quod Omnes Tangit: Globalization, Welfare Regimes and Entitlements" (in *The Welfare State in an Era of Globalization* (2003)); "Sustaining Paradox Boundaries: Perspectives on the Internal Affairs in Domestic and International Law" (*European Journal of International Law*); "European Corporate Law and National Divergences: The Case of Takeover Regulation" (*Washington University Global Studies Law Review*); "Beyond Territoriality: The Case of Transnational Human Rights Litigation" (ConWeb Papers 2005 (available online)). His latest book, *Innovation and Pfadabhängigkeit: Das Recht der Unternehmensverfassung in der Wissensgesellschaft*, is in press.

Acknowledgments

Producing this volume has left us with a large debt to many who have contributed in a number of ways. First and foremost, we thank Dean Don Burnett of the University of Idaho College of Law for giving two young scholars his unqualified support as we put together a symposium and later a book. This project would never have happened without his confidence and generous support.

For support of our research on and the development of this project, thanks are also owed to the directors (Prof. Dr. Armin von Bogdandy and Prof. Dr. Rüdiger Wolfrum) of the Max Planck Institute for Comparative Public Law and Public International Law, Heidelberg, Germany, and Dean Kristen Booth Glen and Acting Dean MaryLu Bilek of the CUNY School of Law.

We owe a special word of thanks to three individuals: Finola O'Sullivan, our editor at Cambridge University Press, who extended us valuable support, patience, and good will in measures above-and-beyond the call of duty; Jane Edwards, Head of Research Services at Michigan State University College of Law, who put her research staff at our disposal; and Jennifer Peavey-Joanis, whose dedication, initiative, and persistence were vital in organizing the symposium that made this volume possible. We owe much to the reviewers of the original book proposal, whose comments enriched our thinking. Many friends, colleagues, and students have helped us with the project that became this book, listening patiently to our ideas and providing constant inspiration as we pursued this project.

Finally, we acknowledge the understanding, patience, and support of our families as we labored in the editing of this book.

With gratitude, we acknowledge the permission granted to republish excerpts of the following articles:

John E. Read, *The Trail Smelter Dispute*, 1 CANADIAN YEARBOOK OF INTERNATIONAL LAW 213 (1963). Reprinted with permission of the Publisher from *The Canadian Yearbook of International Law, Volume 1* edited by C.B. Bourne © University of British Columbia Press 1963. All rights reserved by the Publisher.

Karin Mickelson, *Rereading Trail Smelter*, 31 CANADIAN YEARBOOK OF INTERNATIONAL LAW 219 (1993). Reprinted with permission of the Publisher from *The Canadian Yearbook of International Law, Volume 31* edited by D.M. McRae © University of British Columbia Press 1993. All rights reserved by the Publisher.

Alfred P. Rubin, *Pollution by Analogy: The Trail Smelter Arbitration*, 50 OREGON LAW REVIEW 259–282 (1971). Publication by permission of Oregon Law Review.

The contributors wish to make the following acknowledgments and thanks for support with their individual chapters.

Rebecca Bratspies – I am grateful for research assistance by Vivian Villegas, for research support from Jonathan Saxon, and for comments and suggestions from B. Allen Schulz and Judith Wise.

Neil Craik – An earlier draft of this paper was presented at the first annual *Journal of Environmental Law and Practice* Conference held in Saskatoon, Saskatchewan, June 4–5, 2004.

Mark Drumbl – I thank the Frances Lewis Law Center, Washington & Lee University, for its support.

Jaye Ellis – I am grateful for the research assistance of Alison Fitzgerald and Jared Will, and for comments and suggestions from Jared Will and John Knox.

Russell Miller – Warm thanks to Bradley Richardson (University of Idaho College of Law Class of 2004) for excellent research assistance.

Austen Parrish – I am grateful to James A. Kushner and Robert E. Lutz for their guidance and thoughtful comments on earlier drafts. Thanks also to Andrea G. Duckworth for her research assistance.

Nicola Vennemann – My thanks to Dr. Anja Seibert-Fohr and Prof. Dr. Christian Walter, who critically reviewed earlier drafts of this article and offered helpful comments and suggestions.

Eric Jensen and Judith Wise – We are grateful for the substantial assistance of Daniel K. Sheckler and Vivian Villegas.

Foreword

Some years ago, I began work on a history of international environmental law and policy. A central, iconic event in that history is the *Trail Smelter* arbitration. I decided to visit Trail, British Columbia, and the towns and environs across the border in the United States that were alleged to have been damaged by fumes from the smelter in Trail. The path from Trail down to Northport, Washington, follows the valley of the upper Columbia River. It occurred to me that the trip to the region would be scenic and something the family would enjoy so I consulted a well-known travel guide to the area. It contained a map that highlighted in green the roads in the region that were recommended as particularly scenic. The stretch from Northport to Trail – unlike the roads in adjoining valleys – was not colored green. There, in that absence of color, was proof of the enduring ecological legacy associated with the Trail smelter and the international arbitration it spawned.

It is my pleasure to provide the Foreword to this study of the *Trail Smelter* arbitration: its history, its current relevance to environmental law and policy, and its possible application to transboundary issues beyond the environmental arena. I regret that other matters prevented me from participating in the inaugural Idaho International Law Symposium that is the foundation of this book. It is thus doubly my pleasure that the writing of the Foreword allowed me early access to the richness of this volume.

Any book about an icon, such as the *Trail Smelter* arbitration, runs the great risk that the icon will use the editors and contributors. Icons, by definition, do not reflect objective reality, but are instead people and events that have grown in stature to fill some human need for legend. Perhaps reality has been intentionally appropriated to support an ideological agenda. These forces, often accumulating their own momentum, can overtake even the best-intentioned scholars, leading them to do willing, or sometimes unwitting, service to the agenda of those who regularly polish the icon. But the attraction of wrestling with an icon can be understood. After all, the *Trail Smelter* arbitration became iconic for a reason, and in this case, as is often the case, that reason is that the event was extraordinary. Icons, thus are not only exaggerated or twisted, they often are also a grossly simplified version of what really occurred. That simplification can strip the icon of the complex forces that made the underlying event so extraordinary. These problems represent the great risk that can only be overcome if the book: (1) acknowledges the fact that an event has been transformed into an icon; (2) seeks to recapture the significance of, and choices implicit in, the *event* by returning to its historic details; (3) seeks to identify and critique the power of the *icon* in contemporary events; and, finally, (4) seeks to blend an appreciation of the complexity and contingency of the event into the continuing influence of the icon.

This volume, even confronting an icon as powerful as the *Trail Smelter* arbitration, marvelously fulfills each of these mandates. In my estimation, this book makes a major contribution to our understanding of the events surrounding the Trail smelter in the early 1900s and what those events, and the icon they spawned, might mean today.

The *Trail Smelter* arbitration did not become an icon immediately. It certainly was an important arbitration. But its iconic status came only later in the 1960s and 1970s with the birth of the international environmental movement. The statement of the tribunal in its 1941 award that "no State has the right to use or permit the use of its territory in such a manner as to cause injury by fumes in or to the territory of another or the properties or persons therein" provided the authority and pedigree necessary for the legitimacy of a central aspect of the international environmental movement – the duty of states to respect the environment. Jaye Ellis and John Knox, in their chapters in the book, very ably challenge the precedential authority of this statement, whereas Steve McCaffrey and Günther Handl rise to its defense.

What is the twisting involved in this iconic creation? The *Trail Smelter* arbitration was not about the environment, as James Allum so vigorously relates in his contribution to the book. There is no doubt that Canada and the United States could have, by legislation, allowed damage by fumes to persons and property in their own territory. Indeed, they did. There is no doubt that the United States could have agreed with Canada to mutually allow each other to pollute the other to some degree. Indeed, they implicitly did. Rather, the *Trail Smelter* arbitration was about financial responsibility for damage to property where the vector for the infliction of the harm was transport through the air of a noxious fume and the measure of damage was the commercial value of the damaged property.

Recently, I served as a Commissioner with the United Nations Compensation Commission (UNCC) for claims arising out of the 1990 Gulf War. That Commission helps us reflect on what is truly a claim involving the environment. On the one hand, the oil spills and oil well fires raise a number of public claims for the monies necessary to restore the health of aspects of the local environment. On the other hand, other claims brought by individuals or corporations sought, for example, the costs of repainting a building soiled by oily smoke from the fires. At its core, the *Trail Smelter* claims are much more like the latter case. Yet, paradoxically, the fact that true environmental claims also were part of the docket of the UNCC must be seen as a part of the legacy of the international environmental movement and the *Trail Smelter* icon which is at its center.

What of the exaggeration involved in creating this icon? Should we take this event as a success for this set of claimants, or is it iconic in the sense that we should aspire to such dispute resolution generally? Extending the observation about true environmental claims noted earlier, there also are few controversies that are truly international disputes; that is, disputes actually between two states. The vast majority of disputes, rather, are between individuals, and they often become international because a boundary is inserted in the mix. The problem in the *Trail Smelter* incident was that the boundary between Canada and the United States was not as porous to private litigation as it was to the winds that carried the fumes. Both nations had dealt with the local controversies internally. Local claims in British Columbia against the smelter in Trail were resolved. Local claims in Washington State against a smaller smelter in Northport (closed down decades before the arbitration) also were resolved. The controversies not resolvable (meaning other than via dismissal) at the time were the transboundary claims. Thus, today, after decades of improving cross-border judicial cooperation, one should not expect the current further row over the pollution by the smelter of the Columbia River to give rise to yet another interstate arbitration. Rather, as Neil Craik analyzes in a chapter in the book, it seems destined to proceed in the national courts of one or both of the countries. In this sense, an interstate arbitration *à la Trail Smelter* might be seen as a failure of more efficient private transnational litigation arrangements. Yet, paradoxically, *Trail Smelter* also should be seen as a success in terms of allocating responsibility on the basis of legal principles rather than the all-too-common international response of letting such harm rest simply where it is suffered. Indeed, it is the relevance of the *Trail Smelter* arbitration to a wider variety of transboundary environmental issues that is at the core of a number of contributions to this volume.

Event and icon, decision and precedent, responsibility and complicity – the *Trail Smelter* arbitration raises all these possibilities and complexities. It is a lens through which many of the issues confronting the world of boundaries may be viewed. The contributors look honestly through this lens and in doing so make a singularly significant contribution to our understanding of the event, the icon, and the continuing relevance of both.

David D. Caron
C. William Maxeiner Distinguished Professor of International Law
University of California at Berkeley
Berkeley – September 1, 2005

TRANSBOUNDARY HARM IN INTERNATIONAL LAW

INTRODUCTION

Transboundary Harm in International Law: Lessons from the *Trail Smelter* Arbitration

Rebecca M. Bratspies and Russell A. Miller

PERSPECTIVE

If you go to Trail, British Columbia, as most of the contributors to this volume did in March 2003, you can still see one of the two 409-foot smokestack built there by the Consolidated Mining and Smelting Company in the mid-1920s. It was this smokestack that accelerated a chain of events that ultimately produced the *Trail Smelter* arbitration and etched the name of this tiny Canadian town into the annals of international law.[1] Nestled in an alcove along the shores of the remote but majestic Columbia River, Trail seems an unlikely setting for a case that would assume a prominent role in the law of nations. But viewing the fateful smokestack, which seems somewhat diminished by the combined effect of the smelter's much expanded facilities and the surrounding peaks of the Canadian Rockies, one contributor to this book was moved to exclaim "arbitration works – the arbitration worked." It was a rare, unequivocal endorsement of international law, especially in such an improbable context.

Certainly, the Columbia River Valley, from northeastern Washington state upstream to Trail, is no longer routinely bathed in toxic fumes from the smelter. Gone are the plumes of sulfur dioxide, nitrous oxide, and particulate matter that cut a swath of damage in those earlier years, even while Trail continues as one of the world's most significant centers for mining and smelting. To this extent, the arbitration was undoubtedly a success. The name of the local hockey team, the "Smoke Eaters," now seems a quaint throwback to another time, although James Allum, in his contribution to this volume, puts the team's name to good use in his critical examination of the historical class structures operating in the *Trail Smelter* dispute. Cleaning up the smelter, and thus improving life in the local communities and ecosystems on two sides of an international border, if true,

[1] *See Trail Smelter Arbitral Decision*, 33 AMERICAN JOURNAL OF INTERNATIONAL LAW 182 (1939) [hereinafter "*Trail Smelter* (1939)"]; *Trail Smelter Arbitral Decision*, 35 AMERICAN JOURNAL OF INTERNATIONAL LAW 684 (1941) [hereinafter "*Trail Smelter* (1941)"]. *See* Annex to this volume.

would be no small matter. On this basis alone, the *Trail Smelter* arbitration would undoubtedly fall in the asset column of the ledger of international environmental accounting.

But how far-reaching was the success wrought by the investigation, litigation, decisional reasoning, and monitoring regime to which we refer throughout this volume as the *Trail Smelter* arbitration? With regard to the smelter itself, there are ample grounds for skepticism. As Neil Craik outlines in his contribution to Part One of this book, the beginning of the twenty-first century has seen the reemergence of environmental tensions along the border in the Columbia River valley. Current concerns surround the transboundary environmental damage the smelter has inflicted on the Columbia River itself. There were attempts during the *Trail Smelter* arbitration to bring the damage done to the Columbia River to the Tribunal's attention,[2] but those efforts were unsuccessful, and the smelter's harm to the transboundary Columbia River watershed remains unaddressed.

Looking beyond the smelter and its immediate environs, are there international environmental successes that can trace their origin back to the *Trail Smelter* arbitration? What, if any at all, has been the influence of the *Trail Smelter* arbitration on the approach of international law to transboundary harm more generally?

It was to explore these questions, with the benefit of the half century that had passed since the final decision of the Tribunal (and the benefit of proximity to the smelter itself) that we organized the 2003 Annual Idaho International Law Symposium, held in Coeur d'Alene, Idaho. This book is a product of the dialogue that began among the contributors at the symposium. It collects the commentary of a distinguished set of scholars who were asked to participate in a rigorous reflection on the *Trail Smelter* arbitration, and transboundary harm more generally, from three distinct perspectives. These perspectives form the three parts of this book:

- Part One: *Trail Smelter's* legal and historical foundations and its jurisprudential legacy in international environmental law;
- Part Two: *Trail Smelter's* significance in the normative framework for responding to transboundary environmental challenges, including some of the most pressing environmental problems confronting the international community today; and, most radically,
- Part Three: *Trail Smelter's* resonance in international responses to nonenvironmental transboundary harm.

PART ONE: HISTORY AND LEGACY OF THE *TRAIL SMELTER* ARBITRATION

The *Trail Smelter* arbitration is familiar to any student of international or environmental law. It is the first and, to this day, one of only a handful of international

[2] John D. Wirth, Smelter Smoke in America 101–03 (2000).

environmental law decisions. More specifically, it is usually the only case cited in which "transboundary damage was settled by the application of the general principles of international law on State liability for cross-border damage . . . "[3] Thus, the dispute between Canada and the United States required the Tribunal to decide, for the first (and, for an adjudicatory body addressing an environmental dispute, perhaps last) time, the limits of the fundamental legal concept of the sovereign equality of states. Where Canada's sovereignty implied the right to exploit its natural resources as it willed, that same sovereign norm protected the United States' right to the inviolability of its national territory. The activities of Consolidated Mining and Smelting in Trail, by virtue of climatic conditions that sent its emissions downstream and into the United States,[4] implicated both sovereign rights at the same time.

The *Trail Smelter* Tribunal navigated this clash of sovereignties by articulating what have come to be known as the *Trail Smelter* principles: (1) the state has a duty to prevent transboundary harm, which is commonly expressed in the Latin maxim *sic utere tuo ut alilenum non laedas* ("one should use one's own property so an not to injure another"); and (2) the "polluter pays" principle, which holds that the polluting state should pay compensation for the transboundary harm it has caused.[5] Both of these principles were first announced by the *Trail Smelter* Tribunal in 1941.[6]

The ensuing half century has seen expansive, almost mythological status attributed to the *Trail Smelter* Tribunal and these principles. Having solved the contradiction at the core of sovereign equality, so the reasoning goes, the Tribunal's decisions represent a triumph of international law and diplomacy. *Trail Smelter* has been proclaimed the *locus classicus*[7] and the *fons et origo*[8] of international law on transboundary environmental harm. Indeed, many multilateral environmental treaties endorse the normative quality of the *Trail Smelter* principles. This celebration of the arbitration's success is convincingly advanced in Part One of this book in a contribution from Stephen McCaffrey and a republished excerpt of an article written by John Read, the Canadian Agent in the arbitration and later a judge at the International Court of Justice.

[3] Xue Hanqin, Transboundary Damage in International Law 269 (2003).

[4] *Trail Smelter* (1939), *supra* note 1 at 194–98.

[5] *Trail Smelter* (1941), *supra* note 1, at 716–17. *See* Alexandre Kiss and Dinah Shelton, International Environmental Law 107 (1991).

[6] Cristina Hoss and Pierre-Marie Dupuy argue in their contribution to this volume that "invented" better describes the work of the Tribunal as regards these principles.

[7] Günther Handl, *Territorial Sovereignty and the Problem of Transnational Pollution*, 69 American Journal of International Law 50, 60 (1975).

[8] *See* Alfred P. Rubin, *Pollution by Analogy: The Trail Smelter Arbitration*, 50 Oregon Law Review 259 (1971). Republished in this volume. *See also* Robert Q. Quentin-Baxter, Second Report on International Liability for Injurious Consequences Arising out of Acts not Prohibited by International Law, UN Doc. A/CN.4/346 and Add.1 & 2, *reprinted in* 2(1) Yearbook of the International Law Commission 103, 108–12 (1981).

However, despite the arbitration's ubiquity, there is surprisingly little depth to most invocations of *Trail Smelter*. The dispute's rich factual tapestry remains largely ignored, a criticism thoughtfully explored from various perspectives by James Allum, Jaye Ellis, and John Knox in their contributions to Part One of this volume. It is also a theme raised in articles by Karin Mickelson and Alfred Rubin, which are excerpted and republished here. All raise objections to ritual incantations of the *Trail Smelter* principles, challenging the rhetoric surrounding the *Trail Smelter* arbitration, and reconsidering the Tribunal's mandate, its decisions and their precedential weight.[9] *Trail's* champions portray the arbitration as an expansive declaration of state responsibility and liability, with environmental principles and international law triumphant, but its critics point to the extraordinary narrowness of that victory. After all, under the Tribunal's reasoning, states are responsible for transboundary air pollution only when the resulting harm is "of serious economic consequence"[10] and established by clear and convincing evidence. Without proof of such harm, as Rubin has observed, "there appears to be no international responsibility at all [under the *Trail Smelter* Tribunal's reasoning] for acts of pollution."[11] In Part Two, Phoebe Okowa and Günther Handl take vigorous exception to this criticism of *Trail Smelter.*

Rounding out the contributions to Part One, Mark Drumbl and Mark Anderson explore *Trail Smelter's* relationship to traditional and contemporary, domestic and international jurisprudence on questions of responsibility, liability, and indemnification for harm. These matters were fundamental to the *Trail Smelter* dispute, and in many ways define the complex of interests affected by the Tribunal's resolution of the conundrum of conflicting sovereignties. In particular, Mark Drumbl considers *Trail Smelter's* significance for the International Law Commission's ongoing project of defining and codifying state responsibility (for wrongful acts) and state liability (for non–wrongful acts) in international law.

PART TWO: *TRAIL SMELTER* AND CONTEMPORARY TRANSBOUNDARY HARM

It is not mundane to remark, in fact *Trail Smelter* demands no less, that a boundary lies at the heart of every trans*boundary* harm.[12] An extensive body of literature grapples with the role boundaries play in many global environmental problems,

[9] *See, e.g.*, Samuel Bleicher, *An Overview of International Environmental Regulation*, 2 ECOLOGY LAW QUARTERLY 1 (1972); Rubin, *supra* note 8; Günther Handl, *supra* note 7; Quinten-Baxter, *supra* note 8.

[10] *Trail Smelter* (1941), *supra* note 1, at 716.

[11] Rubin, *supra* note 8, at 273.

[12] "With national boundaries in mind, the term 'transboundary' stresses the element of boundary-crossing in terms of the direct or immediate consequences of the act for which the source State is held responsible. It is the act of boundary-crossing which subjects the consequent damage to international remedy and initiates the application of international rules." HANQIN, *supra* note 3, at 9.

often contributing to the creation of these problems and at the same time frustrating attempts to resolve them.[13] The contributors in Part Two of this book confront the constraints that sovereign boundaries (however sovereignty may be delimited and defined) play in resolving transboundary harms. With regard to this point, one particular lesson repeatedly emerges: the distinct character of the border at issue in the *Trail Smelter* dispute limits the precedential significance of the case.

The *Trail Smelter* transboundary dispute and adjudication occurred across a border, which, throughout its history, has been most distinctively characterized by American and Canadian efforts to downplay its functional significance. The point made by Phoebe Okowa and others is that the history of amicability and cooperation along the 49th parallel in North America made an adjudicatory resolution of the dispute possible.[14] But that amicability and cooperation undermine the relevance of the case for other, more complex transboundary situations. Borrowing from the title of John Knox's contribution to this book, one might be inclined to conclude that *Trail Smelter* involved the "wrong border" for establishing generally applicable principles of international law regarding transboundary harm.

Trail Smelter's relevance to contemporary transboundary environmental harm is further complicated because the case reflects a distinct, historical view of state boundaries. Territorial borders, generally speaking, "delineate areas within which different sets of legal rules apply. There has been, until now, a general correspondence between borders drawn in physical space . . . and borders in 'law space'."[15] The *Trail Smelter* Tribunal worked from a presumption that Canada not only ought to, but could, exert control over its territory. That presumption no longer rings true. Many contemporary environmental threats strain the traditional concept of sovereignty, defined as states' control over defined territories. Pollution, global warming, and loss of ecosystem services defy borders. Indeed, these contemporary problems exploit the limitations imposed by clearly demarcated boundaries of state authority, creating harms over which individual states have little control and few tools to combat. States face new dilemmas of shared risk – problems that cross borders, and issues that no single government can control. The challenge posed by transboundary harm thus represents the dark underside of the reshaped

13 *See, e.g.*, Jutta Brunnée, *The United States and International Environmental Law: Living with an Elephant*, 15 European Journal of International Law 617 (2004). Bradley Karkkainan, *Marine Ecosystem Management & A "Post-Sovereign" Transboundary Governance*, 6 San Diego International Law Journal 113 (2004); Jutta Brunnée, *Of Sense and Sensibility: Reflections on International Liability Regimes as Tools for Environmental Protection*, 53 International and Comparative Law Quarterly 351 (2004).

14 Perhaps for precisely these reasons, the U.S. and Canadian border has generated a rich body of international law. *See, e.g.*, Delimitation of the Maritime Boundary in the Gulf of Maine Area (Canada/U.S) 1984 I.C.J. 246 (Oct. 12).

15 David R. Johnson and David Post, *Law and Borders – The Rise of Law in Cyberspace*, 48 Stanford Law Review 1367, 1368 (1996).

relationship between states that the advances in technology, transport, and communications have produced.[16]

Conscious of the limits imposed by the unique characteristics of the boundary at the center of the *Trail Smelter* dispute, the contributors in Part Two of this book explore *Trail Smelter's* significance to some of today's most pressing transboundary environmental problems. They discover a diverse array of transboundary environmental issues converging in the shadow cast by the *Trail Smelter* arbitration. The Stockholm Declaration's Principle 21 and the Rio Declaration's Principle 2 trace their origins, more or less directly, back to *Trail Smelter*.[17] Many existing multilateral environmental treaties endorse the normative quality of the *Trail Smelter* principles. Encoded within the Tribunal's decisions were the basics of prevention, mitigation, and reparation by which transboundary pollution has since been understood and regulated. The *Trail Smelter* Tribunal, like contemporary international environmental regimes, had to respond to the competing imperatives of science, economics, politics, and environmental protection. In our own contributions to the book we explore, as does Phoebe Okowa, how the Tribunal struck this balance. We reach related but different conclusions about how *Trail Smelter* might speak to the use of science in resolving current environmental problems.

As Günther Handl explains, the problematic concepts of harm, responsibility, and due diligence, central to contemporary international environmental issues, also played out in the context of the *Trail Smelter* arbitration. Where Handl praises the arbitration's engagement with due diligence in his exploration of transboundary nuclear energy issues, Austen Parrish offers a more cautions perspective on *Trail Smelter's* legacy for contemporary hazardous waste issues. In the context of the law of the sea, Dean Stuart Kaye explores the limits of *Trail Smelter's* legacy when environmental harms cross the border between a sovereign state and the global commons.[18] James Jacobsen uses a comparison to the *Gabčíkovo-Nagymaros Project Case* to consider how the *Trail Smelter* principles interact with modern expectations about sustainable development.

16 *See* David Held, *Democracy and Globalization*, 3 Global Governance 251, 257 (1997); Richard Dosecrance, *The Rise of the Virtual State*, FOREIGN AFFAIRS 59–61 (July/August 1996); "In the modern world, this reciprocal relationship between States is further enhanced by the increasing interdependence of States facilitated by the advancement of technology and communication." HANQIN, *supra* note 3, at 289.

17 *See, e.g.*, Report of the Stockholm Conference, U.N. Doc. A/CONF.48/14, princ. 21, *reprinted in* 11 INTERNATIONAL LEGAL MATERIALS 1416, at 1420 (1972); Rio Declaration on Environment and Development, June 14, 1992, Annex I, princ. 2, *reprinted in* 31 INTERNATIONAL LEGAL MATERIALS 874, 879 (1992); Framework Convention on Climate Change, May 9, 1992, Preamble, *reprinted in* 31 INTERNATIONAL LEGAL MATERIALS 849 (1992).

18 In this volume, we employ the term "transboundary" broadly, including within its scope harms that cross a single state boundary, several sovereign boundaries, as well as the boundary between state territory and the global commons beyond national jurisdiction or control. The breadth of the definition employed does not, however, detract from the conceptual work regarding borders achieved by the contributors to Part Two. It is a challenge *Trail* does not easily allow one to evade.

PART THREE: *TRAIL SMELTER* AND TRANSBOUNDARY HARM BEYOND THE ENVIRONMENT

Transboundary harm is a term of art that international law reserves almost exclusively for environmental issues. Implied in the use of the term is a relatively direct line of causation from activity to physical consequences.[19] Scholars typically use the terms cross-border or transnational to refer to less tangible impacts that arise from, for example, economic or political activities that cross sovereign boundaries. We deliberately ignore this distinction. In breaking with scholarly convention on this point, we hope to provoke new thinking about what constitutes "harm." Defining "harm" or "damage," as the *Trail Smelter* Tribunal learned, may be the most confounding facet of forming a legal response to transboundary harm, but the simplicity and logic of the *Trail Smelter* principles invite consideration of their applicability to a broader conception of harm.

In its Draft Articles on State Duties to Prevent Transboundary Harm, the International Law Commission (ILC) accepted a distinction between physical and more inchoate harms when it defined transboundary harm to include a component of physical manifestations.[20] The contributors in Part Three of this book explore the limits of this definition by subjecting nonenvironmental harms to *Trail Smelter's* transboundary lens. This conceptual move responds to the ILC's conclusion that only physical consequences trigger a state's duty to prevent transboundary harm, which seems an artificial formalism that neglects modern international environmental law's consciousness of social and ecological interdependencies. After all, environmental scholars have long recognized that "discriminatory trade practices" or "currency policies" are also likely to have "physical" and particularly "environmental" consequences.

Judith Wise/Eric Jensen and Jennifer Peavey Joanis, in particular among the contributors to Part Three, point to the indeterminacy of notions of harm. They echo Okowa's point that the *Trail Smelter* Tribunal's reasoning is intimately tied to physical manifestations of harm. Other contributors in Part Three reinforce Drumbl's claim that traces of the *Trail Smelter* Tribunal's struggle to define harm have been confronted and refined by the International Law Commission's decision to limit state liability for transboundary harm to those physical harms susceptible to relatively high levels of proof.[21]

[19] In her survey of the field, *Transboundary Damage in International Law*, Xue Hanqin eloquently makes this point. HANQIN, *supra* note 3, at 1, 5.

[20] Draft Articles on the Prevention of Transboundary Harm from Hazardous Activities, together with Commentaries, Article 1, Commentary (2), Report of the International Law Commission on the Work of Its Fifty-third Session, UN GAOR, 56th Sess., Supp. No. 10, at V.E.1, UN Doc. A/56/10 (2001).

[21] Draft Articles on the Prevention of Transboundary Harm from Hazardous Activities, together with Commentaries, Article 1, Commentary (2), Report of the International Law Commission on the Work of Its Fifty-third Session, UN GAOR, 56th Sess., Supp. No. 10, at V.E.1, UN Doc. A/56/10 (2001).

Redefining "harm" also means confronting new actors and new victims. Although certainly a product of its time, *Trail Smelter* is nonetheless a surprisingly modern dispute. In a world shaped by multinational enterprises, international organizations, and the Internet, globalization has forced scholars and policy makers to grapple anew with the definition of transboundary harm. Nominally a dispute between two states, *Trail Smelter* also confronted this question. The arbitration bore all the ambiguities created by the contemporary involvement of multinational industrial interests and civil society in the global political economy. Thus, the situation that gave rise to the *Trail Smelter* arbitration has more in common than one might expect with many of the transboundary issues that arise from globalization. As one of the very few international law decisions squarely confronting the conflicting imperatives of sovereign equality and mutual dependence, *Trail Smelter* may offer lessons beyond its environmental roots.

Terrorism, Drugs, Refugees, Corporate Responsibility, and Human Rights: these are some of the most consuming issues of the twenty-first century. All can be construed as raising issues of transboundary harm. The contributors to Part Three of this volume engage with these issues and, with a glance at the *Trail Smelter* arbitration, join the ongoing debate over how diminished state control over territory, and the rise of new actors, shapes responses to transboundary harm. In doing so, they join the growing scholarly exploration of transboundary and cross-border issues.

Many of the contributors grapple with the lessons of the *Trail Smelter* arbitration as regards current debates over the proper balance between state duties of prevention, mitigation, and compensation. Cristina Hoss/Pierre-Marie Dupuy caution against an overbroad reading of what they term *Trail Smelter's* "reactionary" brand of state responsibility. Judith Wise/Eric Jensen, Nicola Venemann, and Jennifer Peavey Joanis join Hoss/Dupuy in expressing concern about the Tribunal's willingness to embrace, without remarking, Canada's voluntary adoption of the private smelter's actions for purposes of liability. This concern echoes the questions posed by Drumbl and Anderson in Part One of the book. Peer Zumbansen, on the other hand, seems more comfortable with the "attribution" question, and he sees a broader influence for *Trail Smelter* than do the other contributors to Part Three.

Where Zumbansen hears implicit echos of *Trail Smelter's* "contemplative legacy" in developing regimes of corporate social responsibility, Hoss/Dupuy are much less sanguine about the arbitration's influence on global responses to terrorism. In their analysis, they draw a strikingly different portrayal of *Trail Smelter's* approach to due diligence than did Handl in Part Two. Wise/Jensen flatly reject *Trail Smelter's* applicability to the myriad transboundary harms they identify as stemming from the international drug trade. Venemann's meditation on jurisdiction recognizes an inspirational resonance of *Trail Smelter* in the realm of extraterritorial application of international human rights regimes, while Peavey Joanis warns of the dangers inherent in applying *Trail Smelter* too readily to situations that produce international refugee populations. Holger Hestermeyer offers perhaps the most innovative analysis – considering *Trail Smelter's* relevance in the borderless world of the Internet.

In general, the authors conclude that the disadvantages of the *Trail Smelter* paradigm outweigh the advantages with regard to these nonenvironmental transboundary harms. However, many of them draw inspiration from the perspectives and ideas imbedded in the arbitration, even as they reject any doctrinal force in their respective fields for the *Trail Smelter* principles. In measuring *Trail Smelter* against some of the most pressing contemporary harms that cross borders, these chapters make for fascinating reading. Their conclusions reinforce the limitations and strengths of the *Trail Smelter* arbitration also present in the earlier sections of the book.

RÉSUMÉ: *TRAIL SMELTER* AS MECHANISM FOR CONCEPTUALIZING TRANSBOUNDARY HARM

The book underscores that any attempt at conceptualizing transboundary harm and international law's responses thereto must give consideration to the changing international economic and political order, and the wide range of actors vying to determine its content. In this respect, each contributor to this book responds in some way to the phenomenon of globalization and the consequent erosion of the self-contained state. Where the *Trail Smelter* Tribunal could presuppose, both politically and theoretically, "state control of space," or what Ulrech Beck has called "the container theory of society,"[22] such an idea is anathema to the postmodern thinker. The measure of control the *Trail Smelter* Tribunal attributed to the Dominion of Canada over the private smelter operating within its territory no longer rings true in the age of multinational corporations. Whether such an assumption was ever very accurate is beside the point; it was essential to the Tribunal's determination of state responsibility, and, more broadly, to the project of transforming Westphalian notions of "equality among states into the complex treaty-based system at the heart of modern international law."[23]

This volume also focuses attention on the inherent tensions between international liability regimes, which presuppose that harmful conduct will continue, and international prevention regimes, which seek the cessation of harmful activities. Measuring the arbitration against current social, political, and scientific conditions, the authors consider whether the hybrid liability and prevention regime crafted by the *Trail Smelter* Tribunal offers useful guidance for resolving questions of transboundary harm.

Given the diversity of views contained within these chapters, no *a priori* effort has been made to channel them into a single interpretive framework, theoretical

[22] Ulrech Beck, What is Globalization? 23 (Patrick Camiller trans., 2000).

[23] *See* S. S. Lotus Case (Fr. v. Turk.), 1927 P.C.I.J. (ser. A) No. 10 (Sept. 7).

tradition, or consensus conclusion. Rather, the common foundation has been each contributor's engagement with the *Trail Smelter* opinions as a vehicle for reconsidering current debates over transboundary harm. The result is a rich menu of perspectives that reflects the debate, the uncertainty and the intellectual passion swirling around these questions.

To fully explore these transboundary issues, the authors view the *Trail Smelter* arbitration through many different lenses: jurisprudential, environmental, and geopolitical. Each chapter singles out a unique aspect of the *Trail Smelter* arbitration for further study, and together the chapters build a thick theoretical framework for exploring the decisions' many facets. The conclusions differ widely, and make for provocative reading. Although some authors draw substantive and procedural lessons from the *Trail Smelter* arbitration, others warn against the dangers of blindly, or too broadly, applying *Trail Smelter's* vision of state accountability. All agree that extrapolating too freely from *Trail Smelter* can become a perilous enterprise.

More than just an historical accounting of the *Trail Smelter* arbitration, this book seeks to reengage with the *Trail Smelter* arbitration and to reinvigorate discussions of its influence on international law. We were resolved to test *Trail Smelter's* legacy against today's transboundary challenges, fully embracing the possibility that doing so might unravel the arbitration's mythological hold over international environmental law. The project has made two things clear. First, *Trail Smelter* still has much to say as regards sovereignty, boundaries, and harm, the essential elements of transboundary harm. Second, there are contextual as well as conceptual limits to the relevance of *Trail Smelter*, with respect to both environmental and nonenvironmental transboundary harm.

With border-crossing conflicts multiplying and intensifying, approaches to resolving these conflicts have acquired new significance. The time is ripe to revisit *Trail Smelter* and to take its measure against this radically changed world. There are important lessons to be learned from a modern engagement with *Trail Smelter* – including both novel applications of the arbitration and a real sense of its limitations.

Big claims, indeed, for a little town and a pair of solitary smokestacks in the Canadian Rockies.

PART ONE

The *Trail Smelter* Arbitration – History, Legacy, and Revival

1 "An Outcrop of Hell": History, Environment, and the Politics of the *Trail Smelter* Dispute

James R. Allum

SMOKE EATERS

One of the great moments in Canadian hockey history belongs to the small town of Trail, British Columbia. In 1961, a local amateur team from Trail captured the gold medal at the World Hockey Championships, a victory that, at the time, appeared to reaffirm Canada's dominance in international hockey. It turned out to be the end of an era. Thirty-three years passed before another Canadian team – this time composed of elite professionals from the National Hockey League – won the prestigious global tournament. As Canada marched toward victory at the 1994 event, the national media wistfully described the team's quest as the "Trail to Gold," thereby linking past glories with the glittering promise of present opportunities.[1]

What interests me here, however, is not Canada's international hockey reputation, but the Trail team's unlikely nickname: the Smoke Eaters. In truth, the name Smoke Eaters accurately reflected the reality of life in that community. To live in Trail was literally to be an eater of smoke, a consumer of the relentless emissions that poured from the stacks of the local smelter. Built by American mining promoters in 1896, the Trail smelter was consolidated with several mines in nearby Rossland under the ownership of Canadian Pacific in 1906. After the Great War, the Consolidated Mining and Smelting Company perfected the metallurgical process for the refining of low-grade zinc ores, a technological innovation that reflected Trail's transformation from an unstable mining frontier to the smelting

[1] Canada has since gone on to win the World Championships, not to mention the Olympic Gold medal, three times. For an excellent account of the 1961 event, *see* SCOTT YOUNG, WAR ON ICE: CANADA IN INTERNATIONAL HOCKEY 95–112 (1976). Much of this paper is taken from the author's Ph.D. dissertation. JAMES R. ALLUM, SMOKE ACROSS THE BORDER: THE ENVIRONMENTAL POLITICS OF THE TRAIL SMELTER INVESTIGATION (1995) (unpublished Ph.D. dissertation, Queen's University) (on file with author). Readers should also consult JOHN D. WIRTH, SMELTER SMOKE IN NORTH AMERICA (2000).

capital of British Columbia.[2] In the interim, a modern industrial complex was forged out of an allegedly uncharted wilderness, bringing progress and prosperity to a barren wasteland.

As the Trail Smoke Eaters took to the ice in 1961, then, they were more than simply hockey players: they also were ambassadors for modern industrialism. Sponsored primarily by Consolidated Mining, which employed all but a few of the players, the Smoke Eaters evoked the smelter company's peculiar territorial grip on the surrounding countryside. A rural industrial outpost on otherwise forsaken ground, Consolidated's control of the landscape extended beyond mere ownership and into the realm of environmental domination. Guided by the steady hand of state mining policy, the wealth generated by Consolidated Mining transformed the Columbia valley in its own image, subordinating land, timber, and water development to industrial production. In time, the town of Trail, the smelter company, and the landscape became indistinguishable, all united under the enormous smokestacks that towered above the town. Smoke from the smelter emerged as a powerful agent of the mining industry's control of the land, and by extension all who lived upon it. The "thicker the smoke ascending from Smelter Hill the greater is Trail's prosperity" went the local slogan, making smoke eating not merely a condition of residence but also a source of community pride.[3]

It was not always this way. As Frances MacNab, a British travel writer who passed through Trail in the 1890s observed, "[t]he great furnace upon the hill looks like an outcrop of hell," and the blinding and choking "noxious fumes," left the surrounding landscape looking "mean, sordid, and depraved – a veritable blot on the face of nature."[4] Smelter smoke in Trail has had, not surprisingly, a much more bitterly contested history than is implied by the complacent community image evoked by the world champion Trail Smoke Eaters. Throughout the first half of the twentieth century, smoke was anything but a badge of community honor; it was a symbol neither of wealth nor of progress. Rather, for many who lived in Trail, smelter smoke was the symbol of class subordination rooted in environmental domination. Nestled in an alcove along the shores of the Columbia River just north of the international border, Trail's close proximity to the region's abundant resources originally made it the ideal site for the construction of a smelting operation. But Trail was also situated in a narrow canyon of the river valley whose particular topographical and climatic conditions exposed the land to the relentless presence of the fumes. Contaminated by exceedingly high concentrations of sulfur dioxide, the smoke descended on a highly sensitive complex of natural

[2] Jeremy Mouat, *Creating a New Staple: Capital, Technology and Monopoly in British Columbia's Resource Sector*, 1901–1925, 1 JOURNAL OF THE CANADIAN HISTORY ASSOCIATION 215, 215–237 (1990).

[3] *See* ELSIE G. TURNBULL, TRAIL BETWEEN THE WARS 13 (1980).

[4] FRANCES MACNAB, BRITISH COLUMBIA FOR SETTLERS: ITS MINES, TRADE, AND AGRICULTURE 271–72 (1898).

vegetation, coniferous forests, and cultivated crops, leaving a virtual moonscape in its wake.

The first to encounter the lethal impact of the smoke were those who lived in close proximity to the smelter. When local farmers rallied to protect their land from the toxic character of the smoke, however, they soon encountered the Janus-like face of capitalism. For the better part of eight years, between 1917 and 1924, they engaged in a bitter legal battle with the smelter company to defend the future society had promised them: the right to enjoy their property and pursue their own economic self-interest free from interference. Instead, they were obliged to accept a state-negotiated settlement that provided the smelter company with a virtual license to pollute. Under the terms of binding arbitration imposed by provincial law, a local magistrate awarded $60,000 in damages to sixty local farmers but imposed no regulatory conditions on the continued operation of the smelter.[5]

Now long forgotten amid more sanitized historical accounts, this first conflict over smoke in Trail was the catalyst behind the smelter company's much more serious smoke problems that developed after 1925.[6] As the wastes generated by the smelter grew in direct proportion to zinc production, the company constructed a 409-foot smokestack to deal with the exponential expansion of the smelter's productive capacity. But the stack merely served to funnel the fumes higher into the windstream and therefore further down the valley. Within months of its construction, Consolidated Mining was embroiled in yet another smoke controversy, only this time with a potentially more threatening adversary. Like those before them in Trail, farmers across the U.S.-Canadian border, in Stevens County, Washington, complained that the emissions were causing irreparable damage to their gardens, field crops, grazing lands, orchards, and timber lots. For a time, the company ignored these complaints, but by the summer of 1926, Consolidated Mining was faced with an increasingly hostile neighbor who threatened to turn a local dispute into an international incident.

Thus began the *Trail Smelter* dispute, a fifteen-year odyssey through the high court of international environmental politics. Of the years between 1926 and 1941, over thirteen were spent in the arcane world of diplomatic exchange. After direct negotiations between the smelter company and individual farmers descended into bitter acrimony, the case was turned over to representatives of the Canadian and U.S. governments in 1927. Diplomacy, however, proved no more effective in finding a solution. The best that the two governments could do was to send the matter to the International Joint Commission (IJC), an institution with a

[5] For a description of the terms of settlement, *see Judge Forin Renders Decision in Famous Smoke Case*, TRAIL NEWS, May 2, 1924, at 1.

[6] *See* ELSIE G. TURNBULL, TOPPING'S TRAIL (1964). In a later book, Turnbull pays lip service to local smoke conditions, but the issue is treated as just another obstacle in the making of a great industrial enterprise. *See* TURNBULL, *supra* note 3, at 56–61.

questionable record on pollution issues.[7] Almost four years after the official reference, the IJC finally issued a unanimous report, which, like the earlier ruling in the domestic *Trail Smelter* case, once more offered compensation to the afflicted farmers but failed to regulate production at the smelter.[8] Within a month, the American farmers had summarily dismissed the IJC report as a sop to the smelting industry.

There the matter lingered until further diplomatic discussions produced the Convention of 1935.[9] Under the terms of reference, the smelter company agreed to pay $350,000 in compensation for all damage before January 1, 1932, whereas any complaints accruing thereafter were to be sent to an *ad hoc* international arbitral tribunal for mediation. The *Trail Smelter* Tribunal officially convened for the first time in June of 1937, but another four years elapsed before it reached a final decision in the case. Following the release of an interim report in 1938, granting an additional $78,000 to the farmers, the Tribunal's final report in 1941 imposed a regulatory regime on production at the smelter, which sought to improve smoke conditions by predicting the weather.[10]

SMOKE AND MIRRORS

Surprisingly, the *Trail Smelter* arbitration is not generally considered to be either a farce or a tragedy. Despite 15 years of social conflict, fruitless diplomacy, extensive economic dislocation, and severe environmental degradation, most commentators apparently believe that more good than bad emerged from the process. Many, of course, profess a profound sympathy for the unfortunate victims in the dispute, but there is a general consensus that the good associated with the Tribunal's final ruling on state responsibility for transnational air pollution far outweighed the hardship faced by the residents of Stevens County.[11] Indeed, many observers argue that the Tribunal's primary finding that "no State has the right to use or permit the use of its territory in such a manner as to cause injury by fumes in or to the territory of another," marked a triumph of international law and diplomacy.[12] But a review of the historical record suggests that any credit for this ruling ultimately

7 Before the *Trail Smelter* dispute, the IJC's only experience with pollution litigation was the 1912 reference on the Pollution of Boundary Waters.

8 Trail Smelter (U.S. v. Can.), International Joint Commission, Trail Smelter Reference, Final Report (Ottawa, 1931).

9 Convention on Damages Resulting From Operation of the Smelter at Trail, British Columbia, April 15, 1935, U.S.-Canada, 49 Stat. 245. *See* Annex to this volume.

10 *Trail Smelter Arbitral Decision*, 35 AMERICAN JOURNAL OF INTERNATIONAL LAW 684 (1941) [hereinafter "*Trail Smelter* (1941)"]. *See* Annex to this volume.

11 D. H. Dinwoodie's statement that the "resolution of the case through treaty and arbitration – the first international pollution incident settled in this manner – provided legal precedent as well as ultimate satisfaction to the injured," is typical of the traditional interpretation regarding the resolution of the dispute. D. H. Dinwoodie, *The Politics of International Pollution Control: The Trail Smelter Case*, 72 INTERNATIONAL JOURNAL 219, 219–35 (1972).

12 John E. Read, *The Trail Smelter Dispute*, 1 CANADIAN YEARBOOK OF INTERNATIONAL LAW 213, 213–29 (1963). Republished in this volume.

belonged to the so-called smoke farmers south of the line who endured acute suffering to preserve their right to live in an environment free of the hazards of toxic wastes.[13] Theirs was not a movement inspired by agrarian purity or aesthetic ideals but by the vision of building a progressive rural community whose potential was being suffocated by fumes. Neither premodern nor protoenvironmentalist, the farmers organized themselves into the Citizens Protective Association (CPA) to protect their own modest claim to the benefits of modernity by demanding that, whatever the cost, the smelter company be forced to stop this invasion by smoke.[14] Their collective battle to stop the pollution at its source, in order to preserve the ecological integrity of the landscape, revealed an environmental ethic that transcended immediate economic objectives.[15]

Conversely, neither the smelter company, nor either of the national governments, worked toward the end arrived at by the Tribunal. All were party to a process that obstructed more than facilitated a settlement requiring the state to set emission standards in order to respect foreign territory. For Consolidated Mining, the issue was even more basic: emission standards meant a potential limitation on output, and for the better part of twenty years the smelter company fought every attempt to impose any sort of regulatory regime aimed at production levels. Indeed, only after the company learned to recycle sulfur dioxide for the production of fertilizer did it finally consent to emission standards.[16] The Tribunal's ruling on state responsibility for transnational air pollution was designed as a "solution just to all parties,"[17] but had it not been for the efforts of the farmers, this most enduring legacy of the *Trail Smelter* arbitration would now neither be celebrated, nor idealized.

That said, it would nevertheless be unwise to put too much stock in the Tribunal's affirmation of state responsibility for transnational air pollution. As Arthur P. Rubin has perceptively argued, the Tribunal's ruling was more a product of creative legalism than a serious attempt to deal with the impact of industrial pollution. According to Rubin, the 1935 Convention provided the Tribunal with a number of options ranging from the total prohibition of industrial wastes to limited regulation.[18] When the Tribunal opted for the latter, by ruling that a state could only intervene when "the case is of serious consequence and the injury is

[13] *See* Miller in this volume.

[14] British Columbia Provincial Archives (BCPA), Cominco Papers, Volume 457, Appendices in Behalf of the United States to the Canadian Agent, Articles of Association and Agreement of the Citizen's Association, 1–5, 1928 (August 25, 1928).

[15] Colin A. M. Duncan, *On Identifying a Sound Environmental Ethic in History: Prolegomena to Any Future Environmental History*, 15 ENVIRONMENTAL HISTORY REVIEW 5, 5–30 (1991).

[16] This is one of the most underplayed aspects of the *Trail Smelter* case. Consolidated's entire smoke control program was tied to the production of fertilizer, and the company only proceeded with the process as market conditions allowed. *See* J. N. Robinson, *The History of Sulphur Dioxide Emission Control at Cominco Ltd.*, 7 INTERNATIONAL JOURNAL OF SULPHUR CHEMISTRY, PART B, 51, 51–56 (1972).

[17] *Trail Smelter* (1941), *supra* note 10.

[18] Arthur P. Rubin, *Pollution by Analogy: The Trail Smelter Arbitration*, 50 OREGON LAW REVIEW 259, 259–98 (1971). Republished in this volume.

established by clear and convincing evidence,"[19] it conferred on industrial producers the right to maximize production in the pursuit of profit, subject only to the weakest of legal, political, and environmental constraints.

SMOKE ZONES

To understand the intricate process by which the Tribunal arrived at this solution requires that we move beyond the legalism that is symptomatic of the literature surrounding this obscure but crucial event in the history of North American environmental politics. Indeed, rather than being a triumph of environmental diplomacy or international law, the *Trail Smelter* arbitration was a rather more complex process of systemic coercion that sought to depoliticize a bitter social conflict rooted in environmental degradation. Once channeled into a "neutral" forum of dispute resolution, powerful economic, legal, political, and scientific forces converged around the issue of smelter smoke to transform a highly politicized class dispute into a matter of economic adjustment, scientific management, and technological refinement.[20] In the process, air pollution became an accepted, culturally sanctioned consequence of industrial capitalism, and "smoke eating" a normal part of everyday life.

The narrow emphasis in the literature on the legal implications of the *Trail Smelter* arbitration reflects just how successful this process of depoliticization has been. Indeed, the law constitutes only one, albeit important, part of a much larger, and more compelling story. In the *Trail Smelter* case, wealth, power, and status sit uncomfortably beside poverty, subordination, and anonymity. The division between these two poles was often quite literally the 49th parallel, but the international border obscured a more basic class divide. On the Canadian side sat the jewel of the British Columbian mining industry, whose ability to generate enormous wealth made it the locus of regional growth, a pillar of provincial development, and an instrument of Canadian economic policy. South of the border lived several hundred pioneer homesteaders, coalesced around the town of Northport, who toiled in virtual obscurity on the difficult terrain of the Columbia River valley. Both sides were born of the same mining boom that had enveloped the region in the 1890s, but with the collapse of the industry south of the border after 1905, Trail and Northport had come to coexist in almost splendid isolation of each other. As the smoke pushed ever further down the valley, however, the small agrarian community of Northport collided with an emerging industrial empire. As a result, a fifty-mile stretch of land between Trail, British Columbia, and Kettle Falls, Washington, was soon transformed into an arena of intense class struggle.[21]

[19] *Trail Smelter* (1941), *supra* note 10.

[20] *See* Miller in this volume.

[21] Whether the term "class" is appropriate here is of course a matter of interpretation. Although the farmers were not a class in a strict structural sense – they were in effect small businessmen – their class position was conditioned by the environmental consequences of capitalist development.

Such were the environmental boundaries within which the social conflict developed; boundaries shaped not just by political jurisdictions but also by the dynamic relationship between the natural environment of the Columbia River valley and the economic culture of industrial capitalism. Here the dispute centered not on hours, wages, or working conditions, but on conflicting constructions of a shared landscape whose bio-regional properties had triggered the class divisions that transcended national borders. Natural processes working in conjunction with human production practices created a bio-regional smoke zone whose parameters were established by the toxic range of the smoke. Initially, the circumference of these toxic borders included only lands adjacent to Trail. As production increased, however, the smoke zone increased as well. After 1925, the contested terrain extended across the international boundary, bringing American farmers into the orbit of a Canadian industry. United by environment, yet divided by class interests, an assortment of strange bedfellows struggled to reorganize the environmental terms on which the social order would develop.

The *Trail Smelter* arbitration was only one of the many smoke disputes that erupted in North America in the first third of the 20th century.[22] Consequently, there had already developed a standard legal process for resolving such smoke disputes across the continent. Sometimes they were resolved voluntarily, but more often local residents were compelled to rely on the common law to protect their interests from corporate cupidity and neglect. In Canada, as Nedelsky has argued, the "pre-industrial doctrines of property law, nuisance, and riparian rights, permitted small, comparatively unproductive landholders to enjoin or exact damages from industries interfering with traditional rights."[23] But with the rise of industrialization, Nedelsky observes, the courts "radically changed the substance of the common law in order to facilitate and, in effect, subsidize development."[24] The erosion of liberalism's traditional commitment to the individual's right of absolute dominion over private property all but eliminated an important tool of environmental resistance in an era of rapacious resource exploitation.[25]

With the transformation of property law across North America during the nineteenth century, smoke disputes were resolved through the legal test of "beneficial use," whereby environmental damage was assessed against the productive value of the offending smelter.[26] The concept of beneficial use privileged industrial

[22] WIRTH, *supra* note 1.

[23] Jennifer Nedelsky, *Judicial Conservatism in an Age of Innovation: Comparative Perspectives on Canadian Nuisance Law 1880–1930*, *in* 1 ESSAYS IN THE HISTORY OF CANADIAN LAW 281, 284 (David H. Flaherty ed., 1981).

[24] *Id.*

[25] *See* Ted Schrecker, *Of Invisible Beasts and the Public Interest: Environmental Cases and the Judicial System*, *in* CANADIAN ENVIRONMENTAL POLICY: ECOSYSTEMS, POLITICS, AND PROCESS 83–108 (Robert Boardman ed., 1992); MORTON HOROWITZ, THE TRANSFORMATION OF AMERICAN LAW, 1780–1860, 63–81 (1977).

[26] On the subject of "beneficial use," *see* ROBERT CAIL, MAN, LAND AND THE LAW: THE DISPOSAL OF CROWN LANDS IN BRITISH COLUMBIA 111–24 (1974); THEODORE STEINBERG, NATURE INCORPORATED: INDUSTRIALIZATION AND THE WATERS OF NEW ENGLAND 135–65 (1991).

resource utilization by linking private wealth creation with the community (or public) interest. The courts were never prepared to overtly countenance outright environmental devastation for the sake of profits but, as one Ontario judge put it, they also were not prepared to "destroy the mining industry even if a few farms are damaged or destroyed."[27] Accordingly, the courts demanded absolute proof of "tangible economic injury" before granting compensation and only reluctantly imposed conditions on the future operations of even the most egregious offenders.[28]

In effect, the law functioned to give legal standing to smoke zones. Contemporaries referred to these legal constructs as industrial zones – legally sanctioned areas for the disposal of industrial wastes whose borders were defined by the limit of economic damage suffered by local residents.[29] More than just a contiguous geographical area afflicted by smoke, industrial zones were a powerful ideological tool that granted industrial polluters the power of eminent domain in the region within which they operated. Under this implicit charter, polluters were given unlimited access to adjacent lands to dispose of industrial wastes subject only to modest compensation agreements.[30] This was precisely the basis of the smoke settlement in the earlier, Canadian *Trail Smelter* dispute. The traditional right of absolute dominion over private property was exchanged for "just compensation." In 1924, the arbitrator ordered Consolidated Mining to buy out four of the farmers, pay compensation to several more, and purchase smoke easements from the remainder. In return, the company was granted the right to maintain production free of regulatory restraint and legal action.[31]

The international character of the *Trail Smelter* dispute, however, made it inordinately difficult to construct an industrial zone across the 49th parallel. Washington State law strictly forbade foreign corporations from owning or even having an interest in State lands[32] – a condition that Consolidated Mining considered essential to a permanent settlement.[33] Moreover, although it was impossible to try a Canadian corporation in a Washington State court at the time, the farmers in the CPA considered British Columbia's arbitration system to be unfriendly at best. Having observed that their agrarian counterparts in Trail had received only money damages with no guarantee that the nuisance would stop, the CPA rejected this option.[34] Thus, legal constraints and political jurisdictions combined to render traditional remedies inoperative.

[27] Black v. Canadian Copper [1917], 12 ONTARIO WEEKLY NOTES 243–44 (1917).

[28] Cairns v. Canada Refining and Smelting Company, ONTARIO WEEKLY NOTES 423–6 (1913–14).

[29] The term "Industrial Zone" was used by experts deployed by the IJC in a preliminary report to the Commission. *See* Trail Smelter, Document Series A, Appendix A1, *The Deans' Report*, 5–39.

[30] *See* HOROWITZ, *supra* note 25, at 81.

[31] *See* TRAIL NEWS, *supra* note 5, at 1.

[32] Washington State University Archives (WSUA), Papers of W. Lon Johnson, Cage 195, Vol. 68, R. C. Crowe to J. J. Warren, 1927, (March 27, 1927).

[33] BCPA, Cominco Papers, Technical File 227, S. B. Blaylock to J. J. Warren, (January 13, 1926).

[34] *See* Read, *supra* note 12, at 222

Hence, two national states were drawn into a local, albeit transboundary, environmental conflict both as actors and conciliators. Diplomacy acted as the means of mediation, but behind the facade of diplomatic decorum stood the more difficult task of reconciling transnational class interests. Without recourse to the courts, and unable to resolve the matter diplomatically, the governments turned to quasi-legal institutions as the principal mechanisms of dispute resolution. First the IJC, then the *ad hoc Trail Smelter* Tribunal, were entrusted with the function of finding a balance between industry and agriculture but the terms of reference were not conceived with the intent to allow a full vindication of citizens' rights. Rather, international mediation was designed to realign the balance of class forces in the valley, where the controversy had had such a destabilizing effect. Guided by the legal test of beneficial use, both the IJC and the Tribunal attempted to construct an industrial zone across the 49th parallel that would effectively erase the border for economic purposes. By doing so, international mediators sought to preserve the hierarchy of economic interests that already existed in the Columbia River valley.

Consolidated Mining's superior financial resources, corporate connections, and international reputation ensured that the company and the farmers were never equals in this contrived international setting. Backed by the full resources of the Canadian state, which apparently valued national over human and environmental rights, Prime Ministers King and Bennett, the National Research Council of Canada, and the Department of External Affairs all worked tirelessly to protect a model Canadian enterprise. Similarly, the American mining and smelting industry clearly valued corporate over civic loyalties by reaching across the international border to join forces with Consolidated Mining to wage legal, scientific, and political warfare on fellow Americans.[35] Throughout the *Trail Smelter* dispute, but especially in the years leading up to the hearings before the Tribunal, the activities of the American Smelting and Refining Company laid bare the raw power of corporate capitalism that lurks just below the surface of liberal democracy.

The CPA, by contrast, enjoyed only the fleeting support of the American government whose loyalties were increasingly divided between its individual and corporate citizens. Racked by internal dissension, the American government's commitment to the fight waned at the very moment corporate America began to flex its considerable muscle. Faced with the prospect of offending, and perhaps even compromising, powerful industrial interests at home, the American government ultimately retreated from a fight it preferred not to win. When the Tribunal convened for the first time in the Spring of 1937, the CPA had been practically abandoned by its own government.[36] The quasi-judicial proceedings of both the

[35] *See* Miller in this volume.

[36] *Trail Smelter Fumes*, Congressional Record Vol. 76 Part 1, 72 Congress 839 (1933). As early as 1931, the congressional appropriation of $40,000 for American scientific investigations (in place since 1928) was cut in half and an additional $29,000 for public health studies was never spent.

IJC and the Tribunal were designed to give the appearance of a fair, rational, and objective process for dispute resolution, but it was Consolidated Mining that owned a virtual monopoly on the resources of power, thereby enabling it to hold an upper hand in the proceedings. Between 1926 and 1938, the smelter company spent over $1.1 million to preserve its hegemonic status in the Columbia River valley.[37] The present-day image of the *Trail Smelter* arbitration as a model of international environmental regulation conceals this historical reality of class politics and corporate privilege.

THE SMOKING GUN

Despite these elements of legal and political coercion, nearly fifteen years elapsed before the matter was deemed officially closed.[38] Many reasons have been offered to account for this long delay, but ultimately the problem can be traced to Consolidated's refusal to reduce pollution through a reduction in its levels of production. This, in turn, provoked an angry response from the farmers who accused the company of negligent conduct. These tensions were exacerbated further by the company's attempt to reach voluntary agreements with individual farmers in an effort to break their organization, and when that failed, to use its considerable resources to influence the political climate south of the border to contain the resistance.

When the issue was channeled into international mediation, Consolidated continued to stubbornly refuse to consider restrictions on production. The IJC settlement failed precisely because it allowed the smelter company to dictate the terms of smoke control measures without establishing minimum standards for safe production.[39] Only the CPA's insistence on a complete cessation of the nuisance, a position from which it never once deviated, finally forced the more difficult environmental questions to the top of the agenda. Determining how climate conditioned the chemical density of the emissions and how sulfur dioxide interacted with the mechanisms of soil formation and plant and tree growth became the environmental key to unlocking the political stalemate.

Such enormously complex questions were further complicated by the conflicting claims of the farmers and the smelter company. Although the former consistently claimed that smoke injury was "immediate, progressive and irreperable,"[40] the latter would only admit that there was some evidence of marginal economic

37 BCPA, Cominco Papers, ADD MSS 2500, Reel A1509, Technical File 231, U.S. Smoke Easements and Expenses, 1931, May 12, 1932, "*Cost of Smoke Control, 1934–1938 Inclusive*" (October 17, 1940).

38 The *Trail Smelter* Tribunal issued its final decision on March 11, 1941.

39 John Raftis and F. M. Turner to the Secretary of State, March 21, 1931, 76 Congressional Record Part 1, 72 Congress 835–38 (December 21, 1932). Lawyers representing the CPA told Henry Stimson, "The attitude of the claimants has always been that they desire these fumes to be stopped."

40 International Joint Commission, Trail Smelter Reference, Brief for the United States Government to the International Joint Commission, 61–62 (Washington, 1930).

damage.[41] Moreover, when the farmers complained that smoke conditions were responsible for a loan embargo imposed by the local banking community, the smelter company countered that the embargo existed because farmland in the smoke zone was of marginal economic utility.[42] And when the farmers complained that the combination of environmental and economic factors had precipitated a regional economic collapse, Consolidated Mining retorted that the region was already an economic backwater, a pitiful remnant of the arrested development of the mining industry in Northport.[43]

To resolve these tensions, both the IJC and the Tribunal turned to scientific expertise to determine the actual conditions in the field.[44] Secure in the objectivity of scientific reasoning, and armed with the instruments of modern empiricism, only the disinterested expert was considered capable of determining the difference between economic and less important forms of smoke damage. In this, the experts were as much economists as scientists, a role that ratified their collective power to shape the terms of settlement.

Interestingly, as the expert emerged as the authentic voice of authority, the farmers suffered a corresponding decline in credibility. The opposite of the expert was, of course, the amateur, the layman, the opinion witness – names which were all frequently applied to the residents of Stevens County. In the absence of expert knowledge, the farmers were deemed to have little understanding of the complex processes that resulted in environmental degradation or its consequent economic impact. As a result, their testimony as to the condition of the land was considered to be unreliable and unworthy of serious consideration. Eventually, the name "smoke farmer" emerged as a kind of umbrella term to deride the alleged ignorance and greed that drove the farmers to hold the smelter company to ransom in the wake of merely marginal damage to their lands.[45] Once cast as a parasite on productive industrial development, the farmers found themselves marginalized in a forum which purported to take their grievances seriously.

Between 1927 and 1937, the smoke zone between Trail and Kettle Falls was put under environmental surveillance. Statistics on production, emissions, wind patterns, temperature, precipitation, and humidity were combined with crop, timber, and plant surveys to determine correlations within an intricate web of

41 Trail Smelter, *Tory Report*, Appendices to the Statement of the Canadian Agent, Document Series A, Appendix A2 at 29 (Ottawa, 1930). As suggested by this report, Canadian investigators struggled "to arrive at some idea of the magnitude of the crop losses from sulphur dioxide" because "with surprisingly few exceptions the task was one rather of finding the evidence of injury than of computing sustained damage."

42 *Id.* at 34.

43 International Joint Commission, Trail Smelter Reference, Hearing at Washington, January 22–February 12, 1930 Vol. 7, 1912–13.

44 *See* Miller in this volume.

45 BCPA, Cominco Papers, Technical File 226, J. W. Blankenship, *Report on Smoke Conditions About Trail* (September, 1918): 1–20.

linkages and variables. These in turn were compared with data on crop production, tax assessments, and land values in an effort to measure the human impact of environmental degradation.

The knowledge generated by these studies ultimately became the foundation for the resolution of the controversy, but scientific expertise proved to be neither reliable nor particularly objective. Field studies and laboratory experiments routinely provided contradictory evidence on the short- and long-term impact of smelter smoke. Moreover, these divisions within the community of experts divided along national lines. Where scientists deployed by the Canadian government supported Consolidated's view that the environmental damage was of minimal economic consequence, their American counterparts supplied evidence establishing severe and extensive injury.[46] During the IJC hearings, experts proved to be little more than neutered partisans who constructed a knowledge around the subject of smelter smoke that was as deeply political as was the conflict it was intended to resolve.

Nowhere was this politicization of science more apparent than in the highly contested issue of smoke injury. Both American and Canadian scientists agreed that the complex of plants, trees, and crops in the region showed symptoms of acute and chronic injury, both of which manifested clear signs of discoloration.[47] Where the scientists disagreed was over the source and severity of visible injury. Although American scientists supported the farmers' claim that smelter smoke was the primary causative agent of discoloration, Canadian investigators maintained that these "burns" or "lesions" were just as easily attributable to any number of causes including drought, winter injury, disease, or insect infestations.[48] The enduring legacy of the IJC settlement, despite its eventual failure, was to establish the credibility of the smelter company's theory of "polycausality," which effectively minimized Consolidated's liability for conditions south of the border.[49] Although the IJC settlement collapsed, the Commission's report nevertheless served to endorse the perception that the farmers were suffering from paranoid delusions. As one expert put it, the claims made by the farmers conformed to his experience that such damage was almost always "90% psychological, 10% physical."[50]

46 Experts received a lot of criticism during the dispute. *See, e.g.*, *Editorial – Mines, Metals, and Men*, The Globe and Mail, August 5, 1937, at 7 ("Experts in every line usually give evidence that is valuable to the side that employs them.").

47 On the nature of sulfur dioxide injury, *see* Michael Treshow & Franklin K. Anderson, Plant Stress from Air Pollution 44–54 (1989).

48 *See* Tory Report, *supra* note 41, at 1–35.

49 "Polycausality" is a term coined by Schramm. *See* E. Schramm, *Experts in the Smelter Smoke Debate*, *in* The Silent Countdown: Essays in European Environmental History 201–04 (P. Brimblecombe & C. Pfister eds., 1990).

50 International Joint Commission, Testimony of George R. Hill, Trail Smelter Reference, Docket 25, 5 Hearing at Washington, January 22–February 12, 1930, at 1206–20 (January, 28 1930).

Having seen official limits placed around the extent and severity of visible injury, attention turned to the more explosive issue of "invisible injury." From the inception of the controversy, the farmers had consistently complained that the land was withering under a relentless smoke attack that robbed the trees, plants, and crops of their vitality. If it could be proven that sulfur dioxide induced physiological changes in plant metabolism without showing any visible signs of damage, then the extent of damage far exceeded that which was visibly apparent and therefore called for strict regulation of the production at the smelter.

Aware that the Americans were focusing almost completely on this issue in the run up to the Tribunal's hearings, the Canadian National Research Council consciously worked to build a case against the theory of invisible injury.[51] Funded entirely by Consolidated Mining,[52] Canadian government scientists paraded masses of experimental data before the Tribunal to show that smoke injury was limited to that which was visible to the naked eye. American experts countered with experimental data of their own, but so sloppy and disorganized was the complainant's case that the American evidence was dismissed as a product of desperation.[53] In its final decision, the Tribunal rejected the theory of invisible injury,[54] paving the way for the construction of an industrial zone across the international border whose parameters were defined by the economic range of visibly apparent damage.

It is perhaps of small consolation to the "smoke farmers" of Stevens County, but their own "amateur" observations on the hidden impact of sulfur dioxide have since become an accepted category of environmental injury. Years of intense research, says the biologist Michael Treshow, have conclusively demonstrated that "significant losses in production and growth can occur without another symptom ever being seen."[55] One need not look for a conspiracy or even the suppression of evidence to understand how this "smoking gun" was overlooked in the *Trail Smelter* arbitration. The difference between the Canadian and American evidence was not so much a matter of science but of economics. The inability of American investigators to quantify the extent of unseen damage stood in stark contrast to the ability of Canadian experts to compute the cost of observable injury in precise, readily understandable terms. Visible, measurable economic damage, in turn, conformed to the legal test of beneficial use that privileged the producers of wealth creation over the victims of environmental degradation. Together the

[51] BCPA, Cominco Papers, Technical File 230, Memorandum on the Present and Future Status of the Trail Smelter Smoke Investigation, by Morris Katz, August 11, 1931.

[52] BCPA, Cominco Papers, Technical File 232, Warren to A. G. McNaughton, 15 October 1935; Technical File 234, *Cost of Smoke Control*, 1934–1938, *supra* n. 37.

[53] *See* ALLUM, *supra* note 1, at 326–60; WIRTH, *supra* note 1, at 98.

[54] Trail Smelter (U.S. v. Can.), International Joint Commission, Trail Smelter Reference, Final Report 193–209 (Ottawa, 1931).

[55] TRESHOW & ANDERSON, *supra* note 47, at 54.

authentic voice of the expert joined with the economic authority of Consolidated Mining, the political authority of state intervention, and the legal authority of international mediation, to forge a powerful consensus around the issue of smoke injury. Out of this cacophony of voices, a language of development emerged that cast the high price of environmental degradation as an acceptable cost of industrial production.[56]

This consensus found its expression in the Tribunal's now famous dictum on state responsibility for transnational air pollution.[57] Encoded within the ruling were the terms in which transboundary environmental degradation was henceforth to be understood and regulated. Having ruled out the prospect of invisible injury, the Tribunal decreed that the state was responsible for transnational air pollution only when one could absolutely prove it was of serious economic consequence. In the absence of such proof, as Rubin has observed, "there appears to be no international responsibility at all for acts of pollution."[58] In fact, the need for absolute proof undermined "the very existence of a legal wrong."[59] Far from being an expansive declaration of state liability, the Tribunal ultimately constructed an extraordinarily narrow doctrine on transnational air pollution that erased national borders to protect the sovereignty of industrial production in North America. And for this, the *Trail Smelter* arbitration's enduring legacy has proven to be more toxic than legal.

56 *See* Jacobson in this volume.

57 *See* Ellis in this volume.

58 Rubin, *supra* note 18, at 273.

59 *Id.* at 274.

2 The *Trail Smelter* Dispute [Abridged]

John E. Read

The *Trail Smelter* Dispute[1] covered a period of thirteen years from 1928 to 1941. It arose out of the operation by the Consolidated Mining and Smelting Company of Canada, Limited, of a smelter at Trail, British Columbia, on the Columbia River about eleven miles from the international boundary. In roasting sulphur-bearing ores, sulphur dioxide gas was wasted into the air. When the air drift was down the valley, the smoke cloud containing the sulphur dioxide crossed the boundary in sufficient strength to cause damage in the State of Washington. In 1925 and 1926, the output of the smelter was increased and more sulphur dioxide was wasted into the air. By reason of the conformation of the valley and the atmospheric conditions prevalent, it was carried across the boundary into the State of Washington. It was common ground that some damage was caused in the years 1926 to 1930.

The U.S. government intervened, through diplomatic channels, in 1927. The subject matter of the dispute did not directly concern the two governments; nor did it involve claims by U.S. citizens against the Canadian government. It did not seem to come within any of the ordinary categories of arbitrable international disputes. It consisted, rather, of claims based on nuisance, alleged to have been committed by a Canadian corporation and to have caused damage to U.S. citizens and property in the State of Washington. Nevertheless, when the United States proposed to refer the questions at issue to the International Joint Commission, the Canadian government concurred; and, on August 7, 1928, it joined in reference

[1] III Reports of International Arbitral Awards (1905–1982).

[Excerpt reprinted with permission of the University of British Columbia Press. The article originally appeared as: John E. Read, *The Trail Smelter Dispute*, 1 THE CANADIAN YEARBOOK OF INTERNATIONAL LAW 213 (1963)] The writer is a former member of the International Court of Justice. He was directly concerned with the *Trail Smelter* dispute at all stages: the settlement of the terms of reference to the International Joint Commission, the negotiation and drafting of the Trail Smelter Convention, and the special problems that arose and the way in which they were dealt with; and he is in a position to discuss some of the matters that do not emerge from examination of the decisions and the records.

under Article 9 of the Boundary Waters Treaty, 1909. This Article provided for investigation and recommendation, but not for decision.

The Commission made an exhaustive investigation. It appointed two scientists, Dean Miller of the University of Idaho and Dean Howes of the University of Alberta, who, with the aid of experts form the public services of the two countries, made field surveys in 1929 covering most aspects of the damage. The Commission heard the interested parties at Northport (Wash.), Washington (D.C.), and Nelson (B.C.). Witnesses were heard, documentary evidence was filed, and counsel presented arguments.

The Commission made a unanimous report on February 28, 1931.[2] It assessed the damage up to the year 1931 at $350,000 and made other recommendations covering the questions referred to it. The Consolidated Mining and Smelting Company had already commenced the construction of remedial works designed to lessen the output of sulphur from the stacks by about one third. The works were to be completed in 1931, and the report was based on the expectation that they would bring about the cessation of damage.

The recommendations of the Commission were satisfactory to the Canadian government but they were rejected by the United States in February 1933. Negotiations that followed led to the submission of the dispute to a three-member tribunal by a Concentration signed on April 15, 1935.

Without making a detailed examination of the Convention,[3] consideration may be given to the following provisions:

Article I required the government of Canada to pay to the United States the sum of $350,000, to cover all damage which occurred before January 1, 1932.

Article II provided for the constitution of the Tribunal to consist of a neutral chairman and two national members.

Article III stated the questions:

1. Whether damage caused by the Trail Smelter in the State of Washington has occurred since the fist day of January, 1932, and, if so, what indemnity should be paid therefore?
2. In the event of the answer to the first part of the preceding Question being in the affirmative, whether the Trail Smelter should be required to refrain from causing damage in the State of Washington in the future, and, if so, to what extent?
3. In the light of the answer to the preceding Question, what measures or regime, if any, should be adopted or maintained by the Trail Smelter?
4. What indemnity or compensation, if any, should be paid on account of any decision or decisions rendered by the Tribunal pursuant to the next two preceding Questions?

[2] *Id.* at 1918–19.

[3] *Id.* at 1907–10.

Article IV stated the law to be applied by the Tribunal. This will be quoted and discussed later.

Articles V to XI were procedural. Among other things, they required the Tribunal to receive and consider evidence, oral or documentary, presented by the governments "or by interested parties;" and it was given power to administer oaths. It was authorized to make investigations.

After the Pleadings had been exchanged, the Tribunal proceeded to an examination of the Questions. It met in Washington, D.C., on June 21 and 22, 1937, for organization, adoption of rules of procedure, and hearing of preliminary statements. From July 1 to July 6, it traveled over and inspected the area involved in the controversy in the northern part of Stevens County in the State of Washington; and it also inspected the smelter at Trail. It held sessions for the reception and consideration of such evidence, oral and documentary, as was presented by the governments or by interested parties in Spokane, Washington, from July 7 to 29, 1937; in Washington, D.C., on August 16, 17, 18, and 19, 1937; in Ottawa, from August 23 to September 18, 1937; and it heard arguments in Ottawa from October 12 to 19, 1937.

When the members of the Tribunal inspected the area involved in the controversy, it was no casual survey. They made a close examination of many of the farms and orchards and cruised the forests. When they heard the evidence, the witnesses and experts were talking about places and conditions familiar to them.

After examining the voluminous record, the Tribunal notified the Agents that, unless the time limit was extended, it would be compelled to give a permanent decision on the basis of data that it considered inadequate and unsatisfactory. Accordingly, the governments extended the period for final decision until late in 1940 so as to leave three growing seasons for testing a temporary regime and for further scientific investigation.

On April 16, 1938, the Tribunal reported its "final decision" on Question No. 1, as well as its temporary decisions on Questions No. 2 and No. 3, and provided for a temporary regime thereunder. There may be some confusion in terminology. The ultimate decision of the Tribunal was rendered on March 11, 1941, and it was very definitely *final*. The Tribunal there referred to the earlier decision as the "previous decision," an expression which is somewhat awkward for the purposes of this article. Nevertheless, confusion can be avoided by using it and by putting the words "final" or "final decision" in quotation marks whenever it is necessary to discuss the finality of any part of the "previous decision."

As regards Question No. 1, the Agent for the government of the United States presented claims for damages of $1,849,156.16 with interest of $205,855.01, divided into seven categories:

(a) cleared land and improvements;
(b) uncleared land and improvements;

(c) livestock;
(d) property in the town of Northport;
(e) wrong done the United States in violation of sovereignty, measured by cost of investigation from January 1, 1932, to June 30, 1936;
(f) interest on $350,000 accepted in satisfaction of damage to January 1, 1932, but not paid on that date;
(g) business enterprises.

The area claimed to be damaged contained more than 140,000 acres, including the town of Northport.

The tribunal disallowed the claims of the United States under items (c), (d), (e), (f), and (g); but allowed them in part under the remaining items (a) and (b), that is, cleared and uncleared land and improvements.

In conclusion of the "previous decision," the Tribunal reported, as its "final decision" regarding Question I:

> Damage caused by the Trail smelter in the State of Washington has occurred since the first day of January 1932, and up to October 1, 1937, and the indemnity to be paid therefore is seventy-eight thousand dollars ($78,000), and is to be complete and final indemnity and compensation for all damage which occurred between such dates. Interest at the rate of six per centum per year will be allowed on the above sum of seventy-eight thousand dollars ($78,000) from the date of the filing of this report and decision until date of payment. This decision is not subject to alteration by the Tribunal hereafter.
>
> The fact of existence of damage, if any, occurring after October 1, 1937, and the indemnity to be paid therefore, if any, the Tribunal will determine in its final decision.[4]

As regards Questions Nos. 2 and 3, the Tribunal subjected the Trail smelter to a temporary regime and provided for further and "more adequate and intensive study" under the supervision of two "Technical Consultants," the two scientists who had been designated by the governments in pursuance of the provisions of Article II.

For the best part of three years, and over three complete growing seasons, the Tribunal applied the temporary regime and conducted a comprehensive investigation of the behavior of the smoke cloud wasted from the stacks of the smelter. The Tribunal, in its final decision, described this investigation in the following words:

> The investigations made during the past three years on the application of meteorological observations to the solution of this problem at Trail have built up a fund of significant and important facts. This is probably the most thorough study ever made of any area subject to atmospheric pollution by industrial smoke. Some factors, such as atmospheric turbulence and the movement of the upper air currents have been

[4] *Id.* at 1933.

> applied for the first time to the question of smoke control. All factors of possible significance, including wind directions and velocity, atmospheric temperatures, lapse rates, turbulence, geostrophic winds, barometric pressures, sunlight and humidity, along with atmospheric sulphur dioxide concentrations, have been studied. As said above, many observations have been made on the movements and sulphur dioxide concentrations of the air at higher levels by means of pilot and captive balloons and by airplane, by night and by day. Progress has been made in breaking up the long winter fumigations and in reducing their intensity. In carrying finally over to the non-growing areas with a few minor modifications a regime of demonstrated efficiency for the growing season, there is sound basis for confidence that the winter fumigations will be kept under control at a level well below the threshold of possible injury to vegetation. Likewise, for the growing season a regime has been formulated which should throttle at the source the expected diurnal fumigations to a point where they will not yield concentrations below the international boundary sufficient to cause injury to plant life. This is the goal which this Tribunal has set out to accomplish.[5]

The final arguments of counsel were held in Montreal from December 9 to 12, 1940, and the final decision was reported to the governments on March 11, 1941.

* * *

Accordingly, after dealing with these three [preliminary] claims, the Tribunal found:

> Since the Tribunal has, in its previous decision, answered Question No. 1 with respect to the period from the first day of January, 1932, to the first day of October, 1937, it now answers Question No. 1 with respect to the period from the first day of October, 1940, as follows:
>
> 1. No damage caused by the Trail Smelter in the State of Washington has occurred since the first day of October, 1937, and prior to the first day of October, 1940, and hence no indemnity shall be paid therefore.[6]

The second Question was "whether the Trail smelter should be required to refrain from causing damage in the State of Washington in the future and, if so, to what extent?"[7]

After a close examination of the relevant provisions of the Convention, and a comprehensive study of the available precedents, the Tribunal concluded:

> The tribunal, therefore, finds that the above decisions, taken as a whole, constitute an adequate basis for its conclusions, namely, that, under the principles of international law, as well as the law of the United States, no State has the right to use or permit the use of its territory in such a manner as to cause injury by fumes in or to the territory of another or the properties or persons therein, when the case is of serious consequence and the injury is established by clear and convincing evidence.

5 *Id.* at 1973–74. 6 *Id.* at 1962.

7 [In the original, the second note 7 does not have a corresponding note in the footer of the page.]

> The decisions of the Supreme Court of the United States which are the basis of these conclusions are decisions in equity and a solution inspired by them, together with the regime hereinafter prescribed, will, in the opinion of the Tribunal, be "just to all parties concerned," as long, at least, as the present conditions in the Columbia River Valley continue to prevail.
>
> Considering the circumstances of the case, the Tribunal holds that the Dominion of Canada is responsible in international law for the conduct of the Trail Smelter. Apart from the undertakings in the Convention, it is, therefore, the duty of the Government of the Dominion of Canada to see to it that this conduct should be in conformity with the obligation of the Dominion of Canada under international law as herein defined.
>
> The Tribunal, therefore, answers Question No. 2 as follows: (2) So long as the present conditions in the Columbia River Valley prevail, the Trail Smelter shall be required to refrain from causing any damage through fumes in the State of Washington; the damage herein referred to and its extent being such as would be recoverable under the decisions of the courts of the United States in suits between private individuals. The indemnity for such damage should be fixed in such manner as the Governments, acting under Article XI of the Convention, should agree upon.[8]

The third Question was "In the light of the answer to the preceding Question, what measure or regime, if any, should be adopted and maintained by the Trail Smelter?"[9]

The Tribunal imposed a regime of control over the emission of sulphur dioxide fumes from the smelter. This part of the final decision is highly technical and it would take up too much space to deal with it in this article. Nevertheless, the magnitude of the regime may be suggested.

The capital cost of complying with the regime was of the order of $20 million: pre-war costs when a million dollars meant a lot of money. Fortunately for the economic life of the southeastern part of British Columbia, in which the Trail smelter was the most important factor, it was a private enterprise and not a government department. Notwithstanding the enormous capital expenditure and intermittent interruption of the metallurgical operations imposed, in perpetuity, by the regime of control, the Consolidated Mining and Smelting Company succeeded in selling the products of its smoke abatement program for substantial profit. In order to comply with the regime, the Company was compelled to remove from the smoke clouds at the stacks more sulphur dioxide than was taken from the stacks of all other smelters of the North American continent combined.

The tribunal recognized that the regime might prove to be insufficient to abate the nuisance. Accordingly, the Tribunal prescribed machinery for "Amendment or Suspension of the Regime." This consisted of a clause providing for compulsory arbitration by a Commission of Scientists.

[8] *Id.* at 1965–66.

[9] *Id.* at 1966.

The fourth Question was "What indemnity or compensation, if any, should be paid on account of any decision or decisions rendered by the Tribunal pursuant to the next two preceding Questions?"[10]

The Tribunal was "of the opinion that the prescribed regime will probably remove the causes of the present controversy and, as said before, will probably result in preventing any damage of a material nature occurring in the State of Washington in the future."[11] Nevertheless, it decided that an indemnity shall be paid in the event of future damage and that the United States shall be paid the reasonable costs of investigation, up to $7,500 in any one year, as compensation. For the establishment of the occurrence of damage, the Tribunal invoked the provisions of Article XI of the Convention. The obligation to pay the compensation, for costs of investigation, was to arise only upon the establishment of the occurrence of damage; but it was to be regarded as compensation to be paid on account of the answers to Questions No. 2 and No. 3. In this way, the Tribunal avoided the possible suggestion that this requirement was inconsistent with the decisions regarding the claims, I and II(b) under Question No. 1.

This brings to an end the survey of the course of the proceedings, ***

[10] *Id.* at 1980.

[11] [In the original, the second note 11 does not have a corresponding note in the footer of the page.]

3 Of Paradoxes, Precedents, and Progeny: The *Trail Smelter* Arbitration 65 Years Later

Stephen C. McCaffrey

PARADOXES

- A fountainhead of transboundary pollution becomes the fountainhead of law prohibiting transboundary pollution.
- An industrial activity that is synonymous with environmental threats becomes synonymous with environmental protection.
- A small town in Canada becomes known throughout the world for an international arbitration that bears its name.
- For the fact that there was an international arbitration at all we owe thanks to an antique English rule of civil procedure. And,
- The country that "won" the arbitration has more recently been equivocal as to the legal status of the fundamental principle on which the award was based.

On March 11, 1941, just over 65 years ago, the *Trail Smelter* Tribunal, composed of jurists from Canada, the United States and Belgium, delivered an award that ushered in a new era in a field that has become known as international environmental law.[1] The Tribunal held, in essence, that Canada was required to see to it that the smelter at Trail, British Columbia, would "refrain from causing any damage through fumes [to agricultural interests] in the State of Washington; . . ."[2]

I first began studying the *Trail Smelter* arbitration in the early 1970s, which is roughly equidistant in time from both the Tribunal's final award and the publication of this volume. I wondered why this controversy, essentially between a private smelter on one side of the U.S.-Canadian border and private landowners

[1] *Trail Smelter Arbitral Decision*, 33 American Journal of International Law 182 (1939) [hereinafter "*Trail Smelter* (1939)"]; *Trail Smelter Arbitral Decision*, 35 American journal of International Law 684 (1941) [hereinafter "*Trail Smelter* (1941)"]. *See* Annex to this volume. The reference in the text is to *Trail Smelter* (1941).

[2] *Trail Smelter* (1941), *supra* note 1, at 717.

on the other, could not simply have been resolved through national courts. What I discovered was that the old English "local action rule"[3] was then and is still alive and well in British Columbia and prevents courts there from entertaining suits for damage to foreign land.[4] And, of course, Washington did not yet have a long-arm statute that would permit the landowners to sue the smelter in Washington. So, up to the respective governments the dispute went, perhaps unfortunately from the standpoint of efficiency, but fortunately for the field of international environmental law.

PRECEDENT

One really has to admire the courage of the three "*trail*-blazing" arbitrators, if you will pardon the expression. For they were writing on a *tabula rasa*: there were no previous decisions by the World Court[5] or international arbitral tribunals in cases whose facts were remotely analogous. It is true that in the *River Oder* case the Permanent Court of International Justice had found in 1929 that riparian states shared a "community of interest" in a navigable river that formed "the basis of a common legal right…"[6] But the dispute in that case was about navigation rights, not pollution. For their part, the *Trail Smelter* arbitrators declared: "No case of air pollution dealt with by an international tribunal has been brought to the attention of the Tribunal nor does the Tribunal know of any such case."[7] All the arbitrators had to go on were a handful of decisions of the U.S. Supreme Court in American interstate pollution cases, a decision of the Federal Court of Switzerland in a suit between two cantons involving transboundary risks of target practice, and what the tribunal referred to as "[a] great number of… general pronouncements by leading authorities concerning the duty of a State to respect other States and their territory…"[8] These pronouncements were no doubt based in large measure on the famous *Alabama* arbitration of 1872.[9] In that dispute between Britain and the United States, Britain was found to have a duty to prevent

[3] *See* British South Africa Co. v. Companhia de Moçambique, [1893] A.C. 602; *applied in* Canada in Albert v. Fraser Cos. Ltd., [1937] 1 D.L.R. 39, 43 (N.B. 1936). *But see* the opinion of Lord Coke in Bulwer's Case, 7 Coke 1a, 77 Eng. Reprint 411 (1584) (Recognizing an exception to the local action rule where an act in one jurisdiction gives rise to an injury to real property in another.).

[4] *See* Stephen C. McCaffrey, *Trans-boundary Pollution Injuries: Jurisdictional Considerations in Private Litigation between Canada and the United States*, 3 California Western International Law Journal 191, 211 and 224–29 (1973).

[5] At that time, the "World Court" was the Permanent Court of International Justice.

[6] Territorial Jurisdiction of the International Commission of the River Oder (Czechoslovakia, Denmark, France, Germany, Great Britain, and Sweden/Poland), 1929 P.C.I.J. (ser. A) no. 23, p. 5, at 27.

[7] *Trail Smelter* (1941), *supra* note 1, at 714.

[8] *Id.* at 713.

[9] Alabama Claims Arbitration, 1872, Moore, 1 International Arbitration 495. The "Alabama case" is referred to in general terms in the 1941 award of the *Trail Smelter* Tribunal. *Trail Smelter* (1941), *supra* note 1, at 713.

the construction in London of a vessel whose purpose would be to prey on Union shipping during the Civil War. Hardly an overwhelming international lineup of precedents on all fours with the facts of the *Trail Smelter* dispute.

Fortunately, the parties to the arbitration themselves anticipated this authoritative gap,[10] and their agreement submitting the dispute to arbitration provided that: "The Tribunal shall apply the law and practice followed in dealing with cognate questions in the United States of America as well as International Law and Practice, and shall give consideration to the desire of the High Contracting Parties to reach a solution just to all parties concerned."[11] The Tribunal avoided the problem of whether to apply international or U.S. law by finding that the law followed in the United States in dealing with the quasi-sovereign rights of the States of the Union, in the matter of air pollution, while more definite, "is in conformity with the general rules of international law."[12] This conclusion alone was a rather large analytical leap, because, as noted, there does not seem to have been any positive international law on point. In any event, the Tribunal could presumably have had recourse to decisions of the U.S. Supreme Court, as well as those of the courts of other federal states, without explicit permission from the parties, because it was by then accepted that international tribunals could look to national court decisions in analogous cases as evidence of rules of customary international law.[13]

Thus, despite the decisional vacuum from international tribunals, it seems virtually impossible for the *Trail Smelter* Tribunal to have pronounced a *non liquet*, that is, to have declined to decide the case for lack of clearly applicable rules of law, a result that would have been at least highly undesirable for a tribunal to reach at the time, and is still today regarded as not permissible.[14] In any case, Judge John Read, the Canadian Legal Adviser in the dispute, confirms that "both the United States and Canada were determined to avoid the possibility of a finding *non liquet*," in view of the fact that, as he describes the situation at the time, "there was not much international law available dealing with international nuisance."[15]

[10] *See* John E. Read, *The Trail Smelter Dispute*, 1 CANADIAN YEARBOOK OF INTERNATIONAL LAW 213, 227 (1963). Republished in this volume.

[11] Convention on Damages Resulting From Operation of Smelter at Trail, British Columbia, June 5, 1935, U.S.-Canada, art. IV, 49 Stat. 3245.

[12] *Trail Smelter* (1941), *supra* note 1, at 713.

[13] *See* art. 38(4) of the Statute of the Permanent Court of International Justice, available at http://www.worldcourts.com/pcij/eng/laws/law03.htm. [Article 38(4)] provided that the Court "shall apply . . . judicial decisions . . . as subsidiary means for the determination of rules of law." This provision is repeated in statute of the Permanent Court's successor, the International Court of Justice. *See* Statute of the International Court of Justice, June 26, 1945, art. 38(1)(d), 59 Stat. 1055, 33 U.N.T.S. 993.

[14] *See, e.g.*, 1 OPPENHEIM'S INTERNATIONAL LAW 3, 13 (Robert Jennings and Arthur Watts eds., 9th ed. 1992).

[15] Read, *supra* note 10, at 227.

Another most interesting aspect of the parties' instruction to the Tribunal concerning the law it was to apply is the reference to the "desire of the High Contracting Parties to reach a solution just to all parties concerned."[16] The Tribunal highlighted this phrase several times in its award, emphasizing its importance. It seems clear from the text of the award that the Tribunal interpreted the phrase to mean that what the parties were after was a balanced solution, one that neither shut down the smelter nor left the agricultural interests entirely at the mercy of its fumes.[17] This was crucial, for the Tribunal would otherwise have been confronted with the classic "clash of sovereignties": the sovereign right of Canada to use its territory as it pleased, and the equally sovereign right of the United States to be free from outside interference in the use of its territory.[18] Only by curtailing Canada's sovereignty could the Tribunal safeguard that of the United States; but on what basis should the sovereignty of only one of the two states be restricted? How could this Gordian knot possibly be cut?

When confronted with this question in 1895, U.S. Attorney General Judson Harmon had little difficulty. He replied simply that the United States could do whatever it pleased with the waters of the Rio Grande, regardless of the downstream consequences in Mexico.[19] But it was clear to the Tribunal that this would not do in the case of the *Trail Smelter* arbitration. As H. L. Mencken said, for every complex problem there is a solution that is simple, neat, and . . . wrong.[20] So the Tribunal interpreted the questions asked of it by the parties as indicating what it called "a desire and an intention" that its award "would allow the continuance of the operation of the Trail smelter but under such restrictions and limitations as would, as far as foreseeable, prevent damage in the United States, and as would enable indemnity to be obtained, if in spite of such restrictions and limitations, damage should occur in the future in the United States."[21] This is exactly what the Tribunal did. On the basis of in-depth scientific and technical studies, it fashioned a solution that may be said to constitute an equitable, or just, allocation of the shared airshed as between the smelter in Canada and the agricultural community in the United States, and thus between the two countries. The Tribunal thereby recognized that, as Ronald Coase later famously pointed out, "harm" can be a two-way street: granting absolute protection to the farmers would harm the smelter every bit as much as granting absolute protection to the smelter would harm the farmers.[22]

[16] 1935 Convention, *supra* note 11, art. IV.

[17] *See* Jacobson in this volume.

[18] *See* Günther Handl, *Territorial Sovereignty and the Problem of Transnational Pollution*, 69 AMERICAN JOURNAL OF INTERNATIONAL LAW 50 (1975) (discussing whether activities lawful *per se* in one state that give rise to environmental effects in another state may entail state responsibility even where the effects do not constitute material damage).

[19] 21 Op. Att'y Gen. 274 (1895). This opinion has become known as the "*Harmon Doctrine.*"

[20] HENRY LOUIS MENCKEN, PREJUDICES (1919).

[21] *Trail Smelter* (1941), *supra* note 1, at 685.

[22] *See* Ronald Coase, *The Problem of Social Cost*, 3 JOURNAL OF LAW & ECONOMICS 1 (1960). For a critical reading of this perspective on the work of the Tribunal, *see* Allum in this volume.

DAMAGE

Interestingly, it was the very concept of harm, or damage, that brought the dispute to arbitration in the first place. It had initially been taken by the two countries in 1928 to the International Joint Commission (IJC),[23] set up by their 1909 Boundary Waters Treaty,[24] under a provision of the treaty allowing them to refer to the Commission for examination and report not only water-related questions but also "any other question or matters of difference arising between them involving [their rights and interests] along the common frontier..."[25] The IJC made what Judge Read describes as an "exhaustive investigation"[26] and rendered a unanimous report in 1931.[27] The report recommended that the Consolidated Mining and Smelting Company, which operated the Trail smelter, proceed expeditiously with works it had initiated for the extraction of sulfur from the fumes and that it take any other measures necessary to reduce the amount of SO_2 fumes "to a point where [the smelter] will do no damage in the United States."[28] The IJC expected that these works, which were to be completed in 1931, would result in a cessation of damage in Washington[29] and assessed all damage through January 1, 1932, at $350,000.[30] The company would indemnify the interests affected by any damage occurring thereafter. Crucially, the ICJ defined the term "damage" to include "such damage as the Governments of the United States and Canada may deem appreciable, and...shall not include occasional damage that may be caused by SO_2 fumes being carried across the international boundary in air pockets or by reason of unusual atmospheric conditions."[31]

The Commission's recommendations were acceptable to Canada but not to the United States, which objected to the definition of "damage." The U.S. view seems to have been that if "fumigations" sufficient to cause injury continued to occur in Washington, the United States would have grounds for complaint, no matter what remedial works had been installed by the company, and regardless of their effect.[32] Canada nevertheless paid the sum of $350,000 that had been fixed by the IJC to the United States, and indeed the Tribunal was asked to rule only on damage occurring after January 1, 1932. Thus, although the report of the IJC was not accepted as such, it resulted in an effective settlement of the dispute through the first day of January, 1932, assisted significantly in achieving an overall solution to the problem, and provided an important basis for the Tribunal's work. Clearly one important

23 *Trail Smelter* (1941), *supra* note 1, at 693 (stating that the "subject of fumigations and damage claimed to result from them" was referred to the Commission on 7 Aug. 1928).

24 Treaty between Great Britain (Canada) and the United States relating to Boundary Waters and Questions Arising between the United States and Canada, 11 Jan. 1909, 36 Stat. 2448, TS 548, 12 Bevans 319, 102 BFSP 137.

25 *Id.* art. IX.

26 Read, *supra* note 10, at 214.

27 *Trail Smelter* (1941), *supra* note 1, at 693 (giving the date as 28 Feb. 1931).

28 *Id.*

29 Read, *supra* note 10, at 214.

30 *Trail Smelter* (1941), *supra* note 1, at 694.

31 *Id.* at 693.

32 *Id.* at 713.

lesson to be drawn from the *Trail Smelter* arbitration is, therefore, that joint commissions can be extremely helpful in resolving disputes concerning shared natural resources by finding facts and proposing technical solutions for consideration on the political and policy-making levels of the governments concerned.

THE RULE

The level of damage that is considered legally wrongful, which as defined by the IJC caused the United States to reject the Commission's report, figured prominently in the Tribunal's formulation of the principle on which its holding was based. That famous formulation, which has largely inspired the present project, reads, as nearly every student, scholar, and practitioner of international environmental law must know, as follows:

> [U]nder principles of international law, as well as of the law of the United States, no State has the right to use or permit the use of its territory in such a manner as to cause injury by fumes in or to the territory of another or the properties or persons therein, when the case is of serious consequence and the injury is established by clear and convincing evidence.[33]

This ringing declaration, uttered by a Tribunal years ahead of its time, became the wellspring of international environmental law. But in lauding the principle's path-breaking power, we should not lose sight of the fact that a twenty-first-century reader might well find that, as formulated by the Tribunal, it hardly appears radical. In fact, environmentalists today would likely raise objections to the seemingly high standards of proof the tribunal laid down: the case must be of "serious consequence;" and the injury must be established by "clear and convincing evidence."[34] This is hardly an embodiment of the modern precautionary principle, they might point out, which counsels that action to prevent serious environmental harm should be taken even in the absence of clear scientific evidence of a causal link.[35] Especially in a relatively straightforward case of unidirectional transboundary pollution, it could well be asked, why raise the bar above the normal preponderance-of-the-evidence standard? Even in other, less linear cases, should not the law err on the side of protecting the environment, rather than the source of the harm? In addition, why must the law wait until the consequence is "serious" before it steps in? Why not a lower threshold, such as "significant," or even "appreciable," the term used in the IJC's definition? Would such a lower standard not afford greater protection to the environment, especially in view of the difficulty of remediating environmental harm?

The answers to these questions seem to lie both in the fact that this was a case of first impression on the international stage that arose in an earlier and quite different era, and in two additional, but related, considerations: the Tribunal's

[33] *Id.* at 716.
[34] *See* Ellis, Rubin, and Mickelson in this volume.
[35] *See* Bratspies in this volume.

effort to resolve the clash of sovereignties; and its heavy reliance on the only closely analogous authority it was able to find, decisions of the U.S. Supreme Court in interstate air and water pollution cases. For the Supreme Court also had to resolve a clash of sovereignties, or, more accurately, *quasi*-sovereignties. Even in that bygone era, the Court was no doubt sensitive to the hazards of being seen as restricting state sovereignty, recalling, perhaps, its 1832 decision in *Worcester v. Georgia*,[36] restricting that state's authority over the Cherokees, a judgment to which President Andrew Jackson is reported to have responded: "John Marshall has made his decision, now let him enforce it!"[37] For this reason, and perhaps also because it was mindful of the fact that the injunction is, after all, an *equitable* remedy, the Court required that before it would "exercise its extraordinary power ... to control the conduct of one State at the suit of another, the threatened invasion of rights must be of serious magnitude and it must be established by clear and convincing evidence."[38]

This is almost precisely the formula adopted by the Tribunal in its statement of the applicable rule of law. Although this was, to be sure, a high standard of proof, it had to have been of some comfort to Canada, on the one hand, and, on the other hand, it had to be very difficult for the United States to reject, in view of its provenance. Moreover, the Tribunal must have believed the high standard to be necessary in achieving a balanced or "just" resolution to the clash of the sovereignties involved: do not shut down or curtail the operations of the smelter unless it is demonstrably causing some serious damage; that is, do not restrict Canada's sovereignty without clear evidence of a serious interference with the sovereignty of the United States. This idea finds echoes in the European concept of "neighborhood" law, which is similar to the common law of nuisance. Being a good neighbor, whether you are an individual or a state, means tolerating some interference by your neighbor with the enjoyment of your property.[39] In the context of the Tribunal's rule-formulation, the United States would be expected to tolerate minor or insubstantial interferences.[40] Those that went beyond that level would be actionable.

36 31 U.S. 515 (1832).

37 Although well known and widely quoted, there does not seem to be an official source for Jackson's words. Historians often refer to them, however. *See, e.g.*, VERNON L. PARRINGTON, 2 MAIN CURRENTS OF AMERICAN THOUGHT, *available at* http://xroads.virginia.edu/~Hyper/Parrington/vol2/bk01_01_ch03.html.

38 New York v. New Jersey, 256 U.S. 296, 309 (1921) (*quoted in Trail Smelter* (1941), *supra* note 1, at 715).

39 *See generally* Juraj Andrassy, *Les Relations Internationales de Voisinage*, 79 RECUEIL DES COURS (1951–II) 77 (1952).

40 It is true, as John Knox points out, that the U.S. Supreme Court also recognizes that it does not have "quite the same freedom to balance the harm that will be done by an injunction against that of which the plaintiff complains that it would have in deciding between two [private citizens in the same state]." *Trail Smelter* (1941), *supra* note 1, at 716 (*quoting* Georgia v. Tennessee Copper Co., 206 U.S. 230 (1907)). This would certainly have been welcomed by the United States. The injured private parties did not fare so well in the well-known case of Boomer v. Atlantic Cement Co., 26 N.Y.2d 219, 309 N.Y.S.2d 312, 257 N.E.2d 870 (1970), where the New York court granted

THE PROGENY

In any event, the strictness of the Tribunal's statement of the governing principle has been attenuated considerably over the years in the wide array of progeny of the *Trail Smelter* arbitration. The interesting thing one notices when following the arbitration's "family tree," as it were, is that its descendants seem to be arranged in two main branches: the political branch, on the one hand, and the expert and judicial branch, on the other. The "political" side of the family tends to juxtapose a liberalized formulation of the Tribunal's rule with a reaffirmation of the very sovereignty that brought about the clash that confronted the Tribunal in the first place. The expert and judicial side typically omits a reference to sovereignty. A few examples, out of the many members of both sides of this family, will have to suffice.

On the political side, the classic reformulation of the Tribunal's formula is found in Principle 21 of the 1972 Stockholm Declaration on the Human Environment. It reads:

> States have, in accordance with the Charter of the United Nations and the principles of international law, the sovereign right to exploit their own resources pursuant to their own environmental policies, and the responsibility to ensure that activities within their jurisdiction or control do not cause damage to the environment of other States or of areas beyond the limits of national jurisdiction.[41]

This principle was restated, virtually unchanged, twenty years later at the Earth Summit in Rio de Janeiro. The first limb of the principle, the "sovereignty" limb, could have been advanced by either Canada or the United States in the *Trail Smelter* arbitration. The second, or "no-environmental-damage" limb, attempts to cut the Gordian knot. Unlike the Tribunal's rule-formulation, Principle 21 does not, in terms, require the damage to be "serious"; in fact, it uses no adjective at all.[42] Likewise, Principle 21 says nothing about proof, leaving that to be worked out by the states concerned. Finally, a new element, the "environment," appears in this reformulation. However, although at the time of the *Trail Smelter* arbitration that term was not used in the way we use it today, the concept could be considered to be included within the term "territory" in the Tribunal's rule-statement.

Principle 21 has its own progeny. It is incorporated verbatim in Article 3 of the Convention on Biological Diversity[43] and in Principle 1(a) of the Statement of

an injunction against the cement plant but provided that it would be vacated on payment of permanent damages as determined by the trail court.

41 Adopted June 16, 1972, at the United Nations Conference on the Human Environment, Stockholm, *Report of the U.N. Conference on the Human Environment*, U.N. Doc. A/CONF.48/14/Rev. 1 at 3 (1973), U.N. Doc. A/CONF.48/14 at 2 and Corr.1 (1972), *reprinted in* 11International Legal Materials 1416 (1972).

42 It is true, as Günther Handl points out, that the U.N. General Assembly thereafter adopted a "gloss" on Principle 21, explaining that it contemplated "significant" damage. *See* G.A. Res. 2994, U.N. GAOR, 27th Sess., Supp. No. 30, at 42, U.N. Doc A/8730 (1972).

43 Convention on Biological Diversity, June 5, 1992, 1760 U.N.T.S. 143, *reprinted in* 31 International Legal Materials 818 (1992).

Forest Principles.[44] Finally, both the Desertification Convention[45] and the Framework Convention on Climate Change[46] reproduce Principle 2 of the Rio Declaration verbatim in their preambles. All of these texts, negotiated and adopted on the political level, contain both the "sovereignty" limb and the "no-environmental-damage" limb.

On the expert and judicial side of the family, pride of place should be accorded to the International Court of Justice. The Court, being in principle free from the political constraints that influenced the framing of Principle 21 and its progeny, was free to detach the obligation from any mention of sovereignty. First in its Advisory Opinion in *Legality of the Threat or Use of Nuclear Weapons*[47] and later in its judgment in the case concerning the *Gabčíkovo-Nagymaros Project*,[48] the Court stated that: "The existence of the general obligation of States to ensure that activities within their jurisdiction and control respect the environment of other States or of areas beyond national control is now part of the corpus of international law relating to the environment." It seems clear that this general obligation may be traced directly back to the *Trail Smelter* arbitration. There is also little doubt that because its existence was recognized by the International Court of Justice itself,[49] the principle now forms part of customary international law. That the Court applied it in the context of both nuclear weapons and international watercourses speaks volumes about the principle's utility and fundamental nature.[50]

The expert limb of the "expert and judicial" branch of the family is perhaps the most vigorous and fecund. Again, pride of place should perhaps be given to a United Nations expert body, the International Law Commission (ILC). The ILC, whose purpose is to codify and progressively develop international law, prepares drafts on topics that are referred to it by the General Assembly. The draft texts have often formed the basis of treaties. Examples include the four 1958 Geneva Conventions on the Law of the Sea; the Vienna Conventions on Diplomatic and Consular Relations; and the Vienna Convention on the Law of Treaties. The *Trail Smelter* arbitration could actually be said to have spawned one of the topics

44 Statement of Principles for a Global Consensus on the Management, Conservation and Sustainable Development of All Types of Forests, Aug. 14, 1992, 3 Report of the UNCED, UN Doc. A/CONF. 151/26, Annex III (1992) *reprinted in* 31 International Legal Materials 881 (1992).

45 Convention to Combat Desertification in Those Countries Experiencing Serious Drought and/or Desertification, Particularly in Africa, Paris, June 17, 1994, fifteenth preambular paragraph, U.N. Doc. A/AC.241/15/Rev.3, *reprinted in* 33 International Legal Materials 1332 (1994).

46 Framework Convention on Climate Change, May 9, 1992, 1771 U.N.T.S. 107, *reprinted in* 31 International Legal Materials 849 (1992).

47 Nuclear Weapons, 1996 I.C.J. 226, at 241–42, para. 29 (July 1996).

48 Gabčíkovo-Nagymaros Project (Hun./Slov.), 1997 I.C.J. 7, at 41, para. 53 (Sept. 25).

49 *See also, The Corfu Channel* case, in which the I.C.J. referred to the duty of every state "not to allow knowingly its territory to be used for acts contrary to the rights of other states." Corfu Channel (U.K. v. Albania), 1949 I.C.J. 4 (April 1949).

50 This is the challenge taken up by Part Two of this volume, which explores the potential relevance of the *Trail Smelter* principles to a range of transboundary issues not connected with the environment.

the ILC has worked on, or perhaps I should say "labored" on, since it has been before the Commission for well over twenty years. The topic has the rather pedantic title of International Liability for Injurious Consequences Arising Out of Acts Not Prohibited by International Law. Liability for harm caused by a nonprohibited act is a strange concept indeed, at least governments seem to think so. And yet, that is precisely what the *Trail Smelter* Tribunal provided for in Part Five of its award, namely, the possibility that the smelter would pay an indemnity for any damage occurring in Washington, notwithstanding compliance with the regime provided for by the tribunal.[51] The ILC in 2001 adopted draft articles on a portion of the topic, Prevention of Transboundary Harm from Hazardous Activities.[52] Article 3 of that text, which is entitled simply, *Prevention*, reads as follows: "The State of origin [of activities involving a risk of transboundary harm] shall take all appropriate measures to prevent significant transboundary harm or at any event to minimize the risk thereof."[53] This is certainly a refinement of the *Trail Smelter* principle but it flows rather obviously from the rule formulated by the Tribunal.

I had the privilege of serving on the ILC for ten years, during which significant progress was made on the International Liability draft. For five of those years, one of the members of the Commission was Alan Beesley, from Canada. I have checked the reports of the ILC to verify that the recollection I had about Ambassador Beesley's statements on this topic was correct, and indeed it was: virtually every year, in speaking on this topic, Ambassador Beesley would invoke either the *Trail Smelter* arbitration or Principle 21, or both. He would usually refer to these as proof that there was indeed authority, and state practice, bearing upon the topic of Liability for Non-Prohibited Acts.[54] He also referred regularly to the 1982 UN Convention on the Law of the Sea, notably articles 192 and 193, in whose negotiation he was closely involved, which he characterized as implementing Principle 21, in whose drafting Ambassador Beesley also played a role. Article 193 would indeed appear to be a somewhat more elegant version of Principle 21 as applied to the marine environment. It provides that "States have the sovereign right to exploit their natural resources pursuant to their environmental policies and in accordance with their duty to protect and preserve the marine environment."[55]

I could go on at length about the influence of the *Trail Smelter* arbitration on the work of the ILC, but I will confine myself to mentioning a provision of a draft text that is near and dear to my heart, that on the Law of the Non-Navigational

[51] *Trail Smelter* (1941), *supra* note 1, at 733. For further discussion of the ramifications of the *Trail Smelter* arbitration for the state liability project of the ILC, *see* Drumbl in this volume.

[52] 2001 I.L.C Rep. 370 (2001).

[53] *Id.* at 372.

[54] *See, e.g.*, 1 YEARBOOK OF THE INTERNATIONAL LAW COMMISSION 174 (1987).

[55] United Nations Convention on the Law of the Sea, 10 Dec. 1982, art. 193, U.N. Doc. A/CONF/62/122, *reprinted in* 21 INTERNATIONAL LEGAL MATERIALS 1261 (1983).

Uses of International Watercourses. This text was adopted by the ILC in final form in 1994 and formed the basis for the 1997 United Nations Convention on the same subject. Article 7 of both the ILC's draft and the Convention provides that riparian states must "take all appropriate measures to prevent the causing of significant harm to other [riparian] States."[56] Although the context is different, and in the case of a watercourse harm may be caused by taking something out of the medium rather than putting something into it, the influence of the *Trail Smelter* Tribunal's rule-formulation on Article 7 of the UN Watercourses Convention is clear.

A second example of work of the "expert limb" of the "expert and judicial" branch of the family is the final Report of the Brundtland Commission's Experts Group on Environmental Law, published in 1987 in conjunction with the report of the Brundtland Commission itself.[57] As a member of the Experts Group, I can testify to the importance of the *Trail Smelter* arbitration in the Group's deliberations. Article 10 of the Experts Group Report provides that: "States shall . . . prevent or abate any transboundary environmental interference or a significant risk thereof which causes substantial harm, i.e. harm which is not minor or insignificant."[58] As a draft that attempts to codify legal principles for sustainable development, this text represents a significant updating of the *Trail Smelter* principle.[59] Interestingly, it uses the adjective "substantial," which would normally be interpreted to be closer to "serious" than to "significant," despite the negative definition contained in the article.

A final example of work of the "expert" limb of the "expert and judicial" branch of the *Trail Smelter* "family tree" is the International Covenant on Environment and Development, prepared by the Commission on Environmental Law of the World Conservation Union (IUCN), in cooperation with the International Council of Environmental Law. The Covenant, in whose drafting I was privileged to have been involved, was launched at the United Nations Congress on Public International Law, held at United Nations Headquarters in New York in March, 1995. The commentary to the Covenant explains that: "The Duty to prevent harm at the transboundary level has deep roots in customary international law and indeed it finds expression in nearly every provision of the draft Covenant. It is inherent in the *Trail Smelter* arbitral decision . . ."[60] This statement, that the principle for which the *Trail Smelter* arbitration stands runs throughout the entire Covenant, is a testament to the view of major environmental NGOs as to the influence of *Trail Smelter* on modern international environmental law.

[56] Art. 7(1), U.N. Doc. A/RES/51/869, 21 May 1997, *reprinted in* 36 INTERNATIONAL LEGAL MATERIALS 700 (1997).

[57] Brundtland Commission, Environmental Protection and Sustainable Development (1987).

[58] *Id.* at 75.

[59] *See* Jacobson in this volume.

[60] Commission on Environmental Law of the IUCN World Conservation Union, in cooperation with International Council of Environmental Law, International Covenant on Environment and Development 39 (March 1995).

CONCLUSION

The arbitrators of the *Trail Smelter* arbitration were courageous and creative. Still, it is clear why the rule-statement in the *Trail Smelter* arbitration may read a bit strangely through today's lenses. Whether praised as bold and visionary or dismissed as narrow and compromising, even a glimpse of a tiny portion of the rich tapestry of law and legal doctrine that has been woven of the threads of the *Trail Smelter* arbitration confirms this much: This was a remarkable decision indeed.

4 Pollution by Analogy: The *Trail Smelter* Arbitration [Abridged]

Alfred P. Rubin

Where there's muck, there's brass.
– Yorkshire Folksaying.

Every discussion of the general international law relating to pollution starts, and most end, with a mention of the *Trail Smelter* arbitration[1] between the United States and Canada. For example, in the American Law Institute's *Restatement (Second) of the Foreign Relations Law of the United States*, the only precedent cited on the topic of a state's liability to another in connection with pollution is the *Trail Smelter* arbitration.[2] Such heavy reliance on a single precedent breeds overstatement as analysts attempt to reinterpret the case to fit various hypothetical circumstances and new cases. Frequently, the precedent can be applied only by raising it to a level of abstraction far beyond the range of its logic. In the *Restatement* itself, the proposition that the *Trail Smelter* arbitration *is* cited to support is:

> The relation of cause to effect underlies the parallel principle that a state may be held responsible under international law for damage which it causes in the territory of another state. Thus Canada was held responsible to the United States under international law for the production of fumes in Canada which polluted the air in the United States.[3]

Professor of Law, University of Oregon, B. A. (1952), J. D. (1957), Columbia University; M. Litt. (1963), Cambridge University.

[1] *Trail Smelter Arbitral Decision*, 35 AMERICAN JOURNAL OF INTERNATIONAL LAW 684 (1941) [hereinafter "*Trail Smelter* (1941)"].

[2] Restatement (Second) of the Foreign Relations Law of the United States, § Explanatory Note 3 (1965).

[3] *Id.*

[Excerpt reprinted with permission of the author and the University of Oregon Law Review. The article originally appeared as: Originally appeared as Alfred P. Rubin, *Pollution by Analogy: The Trail Smelter Arbitration*, 50 OREGON LAW REVIEW 259 (1971)]

In fact, as will be seen, the arbitration did not hold that polluting the air in the United States was the basis of Canadian liability. But more of that later.

It is clear that the time has come to reexamine the precedential value of the *Trail Smelter* arbitration, to restore it to its rightful place as a landmark case in a continuum of developing law and remove from its shoulders the burdens heaped on it by two generations of publicists.

* * *

THE TRIBUNAL FINDS THE LAW

The Question

The parts of the Tribunal's final decision that consider general pollution problems and that have been subsequently cited by publicists concerned with these problems are contained in the Tribunal's final answers to Questions 2 and 3. The key passages are: the statement that, as a principle of international law, each State has a duty to prevent its territory from being used as a source of pollutants causing damage to the territory, persons, or property of another State; and the portion of the opinion establishing a regime to enforce the duty established by the foregoing statement.

The threshold question was whether the law that the *compromis* directed the Tribunal to apply required a state to refrain from permitting acts within its territory that (1) had *any effect* on foreign territory, *(2) risked doing damage* on foreign territory, or (3) whether *actual damage was* to *be* considered an essential element of the international tort. If the last, then Canada could permit the smelter to continue operations if it kept its emissions below the injurious level. This could be accomplished by lowering the permissible level of air pollution below the level at which injury is caused, to be scientifically determined at some later time, with occasional lapses to be the subject of *ad hoc* settlement or some grievance procedure aimed merely at settling claims rather than alleviating the cause of the claim. If risk were to be avoided, then the entire operation would have had to be reviewed, and might have been closed down. If an absolute prohibition on incursions into the territory of the United States were maintained, the smelter would have been forced to cease operations immediately.

The Answer

As to the substantive issue of law, it was noted earlier that the Tribunal ignored the fact that the two governments had decided on the principle of Canadian liability in agreeing to the *compromis*, which had directed the Tribunal to assess an indemnity to be paid by Canada for whatever "damage" had been done by the Trail smelter in the State of Washington after January 1, 1932. Instead, it seized on the question of whether the smelter should be restrained from causing future damage in the State of Washington and found, as a matter of general international law, that Canada was responsible for the acts of the smelter to the extent fumes emitted therefrom entered U.S. airspace and caused tangible injury for which "damages" were a remedy under U.S. law. In the words of the Tribunal, the nub of the substantive holding was that "under the principles of international law . . . no State has the right to use or permit the use of its territory in such a manner as to cause injury by fumes on or to the territory of another or the properties or persons therein, when the case is of serious consequence and the injury is established by clear and convincing evidence."[4] How did the Tribunal reach that conclusion?

The Tribunal found no international law precedents dealing with air pollution. It found the nearest analogy to be that of water pollution, but found no decision of an international tribunal in that area either. One Swiss Federal Court case was found which dealt with the risks created by one canton maintaining a shooting range near its boundary with another. The Swiss court permitted the range to continue in operation despite the risk to the neighboring canton. But in doing so the court had applied Swiss constitutional law, holding that there was a federal duty to maintain shooting ranges for certain purposes, and that no greater precautions were required with regard to such ranges near the edge of a canton than in the interior. The *Trail Smelter* Tribunal felt that this precedent was not persuasive with regard to the case before it.[5]

The Tribunal found, on no discernable evidence and with no discussion, that

> the law followed in the United States in dealing with the quasi-sovereign rights of the States of the Union, in the matter of air pollution, whilst more definite, is in conformity with the general rules of international law.[6]

On sounder ground, less cavalier with regard to implying knowledge of "the general rules of international law" in the absence of precedent or any examination of state practice, the Tribunal found another reason for applying the precedents of the United States:

> [F]or it is reasonable to follow by analogy, in international cases, precedents established by [the Supreme Court of the United States] in dealing with controversies between States of the Union or with other controversies concerning the quasi-sovereign rights of such States, where no contrary rule prevails in international law and no reason for rejecting such precedents can be adduced from the limitations of sovereignty inherent in the Constitution of the United States.[7]

It is noteworthy that in the entire discussion, the Tribunal did not cite the provision of article IV of the *compromis*, directing it to "apply the law and practice

4 *Trail Smelter* (1941), *supra* note 1.

5 *Id.*

6 *Id.*

7 *Id.*

followed in dealing with cognate questions in the United States of America as well as international law and practice" as a basis for applying U.S. law to the substantive law of international air pollution. Because the Tribunal had ruled that reference to the law and practice of the United States was relevant only to defining the substance of "damage" (but not whether "damages" included intangibles and the costs of preparing the case), and did not purport to apply the law and practice of the United States as such to the substantive issue of Canada's rights to permit air pollutants to escape to neighboring territorial airspace, the Tribunal seems to have acted as an authoritative arbiter of international law, notwithstanding the more restrictive language of the *compromis*. Indeed, after examining the "analogous" precedents of the United States with regard to water pollution,[8] and two interrelated air pollution cases,[9] the Tribunal held "that the Dominion of Canada is responsible in international law for the conduct of the Trail smelter"[10] and that the obligation of Canada extended only to requiring the smelter to refrain from "causing any damage through fumes in the State of Washington."[11] The word "damage" was purportedly defined as "such as would be recoverable under the decisions of the courts of the United States in suits between private individuals,"[12] but it seems clear that the Tribunal was in fact not defining damage at all with this language, but defining "damages" the extent to which there should be monetary recovery for "damage." The importance of this confusion in language, and therefore in logic, cannot be emphasized too strongly.

Canadian responsibility for the smelter in the international arena might have been regarded as settled when Canada agreed to the *compromis*. This proposition might have been implied from the wording of the *compromis*, appearing to limit the U.S. claim to a narrow claim for an indemnity based upon American concepts of "damages" (which might, of course, be considerably less than the full claim the United States might have pressed if it had wanted to vindicate its full rights against Canada in international law). But the Tribunal did not reach its result by reading the *compromis* closely. It reached its result by applying U.S. constitutional law precedents to the international law field by analogy. Thus, the Tribunal apparently ignored the positive state practice of the United States and Canada with regard to the special law applicable on their shared border as reflected in the *compromis*, and instead pronounced on questions of general international law. The substantive rules applied as a matter of general international law were not found by any analysis of international precedents or practice, but by the same evidence that would have been used had the Tribunal been applying its powers more narrowly; the extension

[8] New Jersey v. City of New York, 283 U.S. 473 (1931); New York v. New Jersey, 256 U.S. 296 (1921); Missouri v. Illinois, 200 U.S. 496 (1906); Kansas v. Colorado, 185 U.S. 125 (1902).

[9] Georgia v. Tennessee Copper Co., 237 U.S. 474 (1915); Georgia v. Tennessee Copper Co., 206 U.S. 230 (1907).

[10] *Trail Smelter* (1941), *supra* note 1.

[11] *Id.*

[12] *Id.*

of U.S. precedents to the international law area was done by analogy, intuitively, and without any examination of the fitness of those rules to the essentially different circumstances of international life.

The Analogy in Theory

The circumstances of international life are essentially different from the circumstances of states within a single federal system. Federal systems are characterized by the surrender of "sovereign" rights by constituent states to the central government, while in an international setting; there is no central government to which rights can be surrendered. In the United States, for example, the power to regulate interstate commerce is allocated to the federal government by the Constitution and has been held to include the power to regulate aspects of the manufacturing process having only a tenuous connection with the movement of goods between states."[13] The analogy to the rights and duties of states at international law is far from clear.[14] Such states are not bound to permit goods to flow freely across their borders. Nor are they bound to a procedure for the resolution of disputes among themselves or between themselves and a central authority possessing a defined set of powers.

The analogy between federal and international law becomes even more strained when the peculiar jurisprudence of the U.S. Supreme Court in cases involving disputes between the states is considered. As noted earlier, Canadian legal experts felt U.S. precedents were more favorable to their cause in the *Trail Smelter* dispute than their own. But these precedents were incorrectly interpreted, both by the Canadian experts and by the Tribunal itself. The confusion arose out of a statement by Justice Holmes in *Missouri v. Illinois*:[15] "Before this court ought to intervene [in a dispute between two states] the case should be of serious magnitude . . . "[16] This language was embellished on in *New York v. New Jersey*,[17] where Justice Clarke, speaking for a unanimous Court, indicated that such reticence was not compelled because the comity among states made minor trespasses *de minimis* as a matter of international law or constitutional law, but because the preservation of the delicate balance between state and federal power struck by the Constitution demanded that the Court, as an organ of the federal government, refrain from intervening in disputes between individual states except in extraordinary cases. It should be emphasized that the balance to be preserved was that between "sovereign" states – analogous to the relationship between independent nations at international law. In the words of the Court:

> Before this Court can be moved to exercise its extraordinary power under the Constitution to control the conduct of one state at the suit of another, the threatened invasion of rights must be of serious magnitude . . . [18]

[13] E.g., Wickard v. Filburn, 317 U.S. 111 (1942).

[14] [footnote omitted]

[15] 200 U.S. 496 (1906).

[16] 200 U.S. at 521.

[17] 256 U.S. 296 (1921).

[18] *Id.* at 309.

Nonetheless, the *Trail Smelter* Tribunal quoted this same language in its Final Decision and concluded:

> What the Supreme Court says there of its power under the Constitution equally applies to the extraordinary power granted this Tribunal under the Convention. What is true between States of the Union is, at least, equally true concerning the relations between the United States and the Dominion of Canada.[19]

Yet unlike states of the Union, which are bound by the decisions of the Supreme Court, the United States and Canada could have agreed to ignore the Tribunal's decision. And it is clear that the Court in *New York v. New Jersey* was strongly influenced by the fact that its decision would be binding on the parties before it. It follows that the reasons the U.S. Supreme Court treads gingerly in controversies between individual states do not apply with equal force in international controversies submitted by independent states to *ad hoc* tribunals. The fact that the U.S. Supreme Court feels free to draw on international law precedents when it deems them sufficiently analogous to the situation before it does not, standing alone, justify emulation by international tribunals; before U.S. precedent can be applied to international disputes, it must first be ascertained that the precedent did not hinge upon a characteristic relationship unique to a federal union. The *Trail Smelter* Tribunal, it is obvious, did not even consider the point.

In addition to exaggerating its own power, ignoring the fact that the U.S. Supreme Court is an arm of a federal government rather than an *ad hoc* tribunal, and misconstruing the Court's decisions by failing to see that they rested on a web of federal state relations totally unlike the relations among the Tribunal, the United States and Canada, the Tribunal committed an even more obvious error in applying the decisions of the U.S. Supreme Court. In quoting from the Supreme Court's decision in *Georgia v. Tennessee Copper Co.*, the Tribunal used the elegant prose of Justice Holmes to support its argument that it had the power to grant a limited relief less than complete abatement,[20] but the argument of Justice Holmes was actually going to a different point: Whether the Supreme Court had any authority at all under the U.S. Constitution to grant relief by enjoining state activity. The question of granting a complete abatement of the nuisance was not an issue in the Supreme Court opinions cited. The issue in the parts of the U.S. decisions cited by the Tribunal related to the extent of federal power, not to the rights of individual states against each other.

The "Just" Solution

At any rate, the Tribunal itself accepted the analogy, expressed itself as pronouncing a rule of international law, and proceeded to determine the future regime to be applied to the smelter. In doing so, it implied that a low level of sulfur dioxide flowing from Canada to the United States would not involve violations of Canada's obligations to the United States under general international law. The possibility

[19] *Trail Smelter* (1941), *supra* note 1.

[20] *Id.*

that the permitted operations might indeed cause some damage in the United States in the future was foreseen, and the Tribunal indicated that an indemnity for the damage caused by the smelter ought to be paid in that event.[21]

There can be no doubt that the final decision of the Tribunal embodied practical results sought by the two governments. The *compromis* had explicitly directed the Tribunal to "give consideration to the desire of the high contracting parties to reach a solution just to all parties concerned."[22] This platitude was interpreted by the Tribunal to indicate that the parties would not stand on absolute rights as sovereigns, but were concerned with balancing the interests of, in the Tribunal's phrases, "the agricultural community" with "the interest of industry."[23] The Tribunal saw its role as being to "endeavor to adjust the conflicting interests by some just solution which would allow the continuance of the operation of the Trail smelter but under such restrictions and limitations as would, as far as foreseeable, prevent damage in the United States . . . "[24]

Wilfred Jenks has pointed out that the arbitration decision is "a leading authority concerning the role of equity in international adjudication" as well as being a leading authority in the international law of tort.[25] It would seem from the foregoing review that its role in the law of torts has been far greater than a careful analysis of the case should allow. Its role in applying principles of equity to international disputes was justified by express language in the Tribunal's *compromis* rather than precedent. Thus, its worth as precedent appears questionable.

The former Legal Adviser to the Canadian Ministry of External Affairs and later member of the International Court of Justice, John E. Read, has noted that the real reason for the decision allowing nondamaging fumes to flow from Canada to the United States was that a different decision would have been applicable to activities at Detroit, Buffalo, and Niagara Falls, to the obvious discomfort of the United States.[26] Apparently, the United States did not argue the absolute inviolability of territorial airspace for that reason, and the Tribunal did not raise the point on its own. But these considerations relate to the special circumstance presented by the border between Canada and the United States, and to the extent the Tribunal's decision was affected by that circumstance – or by the tacit agreement of the parties that a compromise solution was more desirable than a full vindication of rights – it is difficult to accord the decision a persuasive weight in general international law.

The Results

Limiting the prohibition against the flow of air pollutants across international boundaries to that which causes tangible injury translatable into provable money damages, the Tribunal's proposed *régime* for the Trail smelter was accepted by

[21] *Id.*
[22] *Id.*
[23] *Id.*
[24] *Id.*
[25] C. Wilfred Jenks, The Prospects of International Adjudication 408 (1964).
[26] John E. Read, *The Trail Smelter Dispute*, 1 Canadian Yearbook of International Law 213, 224–25 (1963). Republished in this volume.

both governments and complied with by the Canadian firm at a cost of some $20,000,000. The Canadian Consolidated Mining and Smelting Company eventually succeeded in selling the products of its smoke abatement program for substantial profit.[27]

Whether or not the opinion of the Tribunal in the *Trail Smelter* arbitration withstands close legal analysis, it has been widely accepted as the foundation of the current general international law of air pollution. One result of this acceptance has been to bring some certainty into the question of polluting activities. It is a certainty that favors the polluter in permitting his polluting activities to continue as long as they do not cause "damage" in the sense of direct injury measurable in money terms to the industrial or agricultural production of a second state. This formulation emphasizing the "damage" aspect of the decision deserves closer analysis than it has yet received.

It often has been noted that the Tribunal applied a rule of strict liability to the action (or inaction) of the state permitting emission from its territory to escape; that is, the United States was not required to prove negligence (much less design) before Canadian responsibility was engaged.[28] But this apparent tightening of the rule of state responsibility must be contrasted with the loosening that resulted from replacing the absolute rule forbidding physical trespass normally applicable in international law with a rule permitting trespasses as long as they do not do injury for which the 1941 federal law of the United States assessed "damages." In a sense, it might be said that the general international law of trespass was replaced by the American law of nuisance.

It is clear that the rules of international law relating to the permissible actions or inactions of a state affecting the territory of another state are extremely complex. Some gross extensions of state authority, such as the sending of military aircraft or troops into foreign territory without the permission of the sovereign whose territorial integrity is being violated, are likely to be viewed as, illegal no matter how "peaceful" the intention or how minimal the "trespass."[29] Other extensions of state authority evoke less certain reactions. For example, is it a violation of international law to support espionage activities in foreign territory? Or to apply legislation to foreign nationals abroad whose activities have a general impact on the economy of a state?[30] Or, to phrase the same issue in a novel way, to permit one's own nationals to act in such a way that their activities affect the general economy of a second state without the permission of that state?

27 *Id.* at 221.

28 Hardy, *International Protection Against Nuclear Risks*, 10 INTERNATIONAL & COMPARATIVE LAW QUARTERLY 739, 751 (1961).

29 E.g., the U-2 incident of 1960. *See* Lissitzyn, *Some Legal Implications of the U-2 and RB-47 Incidents*, 57 AMERICAN JOURNAL OF INTERNATIONAL LAW 135 (1962). This is not to say that some incursions would not be justifiable either in reprisal or self-defense.

30 *See* Jennings, *Extraterritorial Jurisdiction and the United State Antitrust Laws*, 33 BRITISH YEARBOOK OF INTERNATIONAL LAW 146 (1957); Baxter, *So-Called "Unprivileged Belligerency"; Spies, Guerillas and Saboteurs*, 28 BRITISH YEARBOOK OF INTERNATIONAL LAW 323 (1951).

According to the Tribunal in the *Trail Smelter* arbitration and the commentators following its lead, a state would have no hope of successfully presenting a claim with regard to actual physical emissions entering the claimant's territory unless there were actual, provable, substantial, physical damage to a traditional interest. Indeed, the Tribunal explicitly rejected that part of the claim of the United States based on Canadian violations of U.S. sovereignty without tangible damage.[31] Although the basis for this rejection in the first decision was ostensibly a close reading of the arbitration *compromis*, the Tribunal's formulation of its general rules relating to the international law of pollution, which it went to some length to indicate it was declaring in its Final Decision, speaks of responsibility for damage defined in terms of some money recovery; that is, damages, and the need to demonstrate their substantiality. Thus, under the rules pronounced by the Tribunal, there would appear to be no international responsibility at all for acts of pollution not giving rise to tangible monetary injury.

What money damages result from so polluting the air that eagles are endangered as a species? Or sparrows? What is the present value of saplings whose market value can not be determined for another fifty years? Of wild flowers? Must injury be suffered before international responsibility is fixed? That seems to be the rule of the *Trail Smelter* arbitration, which further holds that if the injury cannot be measured in monetary terms there is no "damage" and hence no remedy at international law.

The problems faced by a claimant state in international law seeking to protect its environment from air pollution appear comparable to the problems of "standing" in U.S. courts. In both cases, the claimant must show injury to his legally protected interest.[32] But standing has to do with the claimant's right to be heard, whereas the *Trail Smelter* rule goes to the merits of the controversy. If the injury asserted is not measurable in terms of the productive economy of the United States as of 1941, it seems that the general international law expressed in the *Trail Smelter* arbitration will deny the very existence of a legal wrong. This would appear to be true even when the polluting activity threatens nonproductive, nonmonetary interests of concern to the entire world outside the state permitting the pollution to cross its boundaries. Nonetheless, the need to show damage to legally protected interests under both U.S. and international law seems to reflect similar insensitivity to issues not directly related to traditional property interests.

CONCLUSION

It may be concluded that the rule of the *Trail Smelter* arbitration, forbidding pollution only to the extent the emissions do "damage" in a second state, is founded

[31] *Trail Smelter* (1941), supra note 1.

[32] *See* Davis, *The Liberalized Law of Standing*, 37 UNIVERSITY OF CHICAGO LAW REVIEW 450 (1970); Jaffe, *Standing Again*, 84 HARVARD LAW REVIEW 633 (1971).

on a mistaken analogy to constitutional law practices in the United States, and has results in the international law area that are anomalous. A reexamination of the rule makes it possible to reduce its apparent permissiveness in the light of current information as to the nature of "damage" that pollution wreaks, and in the light of developing techniques to maintain industrial production without the degree of pollution that was felt to be inseparable from production at the time of the *Trail Smelter* arbitration. In view of the likely resistance by many states to any restriction on their industrial development regardless of the pollution involved, negotiation to share the costs of pollution control may be necessary. The first step in bringing about a willingness on the part of industry-poor countries to negotiate may be a greater emphasis on the limitations of the rule of the *Trail Smelter* arbitration as a basis for asserting a right in international law to continue polluting activities up to a threshold of "damage" based on the needs of agriculture of the United States in 1941 and the contemporaneous pollution levels of the Canadian neighbors of the industrial city of Detroit.

The concomitant problem of making industrialized countries willing to help pay for the higher standards of cleanliness sought to be imposed is a political problem that the law cannot help solve beyond pointing out that unless the industrialized countries recognize such an obligation, their views as to the law limiting polluting activities are not likely to be found persuasive by the nonindustrialized members of the world community, to the damage of us all.

5 Has International Law Outgrown *Trail Smelter*?

Jaye Ellis

INTRODUCTION

The *Trail Smelter* arbitration[1] occupies a strange place in international environmental law.[2] It forms part of the canon of the discipline – indeed, it is generally regarded as a foundational case – and reference to it is ubiquitous.[3] Yet there is little agreement on how its holding should be interpreted and applied, and its persuasive value is often called into question. *Trail Smelter* cannot be reconciled with international rules of state responsibility,[4] and the principle for which it appears to stand, namely, that states can be liable for transboundary environmental damage in the absence of a breach of an international obligation, holds very dubious status in international law. International environmental disputes are very rarely litigated; instead, states have invested their efforts in the creation of often very elaborate legal and policy regimes,[5] many of which contain their own internal mechanisms for implementation, compliance, and enforcement. These considerations tend to lead one to the conclusion that international environmental law has moved beyond the *Trail Smelter* arbitration.

Regardless of these difficulties, the influence of the case on subsequent developments in international environmental law is not far to seek. Principle 21 of the Stockholm Declaration, which has itself taken on almost mythical proportions, is generally regarded as a reflection or restatement of the *Trail Smelter* holding, and the case has been relied on heavily in the work of the International

[1] *Trail Smelter Arbitral Decision*, 35 AMERICAN JOURNAL OF INTERNATIONAL LAW 684 (1941) [hereinafter "*Trail Smelter* (1941)"]. *See* Annex to this volume.

[2] Karin Mickelson describes the position of the case as "prominent but somewhat mysterious." Karin Mickelson, *Rereading* Trail Smelter, 31 CANADIAN YEARBOOK OF INTERNATIONAL LAW 219 (1993). Republished in this volume.

[3] Alfred P. Rubin states: "Every discussion of the general international law relating to pollution starts, and must end, with a mention of the *Trail Smelter Arbitration* between the United States and Canada." Alfred P. Rubin, *Pollution by Analogy: The* Trail Smelter *Arbitration*, 50 OREGON LAW REVIEW 259 (1971). Republished in this volume.

[4] *See* Drumbl in this volume.

[5] *See* Miller in this volume.

Law Commission (ILC) concerning liability for harm caused by acts not prohibited by international law.[6] Given these contradictory responses, the question arises whether the *Trail Smelter* arbitration really has anything to teach us about international environmental law, and if so, what lessons we should take from it.

BACKGROUND: THE *TRAIL SMELTER* ARBITRATION

The *Trail Smelter* arbitration arose from smoke and fumes emitted by a smelter owned and operated by the Consolidated Mining and Smelting Company of Canada in Trail, B.C., and causing damage in the state of Washington.[7] An earlier attempt to resolve the dispute through the International Joint Commission was unsuccessful,[8] and the United States and Canada decided to submit the matter to arbitration.[9]

The *compromis* concluded by the two parties indicates that they were well aware of the delicate balance of interests that was at play. The parties did not appear to have in mind a black-and-white appreciation of the situation in terms of Canada's right to allow smelters to operate or the U.S. right to be free of transboundary pollution. Rather, as the wording of the questions makes clear, the parties contemplated the possibility that a pollution control regime would not eliminate transboundary flows of pollutants and that compensation could be an acceptable response to transboundary pollution damage.[10] This is more in line with an approach to the problem grounded in the law of nuisance,[11] which, although it has not won acceptance in international law,[12] does have the advantage of focusing attention on the need to balance the interests of the parties and focus

6 *See* McCaffrey and Drumbl in this volume.

7 *See Trail Smelter Arbitral Decision*, 33 AMERICAN JOURNAL OF INTERNATIONAL LAW 182, 184 (1939) [hereinafter "*Trail Smelter* (1939)"]. *See* Annex to this volume. For an extensive discussion of the factual background to the dispute, *see* JOHN D. WIRTH, SMELTER SMOKE IN NORTH AMERICA (2000).

8 *See Trail Smelter* (1939), *supra* note 7, at 191; John E. Read, *The* Trail Smelter *Dispute*, 1 CANADIAN YEARBOOK OF INTERNATIONAL LAW 213, 214 (1963). Republished in this volume.

9 *See* Allum in this volume.

10 Samuel Bleicher, *An Overview of International Environmental Regulation*, 2 ECOLOGY LAW QUARTERLY 1, 20–1 (1972).

11 Rubin, *supra* note 3, at 273; Read, *supra* note 8, at 214; L. F. E. Goldie, *Liability for Damage and the Progressive Development of International Law*, 14 INTERNATIONAL & COMPARATIVE LAW QUARTERLY 1189, 1227–9 (1965).

12 *See, e.g.*, Goldie, *supra* note 11, at 1228–29. Goldie argues that the *Trail Smelter* decision is not sufficient authority to establish nuisance as a principle of international law, and that, in any event, nuisance is not a good candidate for status as a general principle of international law. Rubin has also cast doubt on the precedential value of the holding. Rubin, *supra* note 3, at 271. Bleicher has noted that the holding goes "well beyond the traditional substantive principle of nuisance, which would allow a balancing of the benefits of the defendant's activities against the damages suffered by the plaintiff." Bleicher, *supra* note 10, at 28–29.

carefully on the facts of the individual case and the context from which it arises.[13] In any event, a conclusion by the Tribunal that transboundary pollution was illegal and would have to be eliminated would not be in the long-term interests of the United States: as one commentator has noted, "[t]he acceptance of the principle of absolute cessation of damage might have shut down the Trail smelter; but it would also have brought Detroit, Buffalo and Niagara Falls to an untimely end."[14]

The questions put to the Tribunal do not contemplate the extent to which Canada should be held responsible in international law for any damage caused by the smelter, a private actor within Canada's borders.[15] Canada accepted responsibility in principle, and the Tribunal was left to resolve the questions whether damage had been caused and what should be done about it.[16] Be that as it may, the Tribunal purported to decide the matter based on international law, which it found to be the same as U.S. law concerning air pollution flowing between states of that federal union.[17] On this point, the Tribunal cited a passage from Eagleton on state responsibility, as follows: "A State owes at all times a duty to protect other States against injurious acts by individuals from within its jurisdiction."[18] The Tribunal then went on to consider what constitutes an injurious act. Because of the absence of international precedent concerning air or water pollution flowing between states, the Tribunal referred to a series of U.S. cases involving air pollution between various U.S. states.[19] The Tribunal cited, and appears to have derived inspiration from, the following passage in one such case:

> It is a fair and reasonable demand on the part of a sovereign that the air over its territory should not be polluted on a great scale by sulphurous acid gas, that the forests on its mountains, be they better or worse, and whatever domestic destruction they may have suffered, should not be further destroyed or threatened by the act of persons beyond its control, that the crops and orchards on its hills should not be endangered from the same source.[20]

13 For a further examination of this important characteristic of the *Trail Smelter* arbitration, namely, the concern for balancing the interests of the parties, *see* Craik in this volume. For a discussion of the principle of sustainable development inherent in the compromise struck by Canada and the United States, *see* Jacobson in this volume.

14 Read, *supra* note 8, at 224. *See* Mickelson, *supra* note 2, at 228.

15 *See* Anderson in this volume.

16 *See, e.g.*, Rubin, *supra* note 3, at 264; Mickelson, *supra* note 2, at 223, n.12.

17 The Tribunal stated:

> The first problem which arises is whether the question should be answered on the basis of the law followed in the United States or on the basis of international law. The Tribunal, however, finds that this problem need not be solved here as the law followed in the United States in dealing with the quasi-sovereign rights of the States of the Union, in the matter of air pollution, whilst more definite, is in conformity with the general rules of international law.

Trail Smelter (1941), *supra* note 1, at 713. *See* Rubin, *supra* note 3.

18 *Id.* (citing Clyde Eagleton, The Responsibility of States in International Law (1928)).

19 *Trail Smelter* (1941), *supra* note 1, at 714. *See* Rubin, *supra* note 3.

20 Georgia v. Tennessee Copper Co., 27 S.Ct 618, 619 (1907 (cited in *Trail Smelter* (1941), *supra* note 1, at 716). Bleicher notes that this case does not in fact impute responsibility to the State of Tennessee,

Then follows the famous *dictum* of the Tribunal:

> [U]nder the principles of international law, as well as of the law of the United States, no State has the right to use or permit the use of its territory in such a manner as to cause injury by fumes in or to the territory of another or the properties or persons therein, when the case is of serious consequence and the injury is established by clear and convincing evidence.[21]

One possible, and perfectly defensible, conclusion is that the *dictum* is *obiter*, in that the question of Canada's responsibility at international law for the damage caused by the smelter was not at issue.[22] However, although this may be a technically correct argument, it does not help us to understand the treatment that the arbitration, and this famous paragraph, have received. In any event, there is no doctrine of precedent in international law;[23] therefore, what ultimately matters is how persuasive the reasoning and conclusion of the Tribunal are found to be by jurists and state representatives seeking to solve similar problems. However, this persuasiveness is compromised by a lack of clarity in the Tribunal's reasoning. In particular, the decision is not based on the international law of state responsibility, and it is extremely difficult to square the decision with that body of law.[24]

A REVIEW OF THE DIFFICULTIES – STATE RESPONSIBILITY

As noted earlier, the Tribunal purported to base its conclusions on international law, and referred to Eagleton, an eminent publicist on state responsibility. However, Eagleton's reasoning is not adopted by the Tribunal, as a closer examination of his text demonstrates. In particular, the Tribunal does not pick up on the necessity identified by Eagleton of determining the basis on which the state is responsible for the harm in question. Eagleton, in the work that the Tribunal cites, is clear on this point:

> But while the responsibility of a state is fundamentally based upon the control which it exercises within its borders, it does not follow that the state may be held responsible for any injury occurring therein. The law of nations does not make the state a guarantor of life and property. It is answerable, under international law, only for those injuries which are internationally illegal in character, and which can be fastened upon the state itself... [T]he responsibility of the state is to be ascertained from the duties

and that the private company that caused the pollution, rather than the State, was ordered to pay compensation. Bleicher, *supra* note 10, at 24.

[21] *Trail Smelter* (1941), *supra* note 1, at 716.

[22] *See, e.g.*, Bleicher, *supra* note 10, at 22.

[23] Article 38(1)(d) of the Statute of the International Court of Justice allows international judicial decisions persuasive authority only, and art. 59 limits the binding authority of decisions of that court to the parties. Statute of the International Court of Justice, June 26, 1945, arts. 38(1)(d) and 59. *See* Ian Brownlie, Principles of Public International Law 5, 19 (6th ed. 2003) ("Article 38 is generally regarded as a complete statement of the sources of international law.").

[24] *See* Drumbl in this volume.

> of control within its territorial limits and over its agents laid down for it by positive international law . . .[25]

The passage cited by the Tribunal is located in a chapter entitled "Acts of Individuals." After reviewing a number of conceptual frameworks that can be used to explain a state's responsibility for acts of its citizens, Eagleton settles on one that he states "is most nearly in harmony with the practice of states to-day," namely, the notion that "the state is never responsible for the act of an individual as such: the act of the individual merely occasions the responsibility of the state by revealing the state in an illegality of its own – an omission to prevent or publish, or positive encouragement of; the act of the individual."[26] Commenting on this conceptual framework, he writes:

> While . . . in a few exceptional cases, the state may be held responsible at once for the act of an individual, it is usually necessary to show an illegality on the part of the state. The state cannot be regarded as an absolute guarantor of the proper conduct of all persons within its bounds. Before its responsibility may be engaged, it is necessary to show an illegality of its own; and this involves simply the question of what duties are laid upon the state with regard to individuals within its boundaries by positive international law.[27]

In this respect, there has been no significant change in the law of state responsibility.[28]

Returning to the fact situation presented in *Trail Smelter*, it is difficult to discover the basis on which Canada would be responsible at international law for the damage caused by the smelter.[29] The Tribunal did not identify the nature of the international legal obligation that Canada breached in this case, and in any event this question was not put to it. The Tribunal's solution was to impute liability for harm caused by a source within a state's borders and affecting the territory of another state, a solution that may be analyzed in terms of strict liability, nuisance, or a failure to meet a due diligence obligation, although the Tribunal did not refer to any of these three principles.[30]

25 Clyde Eagleton, The Responsibility of States in International Law 8–9 (1928).

26 *Id.* at 77. [Footnotes omitted.]. *See* Anderson in this volume.

27 *Id.*

28 *See* International Law Commission, *Draft Articles on Responsibility of States for Internationally Wrongful Acts, 2001*, Report on the 53d Session, UN GAOR, Supp. No. 10, UN Doc. A/56/10 (2001) Ch.IV.E.1: Responsibility, *available at* http://www.un.org/law.ilc [hereinafter *Draft Articles*]. Articles 4–11 address the circumstances under which a wrongful act will be attributable to a state. Exceptions are provided, particularly at arts. 8 and 11, but none of these exceptions would have covered the situation in *Trail Smelter*.

The *Draft Articles* have no binding force in their current form. However, key provisions, including arts. 4, 7, and 8, are well recognised in international law and have likely attained the status of customary international law. Article 11 has also received a significant degree of recognition in international jurisprudence.

29 *See, e.g.*, Bleicher, *supra* note 10, at 21.

30 *See* Anderson in this volume.

Another possibility is to place emphasis on the Tribunal's reference to "serious consequences" and "clear and convincing evidence."[31] Thus, we might argue that, when the consequences are serious enough to affect a state's territorial integrity, the balance between the rights and interests of neighboring states will be disrupted and international recourse may be had. However, we still need a basis on which to impute liability to the *state*. This could be provided by a generally applicable due diligence obligation, such as was articulated in *Corfu Channel*.[32] In this case, it would be necessary to demonstrate that the state had knowledge of the incident and of its consequences to the territorial integrity of a neighbor, and failed to take reasonable steps to prevent or minimize the damage. However, this interpretation of the decision may limit its application to cases of severe and near-catastrophic environmental harm across borders.[33] In the case of the more quotidian, but potentially devastating, consequences of gradual emissions of pollutants over time, the holding may be of little assistance.

ALTERNATIVE APPROACHES TO INTERPRETATION

Equitable Balancing of Interests

As an alternative to arguing that the conclusion in *Trail Smelter* is based on a due diligence obligation, we could argue that the case does what it says it does, namely, imposes an obligation on states to prevent harm to other states originating on their territory. This more literal interpretation of the holding does not appear to have taken root in international environmental law.[34] However, there are other international principles available that point, not to an obligation to prevent any and all transboundary environmental damage but, rather, to a duty on the part of the state of origin of environmental harm to seek to mitigate impacts on its neighbors.

[31] *Trail Smelter* (1941), *supra* note 1, at 716.

[32] Corfu Channel Case (United Kingdom v. Albania), 1949 I.C.J. 4 (April 9).

[33] This will depend, however, in large measure on definitions of serious or significant consequences. For example, the *Draft Articles on Prevention of Transboundary Harm*, which will be discussed in greater detail below, provide a fairly expansive definition of significant harm. Article 2 para. (a) defines "risk of causing significant transboundary harm" as including "risks taking the form of a high probability of causing significant transboundary harm and a low probability of causing disastrous transboundary harm:" Extract from the *Report of the International Law Commission on the work of its Fifty-third Session, Official Records of the General Assembly, Fifty-Sixth Session, Supplement No. 10* (A/56/10), ch. V.E.1.

[34] This is clearly demonstrated by the difficulty that the ILC has had and continues to have with this issue: *see, e.g.*, First Report on the Legal Regime for Allocation of Loss in case of Transboundary Harm arising out of Hazardous Activities, International Law Commission, Fifty-fifth Session, 21 March 2003, A/CN.4/531 at 3–4, paras. 2–3. *See also* Alan Boyle, *Codification of International Environmental Law and the International Law Commission: Injurious Consequences Revisited*, *in* INTERNATIONAL LAW AND SUSTAINABLE DEVELOPMENT: PAST ACHIEVEMENTS AND FUTURE CHALLENGES 61, 74 (Alan Boyle and David Freestone eds., 1999).

It may be argued that the *Trail Smelter* holding is an expression of the *principle of good neighborliness*, or *sic utere tuo ut alienum non laedas*. This topic is often treated as distinct from the principle enunciated in *Trail Smelter*, but, as many authors have recognized,[35] the two are related in many important respects. It has been argued by Special Rapporteur Barboza of the International Law Commission that states have neither the absolute freedom to carry out activities on their territory regardless of the consequences nor the absolute right to be free of impacts from activities of other states. "A legal norm," he states, "cannot be based on an international 'reality' which does not exist, since absolute independence or absolute sovereignty does not exist. In contrast, interdependence has always existed and is becoming more and more prevalent..."[36]

Since 1978, the ILC, in what has been described as "one of the most laboured and confusing studies the Commission has ever undertaken,"[37] has been working on draft articles to address liability based on harm resulting from something other than breach of an obligation.[38] The ILC has adopted a series of *Draft Articles on Prevention of Transboundary Harm from Hazardous Activities* (*Draft Articles*) addressing the topic of prevention,[39] and work on the separate topic of liability for hazardous activities continues under Special Rapporteur Pemmaraju Sreenivasa Rao.[40]

The *Draft Articles* seem very modest, particularly when compared to the ringing and unequivocal language of the *Trail Smelter dictum*. The *Draft Articles* include a series of provisions on prevention of transboundary harm,[41] cooperation to prevent significant harm and reduce risk,[42] the exercise of regulatory control by states of activities on their territory through prior authorizations,[43] environmental impact assessment,[44] notification,[45] and consultation,[46] among other issues. This text does not recognize or create a secondary rule imposing liability for transboundary harm absent breach of an obligation. Nor does it provide a clear answer to the question

35 *See, e.g.*, L.F.E. Goldie, *Concepts of Strict and Absolute Liability and the Ranking of Liability in terms of Relative Exposure to Risk*, 16 NETHERLANDS YEARBOOK OF INTERNATIONAL LAW 175, 213 (1985). This approach was adopted by Special Rapporteur Robert Q. Quentin-Baxter, and is reflected in his balance-of-interest approach to the question of liability for acts not in breach of international law. Robert Q. Quentin-Baxter, *Second Report on International Liability for Injurious Consequences arising out of Acts not Prohibited by International Law* (1981), UN Doc. A/CN.4/346, *passim* but particularly at 112, para. 39; 115, para. 54; 117, para. 61; 122, para. 81. *See* McCaffney in this volume.

36 Julio Barbozo, Second Report on International Liability for Injurious Consequences arising out of Acts not Prohibited by International Law (1986), UN Doc. A/CN.4/402 at 157, para. 54.

37 Boyle, *supra* note 34, at 73.

38 *See* Drumbl in this volume.

39 Draft Articles on Transboundary Harm, *supra* note 28.

40 *Report of the International Law Commission on the work of its Fifty-fourth session*. U.N. GAOR, 56th Sess; Supp. No. 10, U.N. Doc. A/57/10 at 222, para. 438 (2002).

41 Draft Articles on Transboundary Harm, *supra* note 28, art. 3.

42 *Id.* at art. 4.

43 *Id.* at art. 6.

44 *Id.* at art. 7.

45 *Id.* at art. 8.

46 *Id.* at art. 9.

of when a state has breached its obligation to prevent transboundary harm. It does, however, describe a series of procedures through which an answer to that question could be reached. Attention is paid, as in the *Trail Smelter* arbitration, to the need to find a balance between various rights and interests.[47]

One way of understanding this balance would be to focus on the rights and interests of sovereign states. However, when industrial activities cause pollution that harms the health of populations and ecosystems, a solution cast in terms of state territorial integrity seems frustratingly beside the point. The history of the *Trail Smelter* arbitration makes it clear that the actors most directly affected were not sovereign states but farmers, employees, businesspeople – in other words, members of the populations within the transboundary region affected. It is apparent that the balance among interests envisaged in the *Draft Articles* is not limited to the conflicting interests of two states in sovereign territorial integrity. In seeking a solution to an activity creating actual or potential transboundary harm, the parties are to consider a range of issues relating to the costs and benefits of the proposed project, not simply to the states as sovereign entities, but to the populations, economies, societies, and ecosystems of the two states.[48] This approach is, of course, far from perfect. As Alan Boyle has pointed out, a truly equitable balancing of interests may be difficult to achieve in situations of unequal bargaining power.[49] This point is well taken, but we must ask ourselves whether we can really expect to devise a rule of international law that defines precisely the limits of states' rights to cause transboundary environmental harm and their obligations to tolerate it. The task put to international law in such instances is no different from that faced by

47 Alan Boyle, commenting on an earlier version of the *Draft Articles on Transboundary Harm*, stated:

> If, as the ILC draft suggests, the authorization of activities which pose a risk of transboundary harm is permissible only if the parties attempt in good faith to negotiate an equitable solution with neighbouring states then the law will in effect have come close to requiring the equitable utilization of land territory in much the same way that it presently requires states to utilize watercourses or the high seas equitably:

Boyle, *supra* note 34, at 81. *See* Jacobson in this volume.

48 Article 10 ("Factors involved in an equitable balance of interests") refers at para. (a) to the degree of risk of significant transboundary harm and the possibilities of preventing, minimising or repairing such harm – a fairly traditional balancing of state rights and interests. However, it also refers to the risk of significant harm to the environment (para. (c)); to the importance of the activity from social, economic and technical points of view (para. (b)); to the economic viability of the activity in relation to the costs of prevention and to the costs of alternatives to the activity (para. (e)); and, in a nod to common but differentiated responsibility, to the willingness of the state affected to defray prevention costs (para. (d)) and to standards of prevention applied by the affected state and in the region (para. (f)). Thus, the consultations that are envisaged between states would touch on a wide range of questions relating to the activity and the broader national and regional context. *See Draft Articles*, *supra* note 28, at art. 10.

For a contrary reading of the *Trail Smelter* arbitration, *see* Allum in this volume.

49 Boyle, *supra* note 34, at 83. *See also* Daniel Barstow Magraw, *Transboundary Harm: The International Law Commission's Study of International Liability*, 80 American Journal of International Law 305, 327 (1986).

domestic law: this indeterminacy is not so much a feature of a decentralized legal system as it is of a body of formal rules addressing environmental protection.

Developments in Conventional International Law

The body of international environmental law did not exist at the time of the *Trail Smelter* decision, but there has been a veritable explosion of rule-making and regime creation in this field in the last few decades.[50] International environmental conventions pay very little attention to the issue of state liability or responsibility, seeking instead to control or prevent transboundary pollution before damage is caused. The logic behind this approach is obvious. First, if one's objective is to prevent environmental degradation, it is clearly better to develop a preventive regulatory regime rather than to rely on regimes of responsibility and liability.[51] Second, regulatory approaches allow one to bypass completely the thorny problem of attribution of the acts of private actors, such as smelting companies, to states. These regimes impose obligations on states to take positive steps to control pollution, and a failure to meet such obligations would in and of itself constitute a breach of international law. These conventional obligations can also be used to flesh out due diligence obligations. Finally, there is a trend toward creating regimes within regimes for compliance and enforcement,[52] such that a breach of a conventional obligation could be addressed through such mechanisms, thus obviating the need for reference to rules of state responsibility.

CONCLUSIONS: CONTRIBUTIONS OF *TRAIL SMELTER*

Reading *Trail Smelter* is an excellent pedagogical exercise, precisely because it is an awkward fit with other rules of international environmental law, and because it is at the same time a cornerstone of that law. Seeking to understand where the case fits into international environmental law compels us to think analytically

[50] These include Geneva Convention on Long-range Transboundary Air Pollution (and its eight protocols); Canada-United States Agreement On Air Quality; Vienna Convention for the Protection of the Ozone Layer; Montreal Protocol on Substances that Deplete the Ozone Layer; Framework Convention on Climate Change; and Kyoto Protocol to the Framework Convention on Climate Change.

[51] However, as many commentators have pointed out, regimes of responsibility and liability can be designed to be flexible and to offer a wide range of solutions to problems of transboundary pollution, including the creation of pollution control regimes. The *Trail Smelter* arbitration is a perfect example of this. Furthermore, the work of the ILC on liability for transboundary harm emphasises flexibility and, more in particular, focuses on a range of obligations oriented to prevention. *See* Drumbl, Bratspies, and Miller in this volume. On the potential for flexibility within a regime of state responsibility, *see* Alan E. Boyle, *State Responsibility and International Liability for Injurious Consequences of Acts not Prohibited by International Law: A Necessary Distinction?*, 39 International & Comparative Law Quarterly 1 (1990).

[52] *See, in particular*, the Montreal Protocol and the Kyoto Protocol.

and critically about that body of law and about its relation to international law more generally.[53] Furthermore, this intellectual process encourages us to view international law as dynamic, often contradictory, sometimes incoherent, and far from comprehensive – and to view our own role in interpreting international law as striving to give it qualities of comprehensiveness, coherence, and completeness. The question is whether the decision of the *ad hoc* Tribunal, and the principle attributed to it, can be used in any way to reestablish coherence. On one level, the response seems to be "no." However, we should not be too hasty. If we accept that law in general, and international law in particular, is not a perfectly ordered system whose parts relate back to the whole according to the rules of logic, and if we accept further that rules and precedents cannot simply be applied mechanically without interpretation, then we must take responsibility ourselves for lending coherence to the rules and the system with which we have to work. For this reason, it seems that we have a good deal to gain by taking the *Trail Smelter* arbitration seriously and by working hard to understand it and its relationship to the broader body of international environmental law and public international law more generally. Our consideration may be highly critical, but this case gives us an opportunity to gain insights into the problems and opportunities posed by international environmental law as a legal system and as a discipline.

53 Mickelson makes a similar point, arguing that we should read *Trail Smelter* as a "case study," taking into account the context in which the case was heard and decided and the manner in which legal principles were applied in the resolution of this dispute. Mickelson, *supra* note 2, at 232–23.

6 The Flawed *Trail Smelter* Procedure: The Wrong Tribunal, the Wrong Parties, and the Wrong Law

John H. Knox

INTRODUCTION

The *Trail Smelter* arbitration is renowned for its holding: "no State has the right to use or permit the use of its territory in such a manner as to cause injury by fumes in or to the territory of another or the properties or persons therein, when the case is of serious consequence and the injury is established by clear and convincing evidence."[1] Those words have been the subject of an immense amount of scholarly attention. But the procedure that produced the holding is worthy of study in its own right.

The U.S. and Canadian governments created an innovative mechanism to address a common problem in international relations: transboundary environmental harm caused and felt by private actors. The procedure was apparently successful. It resulted in a decision accepted by the governments, payment of damages to the victims of the pollution, and changes in the operation of the polluter to reduce the level of harm. Despite this apparent success, however, the *Trail Smelter* procedure has never been used again to resolve a private dispute over international environmental harm.[2] Why not?

This chapter argues that the refusal to apply the *Trail Smelter* procedure to resolve such conflicts is a result of fundamental flaws in the procedure itself. In establishing the procedure, the U.S. and Canadian governments faced three basic questions: (a) whether to refer the dispute to an international or a domestic tribunal; (b) whether to allow the real parties in interest to appear on their own behalf, or to appear for them; and (c) whether to instruct the tribunal to apply

[1] *Trail Smelter Arbitral Decision*, 35 AMERICAN JOURNAL OF INTERNATIONAL LAW 684, 716 (1941) [hereinafter "*Trail Smelter* (1941)"]. *See* Annex to this volume.

[2] The Lake Ontario Claims Tribunal, created in 1965 to hear claims arising from the Gut Dam in the St. Lawrence River, was also a three-person arbitral tribunal charged with assessing damages for transboundary harm, but the claims it reviewed arose from an action of the Canadian government. *See* Canada-United States Settlement of Gut Dam Claims, Sept. 27, 1968, 8 INTERNATIONAL LEGAL MATERIALS 118 (1969).

domestic or international law. *Trail Smelter's* resolution of each of these issues has proved unworkable. The governments chose the wrong tribunal, the wrong parties, and the wrong law.

The chapter begins by placing the *Trail Smelter* dispute in context. International law has provided few avenues for direct claims by private persons for harm caused by other private actors, and the promise of the *Trail Smelter* procedure is that it offers such an avenue. The next section describes why this promise has been unfulfilled, by examining the practical problems with each of the fundamental aspects of the procedure: the court, the parties, and the law. The final section suggests that procedures with characteristics opposite to *Trail Smelter* on each of these basic points have a far better chance to resolve private international environmental disputes. In short, the *Trail Smelter* procedure is useful primarily as an example of what not to do.

TRAIL SMELTER AS AN ATTEMPT TO ADDRESS PRIVATE INTERNATIONAL ENVIRONMENTAL HARM

The *Trail Smelter* procedure addresses an important problem that international environmental law has yet to solve: finding an effective way to limit transboundary harm resulting from private action. And it does so through a potentially valuable approach that is still rarely used: developing limits based on the private rights of the victims of the harm. In contrast to domestic environmental law, international environmental law has very few mechanisms that allow individuals to seek private remedies for environmental harm. The restrictions it places on private action nearly always result from public limits, publicly enforced.

As a result, international environmental law is likely to be less effective at dealing with certain kinds of problems. Although scholars differ on which method is best suited to limiting which types of private environmental harm, they generally agree that private rights have strengths that public limits do not. Relying on private rights to limit environmental harm may force private actors to compensate those injured by their actions and thereby internalize the actions' environmental costs. In theory, the invisible hand of the market then causes the level of pollution to reach its optimal point without the need for costly regulatory measures that might over- or undercompensate for the harm. Moreover, recognizing private rights to be free from (or to receive compensation for) environmental harm may seem fairer to the individuals concerned than leaving their harm to be addressed only through public regulation. Public regulators may lack accurate information about environmental harm, fail to have full jurisdiction over it, or be captured by special interests.[3] In addition, regulation often makes little or no effort to compensate the individuals harmed.

3 *See* Daniel C. Esty, *Toward Optimal Environmental Governance*, 74 New York University Law Review 1495, 1508–18 (1999).

Public limits do have substantial advantages of their own. They can be imposed before the harm occurs and they can allow the public to make collective decisions about its desired level of environmental protection.[4] And, perhaps most importantly, they may be the only way to address harm that does not lend itself to a private-rights approach. For many types of environmental harm, defining the private rights in question, tracing the causal chain from the actors to the victims, and ensuring that full compensation is paid to everyone who has been or may be harmed are impracticable.[5] The more diffuse the origins and effects of an environmental harm, the less suitable a private-rights approach will be.[6] Trying to apply private rights to limit many international environmental harms, in particular, would be difficult or impossible. One cannot imagine tracing ozone depletion or global warming from its many sources to its many victims.

Not all international environmental problems can be described in this way, however. Some are clearly caused by, and clearly harm, identifiable private parties. *Trail Smelter* was the kind of case that many scholars agree might usefully be addressed through privately enforced private rights: it involved only one polluter, a limited number of individuals alleging harm, and relatively clear causation. In the U.S. and Canada, domestic cases like it were and still are subject to a private-rights approach based on nuisance law, although now each country has added a statutory, public-limit overlay.

Domestic legal remedies do not always translate well when applied to international environmental problems, however. As other chapters in this volume describe, the *Trail Smelter* dispute illustrates common obstacles to the use of domestic remedies to respond to transboundary harm. Potential claimants are likely to be reluctant to sue in the unfamiliar and perhaps unfriendly courts of the actors causing the harm, and defendants will resist appearing in the courts of the victims. Domestic courts may not have personal or subject-matter jurisdiction over such suits in any event, and may be unwilling to exercise jurisdiction even if they have it.

Unable to resolve the dispute under domestic law, the Washington farmers harmed by the Trail smelter turned to the U.S. government, which turned to international diplomacy. But no existing international mechanism was able to resolve the dispute. The U.S. and Canadian governments first referred the case to the International Joint Commission, but its recommendation was non-binding, and when the farmers rejected it (because it failed to ensure that the transboundary pollution would cease), so did the U.S. government. As a result, in April 1935 the governments created an entirely new procedure to resolve the dispute, through an agreement to which they had obtained the prior consent of the farmers and

4 Christopher H. Schroeder, *Lost in the Translation: What Environmental Regulation Does That Tort Cannot Duplicate*, 41 WASHBURN LAW JOURNAL 583, 583–98 (2002).

5 *See* Esty, *supra* note 3, at 1503–08.

6 Schroeder, *supra* note 4, at 598–602.

the Consolidated Mining and Smelting Company ("Cominco"), the owner of the smelter.[7]

The 1935 agreement had two elements: payment of $350,000 for damage up to January 1, 1932, and the creation of a new *ad hoc* international tribunal (composed of one arbitrator from the United States, one from Canada, and one from a neutral country) to decide the remaining issues. In particular, the governments gave the Tribunal jurisdiction to determine four issues: (1) the amount of damage and indemnity for it after 1931; (2) whether the Trail smelter should refrain from causing damage in the State of Washington in the future, and, if so, to what extent; (3) what measures should be adopted by the smelter; and (4) what indemnity should be paid on account of the answers to the second and third issues. The governments agreed to "take such action as may be necessary in order to ensure due performance of the obligations undertaken hereunder, in compliance with the decision of the Tribunal."[8]

By stepping into the shoes of the private parties for the purpose of the arbitration, the governments appeared to transform the private dispute into a public nuisance case. At its root, however, the *Trail Smelter* case remained an effort by private actors, the Washington farmers, to obtain a remedy for environmental harm caused by another private actor, Cominco. Both governments understood that the damages were to be paid, and any other action necessary to comply with the tribunal decision was to be taken, by Cominco for the benefit of the farmers.

If *Trail Smelter* had been followed by similar cases that gradually removed the legal fiction that the governments were the real parties in interest, the effect would have been to create an international equivalent to the domestic law of private nuisance. That such cases did not arise is one of the great roads not taken in the modern history of international law. The absence of an effective way of imposing private limits on private international environmental harm is still felt today. Victims of such harm continue to face obstacles to domestic remedies, just as the Washington farmers did. There is thus a continuing need for the kind of procedure that *Trail Smelter* appeared to pioneer, and a corresponding need to understand why the *Trail Smelter* procedure has not been followed.

THE FLAWS OF THE *TRAIL SMELTER* PROCEDURE

The *Trail Smelter* procedure has not been followed because it has proved to be an impractical mechanism for resolving private environmental disputes. In establishing an adjudicative forum to resolve the *Trail Smelter* dispute, the governments faced three fundamental questions: (a) what tribunal should be given jurisdiction

[7] Convention on Damages Resulting from Operation of Smelter at Trail, British Columbia, June 5, 1935, U.S.-Canada, 49 Stat. 3245. *See* Annex to this volume.

[8] *Id.* arts. II, III, XII.

to hear the claims; (b) which entities should have standing to bring and defend them; and (c) what law should the tribunal be instructed to apply? In answering each of these questions, the governments faced a choice between two opposing alternatives, and chose the one that must have appeared at the time to be the more feasible. Nevertheless, each of the decisions turned out to be mistaken. The governments chose the wrong tribunal, the wrong parties, and the wrong law.

The Wrong Tribunal

The governments first had to decide whether to send the case to a domestic court – by, for example, removing the obstacles to jurisdiction in British Columbia – or to an international one. The governments must have seen the creation of an international tribunal as a much less controversial alternative. Adjusting subfederal jurisdiction through an international treaty might have raised political and legal objections. In addition, neither of the private real parties in interest would have been enthusiastic about the prospect that the courts of the other would resolve the dispute. Creation of an international tribunal avoided these problems and, at the same time, allowed the governments to keep closer control over the litigation by shaping the composition and procedure of the Tribunal.

But the creation of a new international tribunal to resolve a private international dispute turned out to have serious problems. It is far more expensive, both in money and time, than allowing the case to be heard by a domestic court. The costs of the tribunal have to be borne by the governments or the private parties, whereas the marginal costs to a domestic court of hearing an additional case are relatively negligible, and in any event not passed on to the private parties. Moreover, because governments have traditionally taken the position that only they are competent to appear as parties before an international tribunal, deciding to create such a tribunal, even if it is to hear a private case, increases the pressure on governments to appear before it themselves. As a result, governments become involved not only in paying the costs of the tribunal but also in paying the costs of presenting the case to the tribunal. Finally, appearing before the international tribunal themselves creates additional difficulties for governments, which the next section describes.

Peter Sand has written, "Instead of internationalizing a local issue (via an enormous detour to the respective national capitals), a more economic solution in most cases would be to adapt local decision-making processes so that they can handle transfrontier problems like ordinary local ones of comparable size. While this approach requires a number of adjustments between state laws, the social-administrative costs involved would seem to be lower than those of oversized *Trail Smelter* procedures."[9] Governments appear to have agreed. They have not

9 Peter Sand, Transnational Environmental Law: Lessons in Global Change 97–98 (1999) (footnote omitted).

been willing to pay the high costs of creating an international tribunal to resolve essentially private disputes.

The Wrong Parties

In deciding which parties would bring and defend the claim before the Tribunal, the governments had to choose between allowing the private parties to appear and stepping into the private parties' shoes and appearing themselves. The governments again chose the alternative that seemed more reasonable at the time, and that alternative again turned out to be unattractive both to governments and to the private parties.

The U.S. and Canadian governments undoubtedly saw the farmers and Cominco as the real parties in interest in *Trail Smelter*. But allowing them to appear in that capacity before the Tribunal would have violated the long-standing tradition that private persons are not subjects of international law. As a result, the governments took the place of their nationals, "transmuting the claims by individuals against the Trail smelter into claims sounding in international tort by the United States against Canada."[10]

Substituting governmental actors for real parties causes difficulties for both the governments and the private parties, however. By stepping into the shoes of the real parties in interest, governments become subject to the direct and precedential effects of the judgment, whether they like it or not. Thus, even though Canadian and U.S. officials did not want a rule prohibiting and making them responsible for all transboundary pollution,[11] that is essentially what the *Trail Smelter* Tribunal gave them in its famous holding. Moreover, by purporting to ground it in customary international law rather than the 1935 Convention, the Tribunal made the ultimate rule one of general application.[12]

Government involvement also has risks for private parties. Although they may welcome official championing and funding of their case, private parties will dislike the concomitant loss of control, especially as the government will often compromise the claims in ways that are not in the interests of those private parties. *Trail Smelter* illustrates the potential for dilution through espousal. Although the farmers were given an unusually high degree of participation in the *Trail Smelter*

[10] John E. Read, *The Trail Smelter Dispute*, 1 CANADIAN YEARBOOK OF INTERNATIONAL LAW 213, 223 (1963). Republished in this volume.

[11] *See* JOHN D. WIRTH, SMELTER SMOKE IN NORTH AMERICA 88 (2000) ("both governments came to the conclusion that it was expedient to regard Trail as a special discrete case rather than as an opportunity to set broad standards or to break new ground").

[12] *Trail Smelter* (1941), *supra* note 1, at 716–17 ("[T]he Tribunal holds that the Dominion of Canada is responsible in international law for the conduct of the Trail smelter. *Apart from the undertakings in the Convention*, it is, therefore, the duty of the Government of the Dominion of Canada to see to it that this conduct should be in conformity with the obligation of the Dominion under international law as herein determined." (emphasis added)).

arbitration – their attorney was allowed to appear before the Tribunal and in 1938 was even appointed the U.S. agent to the Tribunal[13] – the U.S. government kept control over key elements of the case. Although scientists at the U.S. Department of Agriculture were sympathetic to the farmers and provided evidence to support their claims, engineers at the Bureau of Mines were concerned about the potential effects of the case on U.S. smelter companies.[14] With respect to the crucial issue of naming the scientific advisors to the Tribunal, the State Department sided with the Bureau of Mines, appointing its chief engineer with the understanding that he would "not hold out for some operating requirement being placed on the Trail smelter which might at a later date seriously embarrass the domestic smelters, particularly in a controversy with Mexico where the shoe would be on the other foot."[15] Together with the advisor appointed by the Canadians, who was a consultant to the smelter industry, the U.S. appointee influenced the Tribunal toward placing only those restrictions on the Trail smelter that the smelter industry found acceptable.[16]

Cominco appears to have had much greater control over the conduct of the Canadian side of the case. John Read, the Legal Advisor to the Canadian Department of External Affairs at the time, later wrote that Canadian officials and Cominco worked out the essential terms of the 1935 Convention together, Read prepared filings in consultation with the Cominco general counsel, the two of them divided the argument in the first stage of the arbitration, and, in short, "the Company and the government were yoked together as a team."[17] Cominco may have had so much influence because its chief owner was the Canadian Pacific Railroad, which one Canadian official at the time described as "the most powerful commercial activity in the Dominion,"[18] or because Canadian compliance with the 1935 Convention would depend primarily, if not entirely, on the willingness of Cominco to comply with the terms of the eventual judgment.[19]

Undoubtedly, private parties whose cause is taken up by their government would prefer that the government give them such *de facto* control over the conduct of the case. But this degree of involvement by a private party in a dispute nominally between governments is unusual, if not unique: a government usually tries to keep control of any case that might result in a decision binding on it. Moreover,

[13] Wirth, *supra* note 11, at 39.

[14] *Id.* at 40–41, 92–93.

[15] *Id.* at 40–41.

[16] *See id.* at 98.

[17] Read, *supra* note 10, at 228. If anything, Read understates the company's influence. Cominco secretly funded the Canadian government's scientific research on the case after 1931, and it told the government whom to nominate as one of the two scientific advisors to the tribunal. James Robert Allum, Smoke Across the Border: The Environmental Politics of the Trail Smelter Investigation 301, 326 (1995) (unpublished Ph.D. dissertation, doctoral thesis), *available from* the National Library of Canada, http://www.nlc-bnc.ca/thesescanada/index-e.html. An indication of the extent of the company's involvement in the arbitration and its satisfaction at the result is that after the tribunal's preliminary decision in 1938 Cominco paid $10,000 in bonuses to Canadian government officials, including $2500 to Read himself. *Id.* at 350.

[18] Wirth, *supra* note 11, at 22.

[19] Read, *supra* note 10, at 227–29.

governments will typically feel pressure to balance the conflicting interests of their constituencies, as the U.S. government did in the *Trail Smelter* arbitration, in ways that soften the claims (or defenses) of the nationals whose interests they are purporting to represent. The result of this balancing process is often that the government decides not to espouse the claim of the private party at all. And if governments are reluctant to step into the shoes of private plaintiffs, they are far more reluctant to become defendants, by accepting international responsibility for the actions of their nationals.

The Wrong Law

Finally, the governments had to choose a law for the Tribunal to apply. International law had not defined any private or public right either to be free from or to emit transboundary air pollution. Leaving the Tribunal to resolve the dispute on the basis of existing international law might therefore have resulted in a *non liquet*, which the negotiators of the 1935 Convention wished to avoid.[20] The governments could have created a new international standard through a bilateral treaty but, as noted earlier, they chose not to establish a substantive standard of broad application. Instead, they decided to refer the Tribunal to existing *domestic* law – specifically, "the law and practice followed in dealing with cognate questions in the United States of America."[21] Again, this choice was reasonable, and even innovative. U.S. courts had created a large body of nuisance law, including several decisions of the Supreme Court on transboundary air pollution between states. Why not borrow from it?

The choice turned out to be another poor precedent for the resolution of private international environmental disputes, however. First, the case illustrates the difficulty of applying nuisance law to industrial pollution. Despite the fact that the pollution was understood to come from only one emitter, affected a relatively small area, and had been studied intensively for nearly a decade before the Tribunal was constituted, the Tribunal took another five years to decide how much damage the smelter had caused and to impose a control regime on it. Both sides produced an immense amount of technical evidence, at great expense, but their experts were unable to agree on how much the smelter's emissions were harming the farmers or even on how the emissions were reaching the farms. The Tribunal seemed to rely primarily on the two experts appointed by the governments to assist it, who carried out detailed examinations of their own. And yet, after what was "probably the most thorough study ever made of any area subject to atmospheric pollution by industrial smoke,"[22] the Tribunal underestimated the damage caused

[20] *Id.* at 227. For a different perspective, *see* McCaffrey in this volume.

[21] 1935 Convention, *supra* note 7, at art. IV. The Convention also instructs the Tribunal to apply international law and practice, and to "give consideration to the desire of the high contracting parties to reach a solution just to all parties concerned." *Id.*

[22] *Trail Smelter* (1941), *supra* note 1, at 726.

by sulfur dioxide to crops and timber and entirely ignored its effects on human health.[23]

Some of these problems arise from the nature of nuisance law, which is notoriously unpredictable, as it invites courts to weigh many factors in every case. Governments might avoid those problems by instructing tribunals to apply clearer standards, such as strict liability rules. The difficulty of establishing causation and harm would remain, however. Given the state of scientific knowledge at the time, it is understandable that the Tribunal failed to recognize the full extent of the harm caused by the Trail smelter.[24] But the failure of a tribunal with its financial and technical resources to assess damages accurately should warn against underestimating the difficulty of using private-rights cases to identify and limit complex industrial harm.

However, even with respect to cases of transboundary harm in which the nature of the harm is better understood and the causal chain easier to trace, governments often may be reluctant to ask a tribunal to resolve an international dispute according to domestic law. As any lawyer knows, interpreting a body of domestic law can be difficult even for those who have been trained in it. Attorneys unfamiliar with the legal system that produced a law will often be unable to predict accurately how it will be applied. It is therefore highly risky for any party to agree to the resolution of a dispute according to the domestic law of another country.

According to John Read, Canadian officials were willing to accept U.S. law as the standard in *Trail Smelter* because they believed that Canadian nuisance law would have been more favorable to the plaintiffs.[25] They may have had in mind a 1921 decision by the U.S. Supreme Court in a case involving interstate pollution, in which the Court stated that the "burden upon the state . . . of sustaining the allegations . . . is much greater than that imposed upon a complainant in an ordinary suit between private parties. Before this court can be moved to exercise its extraordinary power under the Constitution to control the conduct of one state at the suit of another, the threatened invasion of rights must be of serious magnitude and it must be established by clear and convincing evidence."[26] The final *Trail Smelter* decision quotes that language shortly before echoing it in its famous holding.[27]

One may wonder, however, whether the Canadians realized that the Supreme Court jurisprudence on interstate pollution contained another element that would help the United States. In 1907, the Court had said that "if the state has a case at all, it is somewhat more certainly entitled to specific relief than a private party might be. . . . This court has not quite the same freedom to balance the

[23] Wirth, *supra* note 11, at 84, 99–102. The Tribunal also ignored the effects of lead emissions and slag pollution in the Columbia River.

[24] *Id.* at 85, 101.

[25] Read, *supra* note 10, at 227.

[26] New York v. New Jersey, 256 U.S. 296, 309 (1921).

[27] *Trail Smelter* (1941), *supra* note 1, at 715, 716.

harm that will be done by an injunction against that of which the plaintiff complains, that it would have in deciding between two subjects of a single political power."[28] This language also influenced the *Trail Smelter* Tribunal, which quotes it at length.[29]

The Canadian attorneys may have felt relatively confident in their ability to construe a body of law similar in many respects to, and with the same roots as, their own, and they may have understood that the U.S. law had elements that would benefit the U.S. as well as the Canadian side. Indeed, they could have supposed that without such elements, the U.S. government would not have agreed to use it. But in most international environmental disputes, the differences in the domestic laws and legal systems of the parties are likely to be greater. Creating a mechanism to resolve disputes according to one of those laws is therefore likely to involve a degree of unpredictability that the governments and the real parties in interest will be unwilling to accept. Moreover, a government may find it politically impossible to choose a standard for resolution of an international dispute that is significantly different from the standard it applies to similar cases arising in its own jurisdiction, because, as a result, it could be accused of treating foreign plaintiffs (or defendants) better than their domestic counterparts.

THE ROAD AWAY FROM *TRAIL SMELTER*

The pitfalls of *Trail Smelter* as a procedural precedent follow from the transboundary nature of the problem, but they are not inevitable results of it. Instead, they result from the specific choices the governments made in deciding how to address the problem: to create an international tribunal, to step into the shoes of the real parties, and to apply domestic law. To fully support this conclusion, more is needed than the above description of the failures of the *Trail Smelter* procedure. The impracticability of that procedure does not mean that an alternative procedure will ensure success merely by reaching opposite answers to the three fundamental questions – that is, by using domestic courts, allowing private parties to bring and defend their own claims, and relying on international legal standards. Although a full analysis of procedures with those characteristics is beyond the scope of this chapter, this concluding section points to two examples suggesting that such procedures are more likely to successfully resolve private international environmental disputes if they avoid all three of the *Trail Smelter* mistakes. In particular, they are more likely to meet the minimum criteria for success: a willingness of governments to create them and of private parties to use them.

The first example is the effort by governments to facilitate private-party access to domestic courts in private international environmental disputes, by removing the

[28] Georgia v. Tennessee Copper Co., 206 U.S. 230, 237–38 (1907).

[29] *Trail Smelter* (1941), *supra* note 1, at 716.

kinds of barriers that kept the Washington plaintiffs from suing Cominco in British Columbia. This approach answers two of the *Trail Smelter* procedural questions differently, by using domestic rather than international tribunals and allowing private rather than governmental parties to appear before them, and gives one question the same answer, by relying on domestic rather than international law. In the early 1980s, uniform law commissioners in the United States and Canada drafted a Uniform Transboundary Pollution Reciprocal Access Act, which has been adopted by seven states and four Canadian provinces.[30] The Act provides that persons in a "reciprocating jurisdiction" who suffer or are threatened with personal or property harm caused by pollution originating in the jurisdiction adopting the Act have the same rights to relief as if the actual or threatened injury occurred in that jurisdiction.[31]

Compared to the *Trail Smelter* procedure, this method of addressing private international environmental harm has the advantages of lower cost to the governments and greater control by the parties directly affected. Nevertheless, the procedure is not very successful, at least in the North American context. The federal governments have not adopted the Uniform Act, they show no interest in revitalizing it or extending it to Mexico, and no state or province has adopted it since 1992. Moreover, plaintiffs do not appear to be using the Uniform Act in the jurisdictions that have adopted it.[32] One reason why the procedure has not been used may be that victims of transboundary harm would prefer to sue in their own jurisdiction rather than in the jurisdiction of the source of the harm. The victims are likely to have less money and less ability to navigate in a strange legal system than the corporations that typically cause the types of harm amenable to private suit.

But the solution to this problem is not simply to establish jurisdiction over private international disputes in the courts of the plaintiffs. Apart from concerns over fairness to the defendant, the judgments of those courts will often be worthless unless governments agree that the judgments are enforceable against the defendant in the defendant's jurisdiction. Moreover, defendants might be unable to comply with the inconsistent judgments that might emerge from a wide variety of plaintiffs' jurisdictions.

[30] The states adopting the Transboundary Pollution Reciprocal Access Act are: Colorado, Col. Rev. Stat. §§ 13–1.5–101 to 13–1.5–109; Connecticut, Conn. Gen. Stat. § 51–351b; Michigan, Mich. Comp. Laws §§ 324.1801 to 324.1808; Montana, Mont. Code §§ 75–16–101 to 75–16–109; New Jersey; N.J. Stat. § 2A:58A–1 to 2A:58A–8; Oregon, Ore. Rev. Stat. §§ 468.076 to 468.089; and Wisconsin, Wisc. Stat. § 299.33. The provinces are: Manitoba, R.S.M. 1985–86, c. 11; Nova Scotia, R.S.N.S. 1994–95, c. 1, part XVI; Ontario, R.S.O., c. T-18; and Prince Edward Island, R.S.P.E.I. 1988, c. T-5.

[31] *See* John H. Knox, *The CEC and Transboundary Pollution*, *in* GREENING NAFTA: THE NORTH AMERICAN COMMISSION FOR ENVIRONMENTAL COOPERATION 80, 87 (David L. Markell & John H. Knox eds., 2003).

[32] *See* Secretariat of the Commission for Environmental Cooperation, *Background Paper on Access to Courts and Administrative Agencies in Transboundary Pollution Matters*, 4 NORTH AMERICAN ENVIRONMENTAL LAW AND POLICY 205, 305 (2000).

An answer to this difficulty is to substitute international legal standards for domestic ones – that is, to answer the third *Trail Smelter* procedural question differently, as well as the first two. Under this approach, governments might allow those harmed to bring suit in their own jurisdiction. Governments could agree that the resulting judgments would be enforceable in the jurisdiction of the defendant with less concern that the judgments would be inconsistent. Moreover, governments could negotiate standards that would minimize the discretion of the court by, for example, specifying exactly what types of harm would be compensable and providing for strict liability. By agreeing on specific, clear rules of international law governing the dispute, governments would reduce the risk of an unpredictable result. In addition, governments could draft the rules to minimize any potential disparity with their domestic laws.

Governments have created procedures for resolving private international environmental disputes that have just these characteristics. They are called agreements on civil liability, and they have been developed in the areas of oil pollution, nuclear accidents, and the transboundary movement of hazardous substances.[33] Although the agreements vary in many details, they all reach answers to fundamental questions about the court, parties, and law that are the opposite of those adopted for the *Trail Smelter* arbitration. They set international standards for liability, provide that domestic courts have jurisdiction over claims, and authorize those harmed to bring suit directly against those that caused the harm. The best-known example is probably the 1969 Convention on Civil Liability for Oil Pollution Damage (CLC), as amended in 1992, which makes owners of oil tankers liable for damage caused by oil pollution; specifies international legal standards for the damage, the resulting liability, the parties liable, and the defenses to liability; provides that suits for compensation may be brought in jurisdictions where the damage has occurred; leaves the suits to be brought and defended by the parties directly involved rather than their governments; and provides that judgments shall be enforceable in the courts of each state that belongs to the Convention.[34]

Compared to the *Trail Smelter* procedure, and even to the Uniform Act, the CLC is remarkably successful. Over one hundred governments have joined, and it provides a framework for the routine resolution of private claims for oil pollution damage. Despite closely resembling the CLC, however, treaties that provide for

[33] *See generally* Jutta Brunnée, *Of Sense and Sensibility: Reflections on International Liability Regimes as Tools for Environmental Protection*, 53 International & Comparative Law Quarterly 351 (2004); United Nations Environmental Programme, Environmental Liability and Compensation Regimes: A Review (2003).

[34] 1992 Protocol to Amend the 1969 International Convention on Civil Liability for Oil Pollution Damage, *reprinted in* Basic Documents on International Law and the Environment 91 (Patricia W. Birnie & Alan E. Boyle eds., 1995). The CLC Convention is accompanied by an agreement establishing a multilateral fund to pay certain claims not fully covered by the Convention. *See* 1992 Protocol to Amend the 1971 International Convention on the Establishment of an International Fund for Compensation for Oil Pollution Damage, *reprinted in id.* at 107.

liability to private actors as a result of other types of marine pollution, nuclear accidents, and transport of hazardous substances have attracted much less governmental support. Some have never even entered into force. Nevertheless, governments continue to draft such agreements and to strengthen existing ones. And at its 2004 session, the International Law Commission adopted on first reading draft principles on civil liability for transboundary harm that would address such harm generally, not just in specific contexts such as oil pollution.[35]

Understanding the degree to which these liability agreements have been successful at attracting government and private support, and why some appear to have been far more effective than others, calls for more attention. No one should conclude, based on the limited and mixed records of these agreements, that every mechanism that includes domestic courts, private party access, and international standards will prove to be practicable and effective. But the relative popularity of the liability agreements, combined with the continuing rejection of the *Trail Smelter* procedure, suggests that while these elements are not *sufficient*, they may well be *necessary*, in the sense that only those mechanisms that contain them will have a chance at successfully addressing private international environmental harm.

[35] Report of the International Law Commission, Fifty-Sixth Session, UN GAOR, 59th Sess., Supp. No. 10, at 143, UN Doc. A/59/10 (2004).

7 Rereading *Trail Smelter*

Karin Mickelson

[. . . M]ore than fifty years after the Tribunal's final decision, the *Trail Smelter* arbitration has come to occupy a prominent but somewhat mysterious position in the legal canon. Although it is one of the best known and most frequently cited international decisions, and is regarded by many scholars as the fountainhead of modern international environmental law,[1] it is more an object of reverence than a subject of analysis. All too often it is invoked as authority by scholars who pause only long enough to mention its name and the principle to which it is said to lend support before moving on. Although the potential dangers involved in overstating either the scope of application or overall significance of the Tribunal's decisions have been noted on numerous occasions,[2] there has been less emphasis on the equally important concern that these passing references constitute reductionist accounts of a highly complex set of circumstances, and as such represent a series of lost opportunities to learn the lessons of *Trail Smelter*.

The numerous warnings about the misuse of *Trail Smelter* tend to focus on its invocation, [. . .] as support for the existence of customary duties to avoid causing

[1] Writing in 1971, one scholar went so far as to say, "Every discussion of the general international law relating to pollution starts, and must end, with a mention of the Trail Smelter Arbitration." Alfred P. Rubin, *Pollution by Analogy: The Trail Smelter Arbitration*, 50 Oregon Law Review 259 (1971) [Republished in this volume]. Although international environmental law has obviously undergone tremendous development since that time, *Trail Smelter* remains a landmark. A recent treatise on international environmental law asserts that "the arbitral judgment . . . in the *Trail Smelter* case is considered as having laid out the foundations of international environmental law, at least regarding transfrontier pollution." A. Kiss and D. Shelton, International Environmental Law 107 (1991).

[2] *See* Rubin, *supra* note 1.

[Excerpt reprinted with permission of the author and the University of British Columbia Press. The article originally appeared as: Karin Mickelson, *Rereading Trail Smelter*, 31 Canadian Yearbook of International Law 219 (1993)] Faculty of Law, University of British Columbia. The author would like to thank Dr. Charles Bourne for his helpful comments on an earlier version of this note.

transboundary environmental damage and to make reparation for such damage should it occur. The applicable principle, referred to as the *sic utere tuo* standard (from the Latin maxim *sic utere tuo ut alienum non laedas*: use your own property so as not to harm that of another), has been characterized as a description "of the other face of the coin of sovereignty"[3] and can be seen as the fundamental building block of a system of international environmental protection.

[...] One statement in the 1941 [*Trail Smelter*] decision has come to be regarded as the classic articulation of the *sic utere tuo* standard as applicable to [transboundary environmental] damage:

> [U]nder the principles of international law ... no State has the right to use or permit the use of its territory in such a manner as to cause injury by fumes in or to the territory of another or the properties or persons therein, when the case is of serious consequence and the injury is established by clear and convincing evidence.[4]

* * *

Although *Trail Smelter* and [the *Corfu Channel* case and *Lake Lanoux* arbitration] are almost invariably cited in the literature dealing specifically with liability for environmental damage, they are usually accompanied by a caveat regarding their limited usefulness as support for a general liability standard applicable to environmental damage in particular. The problems associated with using these cases as support for such a standard are well documented. For one thing, each of these variations on the *sic utere tuo* theme has clear limitations, as even a cursory reading of the circumscribed statements of principle reveals.[5] What is even more problematic is that *Trail Smelter*, the only decision that is directly on point in terms of environmental damage, has limited "precedential value" because of the unique circumstances that surround both the decision to submit the dispute to arbitration and the decision of the Tribunal.[6] Scholars have repeatedly pointed out that the significance of these has been exaggerated,[7] and they should be approached with a certain degree of scepticism [...]

* * *

3 R. Q. Quentin Baxter, *Preliminary Report on International Liability for Injurious Consequences Arising out of Acts not Prohibited by International Law*, 2:1 YEARBOOK OF THE INTERNATIONAL LAW COMMISSION – 1980 247, 258 (1982).

4 *Trail Smelter Arbitral Decision*, 35 AMERICAN JOURNAL OF INTERNATIONAL LAW 684, 716 (1941) [hereinafter "*Trail Smelter* (1941)"]. *See* Annex to this volume.

5 [footnote omitted]

6 [footnote omitted]

7 *See, e.g.*, P. W. BIRNIE AND A. E. BOYLE, INTERNATIONAL LAW AND THE ENVIRONMENT 145 (1992); C. B. Bourne, *The International Law Commission's Draft Articles on the Law of International Watercourses: Principles and Planned Measures*, 3 COLORADO JOURNAL OF INTERNATIONAL ENVIRONMENTAL LAW & POLICY 65, 84–88 (1992). As one scholar aptly put it, Trail Smelter "has become accepted – and, to a certain extent, mythologized – as a landmark case in international law." D. Munton, *Dependence and Interdependence in Transboundary Environmental Relations*, 36 INTERNATIONAL JOURNAL 139, 140 (1980–81). This characterization may apply equally to the other decisions.

Whatever the formal status of the *sic utere tuo* principle within the international legal system, it is clear that *Trail Smelter* and the other two decisions usually cited as support for a general liability standard applicable to transboundary environmental damage fall well short of the mark. However, this has not prevented the widespread misrepresentation of these decisions as providing support for such a standard, particularly in general discussions of international environmental law in which liability is not the focus.[8] The reasons for this are unclear, although it may be that the perception that a *sic utere tuo* standard is indispensable to international environmental law makes scholars particularly eager to find support for it, and that judicial and arbitral decisions provide a convenient reference point. Frequently, the brief statements of principle or allusions referred to above are extracted from the decisions and lumped together; the fact that the principles as articulated in these decisions are subject to clear limitations is often downplayed or omitted. Any detailed discussion of the cases themselves or the unique circumstances out of which they emerge is rare, and the possibility that the decision maker's articulation of a principle might be tailored to the circumstances in which a dispute arose appears to be overlooked. What is unfortunate about this approach is that a separation of abstract statements of principle from the context in which they were articulated runs the risk of creating a fundamentally inaccurate impression of these decisions, as can be illustrated by a consideration of *Trail Smelter*.

* * *

[. . .] The crucial question, which is seldom considered, is the extent to which the circumstances of the case may have influenced the Tribunal's approach, and how this is reflected in its decisions. It is clear that the Tribunal itself treated the surrounding circumstances as being of importance.[9] [. . .]

Clearly, in attempting to reach a "just" solution, and one that would represent a permanent settlement of the dispute, the Tribunal was guided by considerations that required taking these circumstances into account. An analysis of the decision reveals at least four factors that were of crucial importance, not only in the decision on the part of the Canadian and U.S. governments to submit the dispute to arbitration in the first place but also in the approach taken by the Tribunal. These factors are described below.

Proof of damage and identification of its cause: The importance of there being a clearly identified source of damage as well as clearly established harm is reflected in the Tribunal's language limiting the generality of the *sic utere tuo* standard; that is, that the case be "of serious consequence" and the injury be "established by clear and convincing evidence." From the outset, because of the relative lack of industrial development in the border region, it seemed clear that the source of the alleged damage was the smelter. Originally, however, the question of whether

8 *See, e.g.*, Kiss and Shelton, *supra* note 1, at 107–108.

9 [footnote omitted]

damage had in fact occurred was disputed. The scientists involved in the deliberations of the International Joint Commission split in their findings. Whereas the U.S. experts found evidence of widespread damage resulting from the sulphur dioxide fumigations, the scientists brought in by the Canadian government and Cominco, as well as two scientists employed by the Commission, found the damage to be quite limited.[10] No such uncertainty troubled the Tribunal. Its starting point was that "damage occurred in the State of Washington, resulting from the sulphur dioxide emitted from the Trail Smelter" from 1925 at least until the end of 1931,[11] in accordance with the Commission's finding as formalized in the 1935 Convention. The first question before the Tribunal was limited to whether damage had occurred since the beginning of 1932. Thus, there was no room for doubt that the smelter at Trail was the cause of damage, and given that the smelter's operations had not undergone substantial changes between 1931 and 1932, it was not difficult for the Tribunal to reach the conclusion that compensable damage had occurred.

Perception of importance of interests at stake on both sides of the border: Both states involved perceived significant interests as being involved in the dispute. The economic importance of the smelter appears to have been an important factor in the willingness of the Canadian government to become involved and to attempt to ensure that the outcome would not shut down the operation. The smelter constituted the economic backbone of the economy of Trail and the surrounding region, and Cominco was the largest employer in the area, paying taxes in the range of $1 million per year.[12] On the other side of the border, the perception of the interests at stake was equally clear. Although the land allegedly damaged by the fumes was hardly among the most fertile in the state, it was the basis of the livelihood of the inhabitants of Stevens County, a predominantly agricultural area.[13] Concern regarding the smelter's activities triggered an early example of environmental activism, involving the formation of a group called the "Citizen's Protective Association," which initially lobbied fiercely to have the matter taken

10 Dinwoodie asserts that the Commission was "stymied by contradictory expert opinion." D. H. Dinwoodie, *The Politics of International Pollution Control: The Trail Smelter Case*, 27 INTERNATIONAL JOURNAL 219, 226 (1972). He notes that these experts blamed the failure of the region's agricultural economy on "a variety of causes unrelated to the Trail emissions: earlier smelter operations at Northport, forest fires, insect infestations, inadequate soil composition, and poor farming practices." *Id.* at 226–27.

11 *Trail Smelter Arbitral Decision*, 33 AMERICAN JOURNAL OF INTERNATIONAL LAW 182, 190 (1939) [hereinafter "*Trail Smelter* (1939)"]. *See* Annex to this volume.

12 Dinwoodie, *supra* note 10, at 219.

13 There were indications that it was in fact unproductive. For this reason, Cominco's view during the early stages of the dispute was that the whole affair constituted "an attempt at holdup by farmers in a nearly hopeless section who have come to think that they can get much more out of farming this rich corporation across the boundary than from farming their farms, and who are endeavoring to use the Governments at Washington and Ottawa to threaten a complete cessation of operations and thus force extravagant indemnity." *Id.* at 222, n. 5.

up by the U.S. government and later continued to express its concern directly and through the Washington State representative in Congress.[14]

Reciprocity of risk: The element of reciprocity of risk, by contrast, seems to explain the initial unwillingness of the United States to pursue the matter, given the heavy industrialization on the U.S. side of the border further east and the accompanying concern that a dangerous precedent could be set.[15] In its decision, the Tribunal acknowledged that as between the two states the risk of transboundary damage from industrial sources was reciprocal, and that each had an interest in seeing a decision that would take into consideration the need to balance competing interests. The Tribunal stated:

> As between the two countries involved, each has an equal interest that if a nuisance is proved, the indemnity to damaged parties for proven damage shall be just and adequate and each also has an equal interest that unproven or unwarranted claims shall not be allowed. *For, while the United States' interests may now be claimed to be injured by the operations of a Canadian corporation, it is equally possible that at some time in the future Canadian interests might be claimed to be injured by an American corporation.* As has well been said, "It would not be to the advantage of the two countries concerned that industrial effort should be prevented by exaggerating the interests of the agricultural community. Equally, it would not be to the advantage of the countries that the agricultural community should be oppressed to advance the interest of industry."[16]

The fact that the damage suffered could easily have been on the Canadian side and might be so at some point in the future may well have accounted for the Tribunal's willingness to phrase its decision so as to impose liability on Canada outside the terms of the 1935 Convention.

Broader bilateral relations between the states: By the time the dispute was submitted to arbitration, the United States and Canada had an established history of cooperation on matters that had an "environmental" aspect to them. This relationship is most clearly illustrated by the operation of the International Joint Commission, which, as noted earlier, had been the body initially entrusted with the *Trail Smelter* dispute. Established pursuant to the Boundary Waters Treaty of 1909, the Commission was empowered to deal with issues arising from the common boundary. However, the existence of mechanisms for cooperation should not obscure the fact that the relative imbalance of power between the two countries also had an influence; the decision to submit the dispute to arbitration appears

14 *Id.* at 221.

15 Regarding the question *of why* there was not an insistence on the "absolute cessation" of damage, Read notes, "[t]he acceptance of the principle of absolute cessation of damage might have shut down the Trail smelter, but it would also have brought Detroit, Buffalo and Niagara Falls to an untimely end." Read, *supra* note 1, at 224–25.

16 *Trail Smelter* (1941), *supra* note 8, at 685 (emphasis added).

to have been at least partly the result of overt pressure brought to bear on the Canadian government by the U.S. government.[17]

Although these four factors could be characterized as political/ economic/social factors that have little to do with the legal aspects of the decision, they are crucial to an understanding of that decision. In fact, the legal analysis in *Trail Smelter* is difficult to understand when separated from its political, social, and economic context. If one ignores the enormous complexities surrounding the dispute, it is easy to dismiss the principle articulated by the Tribunal as a "lax rule" at worst,[18] a ringing but hollow statement at best. It is therefore unfortunate that these complexities are rarely explored in the legal literature in which the decision makes occasional brief appearances accompanied by *Corfu Channel* and *Lake Lanoux*. What is perhaps most puzzling about the use of these decisions as support for a general liability standard, apart from being problematic for all the reasons summarized here is that it is essentially unproductive. The ritual invocation of principles that enjoy only dubious status as "customary norms" has little practical use; states are unlikely to accept such hazy concepts as a guide to their behaviour. Nor does it have conceptual value, for it does nothing to clarify the theoretical basis of a general liability standard.

* * *

There is a middle ground between those who would reduce cases to their facts, and those who would uproot general statements of principle without any regard for the context within which they were articulated. For those who would explore that middle ground, *Trail Smelter* [continues] to provide a rich vein to be tapped, one that is worthy of serious attention and analysis.

17 See the description of the circumstances leading up to the 1935 Convention in, Dinwoodie, *supra* note 10, at 227–33.

18 As it is described by Rubin. Rubin, *supra* note 1, at 275.

8 *Trail Smelter* and the International Law Commission's Work on State Responsibility for Internationally Wrongful Acts and State Liability

Mark A. Drumbl

INTRODUCTION

On August 9, 2001, the International Law Commission (ILC) adopted Draft Articles on the Responsibility of States for Internationally Wrongful Acts (Draft Articles) together with detailed Commentaries.[1] The Draft Articles – the product of nearly fifty years' work – represent an important step in the codification and development of international law. In their preface, the Commentaries describe the Draft Articles as comparable in significance to the 1966 Draft Articles on the Law of Treaties that became, with limited changes, the 1969 Vienna Convention on the Law of Treaties.[2]

State responsibility involves the consequences to states of their internationally wrongful activities. Draft Article 1 stipulates that every internationally wrongful act of a state entails the international responsibility of that state. Internationally wrongful acts take two forms: (1) breach of an international obligation of the state; or (2) a serious breach of an obligation under a peremptory norm of general international law.

[1] Draft Articles on Responsibility of States for Internationally Wrongful Acts, in Report of the International Law Commission on the Work of Its Fifty-third Session, UN GAOR, 56th Sess., Supp. No. 10, UN Doc. A/56/10 (2001), *reprinted in* JAMES CRAWFORD, THE INTERNATIONAL LAW COMMISSION'S ARTICLES ON STATE RESPONSIBILITY: INTRODUCTION, TEXT AND COMMENTARIES (2002). "[T]he commentaries provide vital insights. [...] [T]hey make heavy use of international court judgments and arbitration awards." Daniel Bodansky and John R. Crook, *Symposium: The ILC's State Responsibility Articles: Introduction and Overview*, 96 AMERICAN JOURNAL OF INTERNATIONAL LAW 773, 789 (2002).

[2] CRAWFORD, *supra* note 1, at ix. *But see contra*, Pierre-Marie Dupuy, *A General Stocktaking of the Connections between the Multilateral Dimension of Obligations and Codification of the Law of Responsibility*, 13 EUROPEAN JOURNAL OF INTERNATIONAL LAW 1053, 1055 (2002) (concluding that the General Assembly's response to the Draft Articles was much more cautious and discrete than its response to the ILC Draft Articles on the Law of Treaties.)

Professor James Crawford was Special Rapporteur for State Responsibility of the ILC at the time of the second reading of the Draft Articles.[3] On December 12, 2001, the United Nations General Assembly adopted Resolution 56/83, which "commend[s the Draft Articles] to the attention of governments without prejudice to the question of their future adoption or other appropriate action."[4] Assuredly, the Draft Articles already have exerted considerable impact as a subsidiary means of determining the content of international law;[5] they already have been relied on by the International Court of Justice.[6] Their influence would continue to grow were states to sign onto them in treaty form or otherwise declare them to constitute evidence of customary international law. Thus far, however, there is no consensus to push the Draft Articles toward a formalized convention or treaty and, tellingly, nothing further has been done in this regard.

It is essential to underscore that the Draft Articles do not set forth any particular substantive obligations but, instead, determine when an obligation has been breached and the legal consequences flowing from that breach. The Draft Articles do not establish primary obligations that define acceptable or unacceptable standards of conduct. Instead, they establish secondary obligations that flow from a breach of an independent and preexisting primary obligation. The breach of a primary obligation therefore gives rise to a new legal regime, that of state responsibility, that contains its own distinctive set of duties and rights.

The primary rule established in the *Trail Smelter* arbitration is that "no state has the right to use or permit the use of its territory in such a manner as to cause injury by fumes in or to the territory of another or the properties of persons therein, when the case is of serious consequence and the injury is established by clear and convincing evidence."[7] Günther Handl documents, in his contribution to this volume, how this primary rule has moved from its evocation in *Trail Smelter* to a variety of international instruments and decisions, such as Principle 21 (Stockholm Declaration) and the opinion of the International Court of Justice in the *Gabčikovo-Nagymaros* dispute.[8] In earlier writings, Handl characterized *Trail Smelter* as the *locus classicus* of international environmental law.[9]

[3] Crawford had been a member of the ILC since 1992. The previous Special Rapporteurs for State Responsibility were F.V. Garcia Amador (1955–1961), Roberto Ago (1963–1979), Willem Riphagen (1979–1986), and Gaetano Arangio-Ruiz (1987–1996).

[4] G. A. Res 56/83, UNGAOR, 6th Sess., at para. 3, U.N. Doc. A/RES/56/83 (Dec. 12, 2001). Throughout the drafting process, the ILC received considerable feedback from the Sixth Committee (Legal) of the United Nations.

[5] They constitute an example of the teachings of the most highly qualified publicists of the various nations. *See* Statute of the International Court of Justice, 59 Stat. 1055, T.S. No. 993 (June 26, 1945), art. 38(1)(d).

[6] *See, e.g.*, Legal Consequences of the Construction of a Wall in the Occupied Palestinian Territory, 2004 I.C.J. 131 (July 9) at para. 140

[7] *Trail Smelter Arbitral Decision*, 35 American Journal of International Law 684 (1941) [hereinafter "*Trail Smelter* (1941)"]. *See* Annex to this volume.

[8] *See* Handl in this volume.

[9] Günther Handl, *Territorial Sovereignty and the Problem of Transnational Pollution*, 69 American Journal of International Law 50 (1975).

Although John Knox, in his contribution to this volume, is wise to point out that neither Canada nor the United States intended for *Trail Smelter* to become such a vital precedent,[10] the fact that it did demonstrates the mysterious, and at times unpredictable, evolution of international environmental law.

Assuredly, *Trail Smelter* best is remembered for its primary rule. But this is not its only legacy. Although not central to the *Trail Smelter* arbitration, important aspects of state responsibility do arise insofar as Canada was ordered to pay reparations for its infringement of the newly established primary rule. In this regard, I welcome Stephen McCaffrey's characterization in this volume of *Trail Smelter* as an important thread that has woven its way into a number of diffuse corners of the tapestry of international law.[11]

The primary goal of this Chapter is to track the relationship between *Trail Smelter* and the 2001 Draft Articles that codify the secondary obligations of state responsibility. The jurisprudential value of *Trail Smelter* in terms of the obligation not to cause serious transboundary environmental harm does not specifically relate to the Draft Articles. Instead, germane to the Draft Articles are those elements of *Trail Smelter* that deal with determining a breach of a preexisting obligation and assessing the legal consequences that flow from such a breach. All things considered, although *Trail Smelter* has played an important role in influencing the development of international environmental law, its influence on the law of state responsibility – although extant – is less catalytic.

The secondary goal of this Chapter is to consider the interface between *Trail Smelter* and the ILC's work in the collateral yet distinct area of state liability, a topic also currently on the agenda of the ILC and for which articles on the subissue of prevention were drafted in 2001[12] and principles on the subissue of allocation were readied in 2004.[13] State liability differs from state responsibility insofar as it covers situations in which no illegal or unlawful conduct has occurred, although the conduct has triggered harm. *Trail Smelter* has exerted greater influence on the preventative aspects of the law of state liability (where *Trail Smelter's* primary

[10] *See* Knox in this volume. *See also* D. H. Dinwoodie, *The Politics of International Pollution Control: the Trail Smelter Case*, 27 INTERNATIONAL JOURNAL 219, 224 (1972) (reporting that the Canadian government initially decided to refer the Trail Smelter controversy for investigation of injury and compensation as "a gesture of comity").

[11] *See* McCaffrey in this volume.

[12] Draft Articles on the Prevention of Transboundary Harm from Hazardous Activities, together with Commentaries, Report of the International Law Commission on the Work of Its Fifty-third Session, UN GAOR, 56th Sess., Supp. No. 10, at V.E.1, UN Doc. A/56/10 (2001). For the moment, it remains unclear whether the Draft Articles on Prevention will move forward on their own or as part of a broader legal instrument in which liability also is addressed. State liability is connected to state responsibility insofar as failure to perform duties of prevention entails state responsibility.

[13] International Law Commission, Draft Principles on Allocation, International Liability for Injurious Consequences Arising out of Acts Not Prohibited by International Law (International Liability in Case of Loss from Transboundary Harm Arising out of Hazardous Activities), Report on the Fifty-Sixth Session, Chapter VII (2004).

rule has been of guidance) than on the reparative aspects that have very recently crystallized in the principles on allocation.

OVERVIEW OF THE DRAFT ARTICLES ON STATE RESPONSIBILITY

There are fifty-nine Draft Articles. They have general application to all areas of international law and develop a regulatory framework for the obligations states owe to each other. One major innovation of the Draft Articles is that they view state responsibility not just bilaterally – which is the traditional approach and one that informs the *Trail Smelter* arbitration – but also multilaterally. In this sense, they modestly contribute to the construction of some sort of global *ordre public*.

The basic predicate of the Draft Articles is that "conduct not in conformity with an international obligation and attributable to a state equals an internationally wrongful act resulting in state responsibility."[14] Internationally wrongful acts take two forms: "a breach of an international obligation of the state" or "a serious breach of an obligation under a peremptory norm of general international law."[15] The language of this latter category was deliberately chosen over the more controversial term "state crime."[16] Although it may seem self-evident that a breach of a peremptory norm may be more serious than a general breach of an international obligation, the Draft Articles do not differentiate according to the nature of the obligations breached.[17] As such, international responsibility of states, even on central matters of the global *ordre public*, is decriminalized.

The Draft Articles explore – in some cases in great detail, in other cases in great generality – many concepts implicated by this basic predicate. The Draft Articles touch on many themes, including: attribution of conduct to a state; determination of breach; responsibility of a state in connection with the acts of another state; circumstances precluding wrongfulness (such as consent, self-defense, or necessity); cessation; assurances of nonrepetition; reparation and compensation; responsibility owed bilaterally and *erga omnes* (to the international community as a whole); which state may invoke a breach of responsibility and the process of invocation; countermeasures; procedure to be followed before initiating countermeasures; and whether states that are not individually injured can implement countermeasures.[18]

The Draft Articles engage in some progressive development of *inter*-state responsibilities. For example, they give states a right to invoke the responsibility of

[14] Bodansky and Crook, *supra* note 1, at 782.

[15] Draft Articles, *supra* note 1, arts. 2, 40(1). "Serious" is defined as a "gross or systemic failure." *Id.* art. 40(2).

[16] As a matter of very general impression, this constitutes an important semantic change, demonstrating the moderation and conservatism of the Draft Articles.

[17] Dupuy, *supra* note 2, at 1058.

[18] The availability of countermeasures must be balanced insofar as too restricted an availability may unduly crimp national sovereignty whereas too generous an availability may roil order in international affairs.

other states for breaches of obligations owed to the international community as a whole. This could be an important mechanism to cultivate responsibility for state conduct that breaches obligations to protect common concerns of humanity – such as biodiversity and the atmosphere. That said, the Draft Articles do not attempt to "codify, much less progressively develop, [. . .] the growing importance of non-state actors as holders of international rights and obligations."[19] Although the rules of attribution reference the potential responsibility of states for the conduct of nonstate actors, they do so only partially. In this sense, the Draft Articles focus on states and, as a consequence, work within a traditional view of the international legal system.[20]

To reiterate for clarification: the Draft Articles do not establish primary obligations that define standards of conduct. Instead, they establish the secondary obligations that flow from a breach of an independent and preexisting primary obligation and, also, that determine when such a breach occurs. These secondary obligations include "the modalities by which states can legitimately claim redress for violations of 'primary' international norms [. . .]."[21] As James Crawford explains: "[T]he key idea is that a breach of a primary obligation gives rise, immediately by operation of the law of state responsibility, to a secondary obligation or series of such obligations (cessation, reparation . . .)."[22] This distinction between primary and secondary obligations had been pushed by Special Rapporteur Ago (1963–1979).[23] Ago felt that it would be impossible to formulate Draft Articles in the absence of such a distinction. Although Ago characterizes the Draft Articles as dealing with secondary rules (and this characterization is widely accepted by commentators), others prefer different language. David Caron, for example, terms the subject matter of the Draft Articles as "trans-substantive rules."[24] By this he means "a set of rules present in state responsibility independently of the particular substantive obligation in question."[25]

As Caron points out, the Draft Articles seek to establish a general law of state responsibility. In an international legal regime where law is subject to increasing specialization that creates individualized systems of dispute resolution and responsibility, this is an important undertaking insofar as it seeks to promulgate

[19] Bodansky and Crook, *supra* note 1, at 775. In the *Trail Smelter* arbitration, both governments were willing to step in and assume responsibility for the acts of the non-state corporations or farmers in question. *See* Miller in this volume.

[20] Edith Brown Weiss, *Invoking State Responsibility in the Twenty-first Century*, 96 American Journal of International Law 798, 798 (2002).

[21] David Bederman, *Counterintuiting Countermeasures*, 96 American Journal of International Law 817, 823 (2002). *See also* Dupuy, *supra* note 2, at 1059.

[22] James Crawford, *The ILC's Articles on Responsibility of States for Internationally Wrongful Acts: A Retrospect*, 96 American Journal of International Law 874, 876 (2002).

[23] David D. Caron, *The ILC's Articles on State Responsibility: The Paradoxical Relationship Between Form and Authority*, 96 American Journal of International Law 857, 870 (2002); Report of the ILC on the Work of Its Twenty-second Session, 2 Yearbook of the International Law Commission, para. 66(c)(1970).

[24] Caron, *supra* note 23, at 871.

[25] *Id.*

some general guidelines. To be sure, this has prompted some commentators to describe the ILC project as "a bit anachronistic."[26] This observation emanates in particular from common-law lawyers, for whom even national legal systems have no generalized rules of responsibility. Instead, substantive rules are classified by subject matter (e.g., contract law, tort law, constitutional law, family law), "each characterized by its own regime of 'responsibility' with its own remedies, rules of attribution and invocation [. . .]."[27] For the *civiliste*, there is more of a natural compatibility with generalized rules of responsibility, although the fragmented nature of the international legal order does make it difficult to conceive of a generalized approach to responsibility, insofar as there may not be an international "legal system" to speak of. The interaction between the Draft Articles and specialized rules of responsibility is clear: the Draft Articles represent only "default or residual rules;"[28] treaty regimes (for example, the WTO, ICC, NAFTA, or regional human rights systems) or specialized custom can develop their own *lex specialis* regarding responsibility that supplants the Draft Articles.

TRAIL SMELTER AND THE DRAFT ARTICLES ON STATE RESPONSIBILITY

The *Trail Smelter* case is referenced four times in the over three hundred pages of Commentaries to the Draft Articles. This is significantly less than many other important international decisions, such as:

- *Air Services Agreement* (1979, international arbitral award) (referenced eight times);
- *Barcelona Traction* (1970, ICJ) (fourteen times);
- *Corfu Channel* (merits and compensation) (1949, ICJ) (seven times);
- *Factory at Chorzow* (jurisdiction and merits) (1927–1928, PCIJ) (twenty-three times);
- *Gabčikovo-Nagymaros* (1997, ICJ) (thirty-four times);
- *LaGrand* (provisional measures and merits) (2001, ICJ) (twenty times);
- *Loizidou v. Turkey* (preliminary objections and merits) (1995–1996, ECHR) (6 times);
- *Nicaragua v. U.S.* (1986, ICJ) (seventeen times);
- *Rainbow Warrior* (1990, ICJ) (twenty-four times);
- *The M/V Saiga* (1999, ITLOS) (six times);
- *The S.S. Wimbledon* (1923, PCIJ) (eleven times); and
- the various awards and decisions of the Iran-United States Claims Tribunal (twenty-eight awards and decisions cited).

[26] Bodansky and Crook, *supra* note 1, at 774.

[27] *Id.* at 780.

[28] *Id. See also* Draft Articles, *supra* note 1, art. 55.

Many of the sources to which frequent reference is made are ICJ, PCIJ, or ITLOS decisions, which are imbued with considerable international legal authority as subsidiary sources of international law. Within the fairly large peer group of arbitral decisions, *Trail Smelter* clearly is among the more frequently referenced. In the end, it is fair to say that the influence of the *Trail Smelter* decision exceeds its rather modest provenance. This phenomenon can be explained in part by the fact that *Trail Smelter* constitutes one of the first renditions of state responsibility for transboundary environmental harms.[29]

Trail Smelter influences four aspects of the Draft Articles: the continuation of the obligation to prevent a wrongful act; special measures to guarantee nonrepetition, remoteness of harm, and compensation for reduction in property value. This chapter now considers each of these in turn.

The Continuation of the Obligation to Prevent a Wrongful Act

Draft Article 14 (extension in time of the breach of an international obligation) provides in subsection (3) that: "The breach of an international obligation requiring a state to prevent a given event occurs when the event occurs and extends over the entire period during which the event continues and remains not in conformity with that obligation." Draft Article 14(3) thus deals with the temporal dimensions of a subset of international obligations, notably the breach of an obligation to prevent the occurrence of an event. Obligations of prevention usually are construed as best efforts obligations, requiring states to take all reasonable or necessary measures to prevent a given event from occurring, This does not, however, rise to the level of a strict liability warranty that the event in question shall not occur. *Trail Smelter* is cited in Commentary (14) to Draft Article 14(3) in support of the proposition that the breach of an obligation may well be a continuing wrongful act, although the breach only continues if the state is bound by the obligation for the period during which the event continues and remains not in conformity with what is required by the obligation. "For example, the obligation to prevent transboundary damage by air pollution, dealt with in the *Trail Smelter* arbitration, was breached for as long as the pollution continued to be emitted. Indeed, in such cases the breach may be progressively aggravated by the failure to suppress it."[30] Continuation of a wrongful act entails a continuation of responsibility.

Special Measures to Guarantee Nonrepetition

Draft Article 30 (cessation and nonrepetition) provides that: "The State responsible for the internationally wrongful act is under an obligation: (a) to cease

[29] *See* Karin Mickelson, *Rereading Trail Smelter*, 31 CANADIAN YEARBOOK OF INTERNATIONAL LAW 219 (1993). Republished in this volume.

[30] CRAWFORD, *supra* note 1, at 140.

that act, if it is continuing; (b) to offer appropriate assurances and guarantees of non-repetition, if circumstances so require." *Trail Smelter* is cited in a footnote to Commentary (13) on Draft Article 30 in support of the proposition that the injured state may require specific conduct to be taken by the injuring state. In *Trail Smelter*, after all, the Tribunal specified measures to be taken by Canada, including measures designed to "prevent future significant fumigations in the United States."[31] These requests for specific conduct go to the obligation on the part of the responsible state to avoid repetition. To be sure, this may be an exceptional remedy – "if circumstances so require" – although, as *Trail Smelter* may well indicate, transboundary environmental harm could present a compelling circumstance to order such an exceptional remedy.

Remoteness of Harm

Draft Article 31 (reparation) states as follows: "31(1). The responsible state is under an obligation to make full reparation for the injury caused by the internationally wrongful act; (2) Injury includes any damage, whether material or moral, caused by the internationally wrongful act of a State." According to the Draft Articles, breaches of an international obligation carry with them in principle a duty to repair harm caused. *Trail Smelter* is cited in the Commentary (10) in regard to discussion of the allocation of injury or loss to a wrongful act. What link must exist between the wrongful act and the injury in order for the obligation of reparation to arise? The *Trail Smelter* Tribunal held that damage that is "too indirect, remote, and uncertain to be appraised" falls outside the scope of the reparative obligation.[32] To this end, causality-in-fact is one element of the obligation to make reparation but, even assuming causality, the obligation ceases once the injury is too remote or indirect from the initial or ongoing breach. Moreover, even if there is general causality, the obligation ceases when the damage is too uncertain to be appraised.

Compensation for Reduction in Property Value

Draft Article 36 (compensation) stipulates: "36(1) The State responsible for an internationally wrongful act is under an obligation to compensate for the damage caused thereby, insofar as such damage is not made good by restitution; (2) The compensation shall not cover any financially assessable damage including loss of profits insofar as it is established." At this stage, it is important to recall that the Draft Articles, reflecting the famous PCIJ decision in *Chorzow Factory*, prefer restitution in kind (namely making up for the harm caused) over

[31] *Trail Smelter* (1941), *supra* note 7. *See* Bratspies and Miller in this volume.

[32] *Trail Smelter Arbitral Decision*, 33 American Journal of International Law 182, 206 (1939) [hereinafter "*Trail Smelter* (1939)"]. *See* Annex to this volume.

compensation, but require compensation when restitution is impossible, problematic, or only partially satisfactory.[33] There are no punitive or exemplary awards: damages fulfill a purely compensatory purpose.[34] *Trail Smelter* is cited in Commentary (15) to Draft Article 36. It is used to assist in determining the scope of payments when compensation has been awarded or granted following an internationally wrongful act that causes or threatens environmental damage. In particular, *Trail Smelter* creates precedent for directing payments to provide compensation for a reduction in the value of polluted property. After a thorough, and oft-neglected, discussion of international indemnity principles, the *Trail Smelter* panel compensated the United States for damage to land and property caused by the sulfur dioxide emissions that crossed the border. Damages were assessed on the basis of the reduction in value of the affected land. Of course, the *Trail Smelter* principle is narrow, insofar as it does not contemplate environmental damage that cannot be readily quantified in terms of clean-up costs or property devaluation. Commentary (15), by contrast, raises this point. In particular, it focuses on damage to "non-use values," such as biodiversity, which "as a matter of principle, [is] no less real and compensable than damage to property, though it may be difficult to quantify."[35]

TRAIL SMELTER, STATE RESPONSIBILITY, AND INTERNATIONAL ENVIRONMENTAL LAW

No discussion of the relationship between *Trail Smelter* and the Draft Rules is complete without unpacking the relationship between state responsibility and international environmental law generally. Here, I see an important distinction in terms of the currency of *Trail Smelter*'s primary rule (no state has the right to use or permit the use of its territory in such a manner as to cause serious injury established by clear evidence in or to the territory of another), on the one hand, and the currency of *Trail Smelter*'s secondary obligation of reparation and compensation for violation of that primary rule, on the other hand. This distinction arises from Dinah Shelton's astute observation that there has been a shift in the way the international community responds to interstate breaches of primary obligations. This shift inclines toward facilitating compliance instead of seeking compensation for noncompliance. In fact, Shelton argues, "[i]nterstate issues of compliance and breach are increasingly handled through nonconfrontational procedures within international organizations and treaty bodies. The rise

33 Chorzow Factory (Germany v. Poland), Indemnity, 1928 P.C.I.J (ser. A) No. 17 (Sept. 13).

34 Dinah Shelton, *Righting Wrongs: Reparations in the Articles on State Responsibility*, 96 American Journal of International Law 833, 844 (2002).

35 Crawford, *supra* note 1, at 223. In this sense, the Draft Articles evade the "visible" and "invisible" injury debate with which the Trail Smelter Tribunal was confronted. *See* John D. Wirth, Smelter Smoke in North America (2000).

of such nonadversarial compliance procedures seems to have brought a corresponding decline in recourse to the law of state responsibility."[36] Compliance can be facilitated through carrots rather than sticks and often in the context of specialized rules and regimes (for example, the compliance mechanisms envisioned by the Montréal Protocol on Substances that Deplete the Ozone Layer).[37]

In addition to movements toward managing compliance[38] through incentives, reporting, capacity-building, and monitoring, another reason the Draft Articles may have limited influence on elevating the profile of *Trail Smelter* is that the Draft Articles do not address the content of primary rules such as the duty of care, which remain ambiguous. This ambiguity constitutes another factor that dissuades states from bringing claims. Furthermore, even if responsibility is found, providing restitution or compensation is often incredibly expensive and onerous. This is especially so when those responsible are developing nations. Who will pay? If there can be no payment, why sue? This shift from *ex post* enforcement to *ex ante* compliance thus may dissuade the flowering of *Trail Smelter*-style arbitration as a method of dispute resolution.[39]

Secondary rules permitting reparations from states that cause transboundary environmental harm do occupy a vital residual role. They may set the stage for a more effective compliance regime, insofar as the threat of reparations for noncompliance is an important implicit "stick." This, in turn, helps imbue international law with some predictability and accessibility. Moreover, remedies are relevant to some injured parties. Litigation at times may serve a useful purpose in terms of setting precedent, expressing norms, and vindicating rights, even when its actual effectiveness may be more symbolic than tangible.[40]

Nonetheless, despite the ILC's clarification of the rules regarding state responsibility, *Trail Smelter* may continue to lead a somewhat lonely existence. Shelton observes that "the *Trail Smelter* arbitration is almost alone today in being cited for state responsibility and reparations in the field of environmental protection, because virtually no interstate cases have been brought in the decades since it was decided."[41] As such, *Trail Smelter* may have had some influence in terms of defining the content of state responsibility, but may offer little in the way of precedent for "law-in-practice," namely "the law that counts in world politics."[42] As the international order moves toward "promoting changes in state behavior

[36] Shelton, *supra* note 34, at 854.

[37] Montreal Protocol on Substances that Deplete the Ozone Layer, Sept. 16, 1987, 1522 U.N.T.S. 3, *reprinted in* 26 International Legal Materials 1550 (1987), art. 5.

[38] Abram Chayes and Antonia Handler Chayes, The New Sovereignty (1995).

[39] *See* Craik in this volume.

[40] For instance, the invocation of the state responsibility principle in the *Gabčikovo-Nagymaros* litigation arguably furthered each of these goals.

[41] Shelton, *supra* note 34, at 854–55.

[42] International Incidents: The Law That Counts in World Politics (W. Michael Reisman & Andrew R. Willard eds. 1988).

rather than sanctioning breaches of international obligations,"[43] I posit that *Trail Smelter* may remain somewhat solitary as an example of the "law-in-practice." With these observations in mind, then, is it at all surprising that no interstate responsibility claims were brought regarding, for example, the catastrophic transboundary harms occasioned by the Chernobyl incident?[44]

TRAIL SMELTER AND THE COLLATERAL TOPIC OF STATE LIABILITY

Trail Smelter may exert greater influence in terms of "law-in-practice" in the area of state liability, particularly concerning prevention, which also remains an important topic for the ILC. In the seventy pages of Commentaries to the 2001 Draft Articles on Prevention of Transboundary Harm from Hazardous Activities (Prevention Draft Articles), *Trail Smelter* is cited four times, and in a more extensive manner than in the Draft Articles and supplemental commentaries on state responsibility. By way of overview, the Prevention Draft Articles apply to activities not prohibited by international law that involve a risk of causing significant transboundary harm.[45] Central to the Prevention Draft Articles is the obligation of due diligence.[46] The Prevention Draft Articles also require states to authorize hazardous activity before the initiation of that activity. There is thus an important procedural aspect to the Prevention Draft Articles that parallels obligations in domestic law for the undertaking of impact assessments before commencing projects that could have deleterious effects on the environment. The precautionary principle and the polluter-pays principle are noted, although neither is suggested as a strict legal obligation.[47]

Trail Smelter is cited in General Commentary (4) to the Prevention Draft Articles in support of the well-established nature of the principle of prevention, which is viewed as a procedure or duty involving the phase before the situation arises where significant harm or damage might actually occur.[48] Here the Commentaries reference *Trail Smelter* in the same line as they do Principle 21 of the Stockholm Declaration, Principle 2 of the Rio Declaration, and General Assembly Resolution 2995 (XXVII) (December 15, 1972).[49]

[43] Shelton, *supra* note 34, at 855.

[44] Alexandre Kiss and Dinah Shelton, International Environmental Law 551–52 (2d ed. 2000).

[45] Draft Articles on Prevention, *supra* note 12, art. 1.

[46] *Id.* art. 3. The scope of due diligence is that which is generally considered appropriate and proportional to the risk of transboundary harm in the case in question. A state must properly inform itself of factual and legal components of the activities contemplated and, moreover, must take appropriate responsive measures in a timely fashion. A contextual analysis – accounting for a state's economic level – is appropriate in determining whether the obligation of due diligence has been satisfied, although a state's economic level cannot be used to exempt a state from its obligations. Draft Articles on Prevention, *supra* note 12, art. 10.

[47] When this risk assessment suggests the likelihood of significant transboundary harm, the state of origin is to provide notification and information to other states that are likely to be affected. Draft Articles on Prevention, *supra* note 12, arts. 6–13. *See* Bratspies in this volume.

[48] *Id.* commentaries at 378.

[49] *Id.*

Trail Smelter also is cited in Commentary (6) to Article 2 (use of terms). Article 2 defines a number of terms. One such term is the "risk of causing significant transboundary harm," which is defined as including "risks taking the form of a high probability of causing significant transboundary harm and a low probability of causing disastrous transboundary harm."[50] *Trail Smelter* is cited in the Commentaries in support of the "idea of a threshold." Commentary (6) states that the use of the words "serious consequences" in *Trail Smelter* indicates that environmental impacts must bear the risk of serious consequences before the threshold of preventative duties arises.[51]

What is more, *Trail Smelter* is cited in Commentary (2) to Article 6 (authorization), specifically Article 6(1)(a) ("The State of origin shall require its prior authorization for: (a) Any activity within the scope of the present articles carried out in its territory or otherwise under its jurisdiction or control"), as informing the substantive scope of that control.[52] Commentary (2) reads as follows:

> The requirement of authorization [. . .] obliges a State to ascertain whether activities with a possible risk of significant transboundary harm are taking place in its territory or otherwise under its jurisdiction or control and implies that the State should take the measures indicated in these articles. It also requires the State to take a responsible and active role in regulating such activities. The Tribunal in the *Trail Smelter* arbitration held that Canada had 'the duty . . . to see to it that this conduct should be in conformity with the obligation of the Dominion under international law as herein determined'. The Tribunal held that, in particular, 'the Trail smelter shall be required to refrain from causing any damage through fumes in the state of Washington'. Article 6(1)(a) is compatible with this requirement.[53]

Finally, *Trail Smelter* is cited in Commentary (2) to Article 7 (assessment of risk). Article 7 provides that "[a]ny decision in respect of the authorization of an activity within the scope of the present articles shall, in particular, be based on an assessment of the possible transboundary harm caused by that activity, including any environmental impact assessment." Commentary (2) mentions the study undertaken in the *Trail Smelter* case. This study was undertaken "by well-established and known scientists" and was, in the words of the Tribunal, "probably the most thorough [one] ever made of any area subject to atmospheric pollution by industrial smoke."[54] Although the Commentaries observe that the assessment of risk in *Trail Smelter* may not directly relate to liability for risk, *Trail Smelter* is cited in support of the "importance of an assessment of the consequences of an activity causing significant risk."[55]

Prevention aligns more closely with managing compliance than does *ex post* reparative responsibility for internationally wrongful or merely harmful acts. I posit that – looking ahead – this is the area in which *Trail Smelter* may carry greater

[50] *Id.* at 386.
[51] *Id.* at 388.
[52] *Id.* at 399.
[53] *Id.* at 400.
[54] *Id.* at 402.
[55] *Id.*

influence. This should not be surprising, insofar as the approach of the *Trail Smelter* Tribunal was not one of shutting down the smelter to protect the farmers' property interests but, rather, one of managed compliance, namely permitting the smelting to continue (and thereby recognizing the centrality of the smelter to the economic life of Trail), but only if the emissions could be controlled such that they did not adversely affect the agricultural interests of the Washington farmers.[56] The approach was one of balance, not a zero-sum game.

The law of state liability involves multiple subsets. Prevention is only one of these. It goes without saying that, even if states comply with the duty of prevention, harm still could occur.[57] Preventative measures, even if faithfully implemented, could prove inadequate or the particular risk that causes the harm may not have been identified at the time.[58] Consequently, the ILC has considered what the law of state liability might mandate in the event of harm that occurs when the triggering event falls outside of the law of state responsibility. In 2004, with these considerations in mind, the ILC in its fifty-sixth session finalized eight preliminary draft principles on allocation of loss in the case of transboundary harm arising out of hazardous activities, together with commentaries.[59] The ILC has transmitted these draft principles, through the U.N. Secretary-General, to governments for comments and observations. The influence of *Trail Smelter* on allocation is more muted than its influence on prevention. In the Draft principles on allocation and commentaries thereto, *Trail Smelter* is referenced only once. This reference occurs in Commentary (1) to Principle (2)(a), which defines "damage" as significant damage to persons, property, or the environment. Commentary (1) cites *Trail Smelter's* concern with the "serious consequences" of the smelter's operation and, with this as a baseline, states that "it is important to note that damage to be eligible for compensation should acquire a certain threshold and that in turn would trigger the operation of the present draft principles."[60] *Trail Smelter*, along with a number of other treaty and jurisprudential sources, is used to qualify the threshold of harm or damage that would invoke state liability.[61] The fact that *Trail Smelter* is only referred to once in these Commentaries supports the notion that the primary

[56] WIRTH, *supra* note 35, at xv. *See* Jacobson in this volume. For a contrary view, *see* Allum in this volume.

[57] International Law Commission, Report of the Working Group on International Liability for Injurious Consequences Arising Out of Acts Not Prohibited by International Law, 54th Sess., U.N. Doc. A/CN.4/L.627 (2002), para. 3.

[58] The Working Group conceptualized state liability along the lines of managing compliance, in particular as "allocation of loss among different actors involved in the operations, such as, for instance, those authorizing, managing or benefitting from them." *Id.* One possibility is for risk to be shared through specific regimes or insurance mechanisms. The Working Group also posits that in terms of managing the distribution of loss, it may be feasible for the operator to bear primary liability, limited by the ability to pay, with states also playing a role, although that role need to be defined. *Id.* at paras. 10–15.

[59] Draft Principles on Allocation, *supra* note 13.

[60] *Id.* commentaries at 170.

[61] *Id.* at 170–71.

jurisprudential legacy of *Trail Smelter* in international environmental law lies in the area of prevention and not reparation or satisfaction of harm.

CONCLUSION

This chapter tracks the influence that the *Trail Smelter* arbitration has had on the 2001 Draft Articles on State Responsibility. The Draft Articles establish the secondary obligations that flow from a breach of an independent and preexisting primary obligation. The Draft Articles address a number of issues, including the attribution of conduct to a state, justifications for breach, circumstances precluding wrongfulness, reparation, compensation, bilateral and *erga omnes* obligations, and countermeasures. *Trail Smelter* is referenced four times in the Commentaries to the Draft Articles. It has played some part in the formulation of specific articles regarding continuing breach, non-repetition of breach, remoteness of harm, and compensation. On a more general note, however, this Chapter concludes that *Trail Smelter's* influence is limited by shifts in international environmental law toward facilitating compliance with primary rules rather than seeking compensation for breaches of those rules. This explains why *Trail Smelter's* evocation of secondary obligations – and the law of state responsibility more generally – lead somewhat of a lonely existence in terms of the "law-in-practice" regarding international environmental protection. *Trail Smelter* has played a more vivid role in the ILC's work on state liability, in particular the preventative aspects. Here, the primary rule of *Trail Smelter* – namely, the obligation not to cause serious environmental harm – has acquired some currency, exceeding that of *Trail Smelter's* secondary obligation of reparation and compensation for violation of that primary rule. Moreover, this secondary obligation unsurprisingly plays a limited role in the aspects of the law of state liability that parallel the content of the Draft Articles on State Responsibility, namely allocation, thereby supporting the general thesis that the dominant legacy of *Trail Smelter* lies in the obligation not to commit serious environmental harm rather than the separate obligations that flow once harm – whether intentional, negligent, or accidental – has been occasioned.

9 Derivative versus Direct Liability as a Basis for State Liability for Transboundary Harms

Mark Anderson

INTRODUCTION

The *Trail Smelter* arbitration[1] has become synonymous with the "polluter pays" principle[2] – the basic vision of environmental liability that infuses both domestic and international law. Under the "polluter pays" principle, the polluter bears the expense of carrying out pollution prevention, control, and remediation measures.[3] It is somewhat ironic that despite frequent characterizations of *Trail Smelter* as embodying, or even establishing, this principle, Consolidated Mining, the owner and operator of the polluting smelter, was not a party to the arbitration. The arbitration was instead between Canada and the United States, and the Tribunal's award directed Canada, not the smelter, to pay damages for environmental harms suffered in the United States. Given this somewhat paradoxical relationship between the perception of *Trail Smelter* as creating a "polluter pays" principle and the fact that the polluter was not a party to the arbitration, it is worth exploring the alternative theories of liability that could provide a basis for the *Trail Smelter* principle.

I am reminded of a story a friend once told me. While on a research expedition in a remote part of the world, one expedition member's camera was stolen. The

[1] *See Trail Smelter* Arbitral Decision, 33 AMERICAN JOURNAL OF INTERNATIONAL LAW 182 (1939) [hereinafter "*Trail Smelter* (1939)"]; *Trail Smelter Arbitral Decision*, 35 AMERICAN JOURNAL OF INTERNALTIONAL LAW 684, 716 (1941) [hereinafter "*Trail Smelter* (1941)"]. *See* Annex to this volume.

[2] *See, e.g.*, Daniel Esty, *Towards Optimal Environmental Governance*, 74 NEW YORK UNIVERSITY LAW REVIEW 1495 (1999) (describing the *Trail Smelter* arbitration as enforcing the polluter pays principle); *see also*, Günther Handl, *Territorial Sovereignty and the Problem of Transnational Pollution*, 69 AMERICAN JOURNAL OF INTERNATIONAL LAW 50, 60–61 (1975). The citations to the *Trail Smelter* arbitration for this point are far too numerous to reproduce here, but readers looking for a typical example should review Hyun S. Lee, *Post Trusteeship Environmental Accountability: Case of PCB Contamination on the Marshall Islands*, 26 DENVER JOURNAL OF INTERNATIONAL LAW AND POLICY 399, 413–414 (1998).

[3] *See* Organization for Economic Co-Operation and Development: Council Recommendation on the Application of the Polluter-Pays Principle to Accidental Pollution, July 7, 1989, *reprinted in* 28 INTERNATIONAL LEGAL MATERIALS 1320 (1989).

expedition members contacted the local police. Several police officers went to the nearby village and summoned the villagers. The officers identified the village leader and told the villagers that the leader would be beaten until the camera was returned. The officers proceeded to beat the village leader until eventually the thief returned the camera.

This possibly apocryphal story contains a fundamental question that often arises in the law, and plainly arose in the *Trail Smelter* arbitration. When should one person bear consequences for the conduct of another person? Should the village leader suffer the beating because the thief stole the camera? Although the beating of the village leader was effective in securing the return of the camera, it is shocking since there was no evidence that he either encouraged the thief to take the camera or participated in some other way in the theft. Somewhat analogously, in the *Trail Smelter* arbitration, Canada bears liability for the damage caused by sulfur smoke emitted by the Trail smelter, yet this outcome is not similarly shocking. In both situations, a primary actor caused harm and a secondary actor suffered a consequence in relation to the conduct of the primary actor.

In both *Trail Smelter* and in the case of the stolen camera, the liability of the secondary actor could be based on two different approaches to liability. First, the primary actor could be considered the principal wrongdoer and the secondary actor could be considered so connected to the wrong committed by the primary actor that the secondary actor should be held derivatively liable. Alternatively, the conduct of the primary actor could be considered relatively insignificant in assessing the liability of the secondary actor. Although the primary actor may have physically committed the act which resulted in the harm, the secondary actor could be viewed as the main wrongdoer. Under this approach, the secondary actor is held directly liable for its own wrong rather than derivatively liable for a wrong committed by the primary actor. In short, the derivative liability approach views the secondary actor as liable for a wrong committed by the primary actor, while the direct liability approach views the secondary actor as committing the wrong itself.

This chapter will assess whether the *Trail Smelter* principle is more appropriately analyzed as an implementation of derivative or direct liability as these concepts are implemented under the domestic law of the United States. It will begin by analyzing derivative and direct liability doctrines in the civil and criminal settings. It will then analogize Canada's liability in the *Trail Smelter* arbitration to both derivative and direct liability doctrines in both civil and criminal settings. In doing so it will examine how choosing between derivative and direct liability analysis will alter the scope of liability under the *Trail Smelter* principle.

DERIVATIVE LIABILITY EXPLAINED

Under domestic law in the United States, derivative liability in both the civil and criminal contexts usually depends on a high degree of intentionality on the part of the secondary actor. The most overt form of derivative civil liability is based on express contracts for liability insurance. In exchange for money, the insurer agrees to assume liability for certain claims against the insured. The insurer's liability is controlled by the intention expressed in the insurance contract.

Similarly, a principal may bear derivative liability in contract for the actions of an agent. An agent with actual or apparent authority may bind the principal in contract. This liability is derivative in nature, because it is the agent rather than the principal who enters into the contract on behalf of the principal. The derivative liability of the principal is affected by its intention as actual and apparent authority are based on the manifestations of the principal in regarding the agent.

A principal also may bear derivative liability for the torts of an agent if the principal is a master, the agent is a servant, and the servant is operating within the scope of his or her employment. This liability derives from the master's intentional acquisition of control over how the servant performs services within his or her scope of employment. The principal can avoid derivative liability by allowing his or her agents sufficient discretion with regard to the physical performance of those duties so that the principal avoids becoming a master of a servant agent.

Agency law also limits the liability of secondary actors by allowing them to intentionally avoid becoming principals or masters. Lenders, landlords, licensors, and customers all can benefit from the actions of their borrowers, tenants, licensees, and suppliers without being liable for their conduct. In the *Trail Smelter* arbitration, those who supplied raw materials to the smelter, purchased the metal produced by the smelter, leased equipment or licensed technology for use in the smelter were not derivatively liable for the smelter's conduct that adversely affected its neighbors. Each of these secondary actors benefited from the smelter's conduct, but none stood in a principal-agent relationship with it.

Under criminal law in the United States, derivative liability also usually depends on a high level of intentionality on the part of the secondary actor. Accomplice liability under the Model Penal Code requires that the secondary actor intends to promote the conduct of the primary actor, the principal.[4] If the accomplice intends to promote the conduct of the principal, a lower level of intentionality may be sufficient regarding the results of the principal's conduct.[5] For example, an accomplice may purposely encourage the principal to fire a machine gun into the air. The accomplice may be negligent with regard to whether one of the bullets fired by the principal will strike someone. Because the accomplice is purposeful regarding the conduct of the principal, his or her negligence regarding the result of the conduct will make the accomplice liable for negligent homicide if someone dies from the conduct. Thus, under the Model Penal Code an assessment of the intentionality of an accomplice to a crime requiring a result must proceed both as to the conduct of the principal and as to the result of that

4 Model Penal Code 2.06(3)(1962).

5 Model Penal Code 2.06(4)(1962).

conduct.[6] This two-level inquiry will become important in the analogy to *Trail Smelter*.

DIRECT LIABILITY EXPLAINED

Under domestic law in the United States, direct liability on the part of a secondary actor is also possible in both the civil and criminal contexts. In the contractual context, an agent may refrain from entering into a contract on behalf of the principal. Rather, the agent may present an offer to the principal for its acceptance. If the principal accepts the offer, it incurs direct liability to the third party based on its own acceptance of the offer communicated by the agent. Similarly, the unauthorized contract of an agent may be ratified by a principal. In these situations, the liability of the principal depends on its own conduct in accepting the offer or ratifying the contract, not on the contractual act of the agent.

In the tort context a principal may be directly liable to an injured third party. The principal's direct liability may be based on its negligence in selecting and supervising an agent. When the agent injures a third party, the principal's direct liability depends on the negligence of the principal, not on the imputed negligence of a servant agent. The direct liability of the negligent principal depends on its own breach of a duty owed to the injured party.

In the criminal context, the secondary actor may also become directly liable despite the participation of the primary actor. If the secondary actor is a cause in fact and proximate cause of the harm to the victim, the participation of the primary actor does not preclude direct liability for the secondary actor. For example, if a bomber sends a bomb through a package delivery service, the fact that the package delivery service is the primary actor in handing the bomb to the victim does not prevent the bomber from being directly liable. The bomber is directly liable for his or her own crime. The bomber is not derivatively liable for a crime committed by the package delivery service. The package delivery service committed no crime. In the bomber example, the bomber happens to have a high level of intentionality. However, a high level of intentionality is not necessary for direct criminal liability. If, instead of a bomb, the shipper had negligently sent a hazardous package to the recipient, the negligence of the shipper could be the basis for criminal liability for the death of the recipient. Because the delivery service lacked information about the harmful nature of the package, its participation as a primary actor would not preclude direct criminal liability of the shipper as a secondary actor.

Similarly, the secondary actor may be viewed as a proximate cause of harm when a primary actor lacks volition rather than information. For example, if the

[6] The analysis in the text is based on the Model Penal Code. Some jurisdictions depart from the accomplice liability approach to intentionality advocated by the Code. In some jurisdictions if an accomplice intends to aid one crime, he is liable for foreseeable crimes committed by the principal. *See, e.g.*, People v. Luparello, 187 Cal. App. 3d 410, 231 Cal. Rptr. 832 (1987).

secondary actor is a terrorist who points a gun at the primary actor and directs the primary actor to burn down a building, the volition of the primary actor is impaired and the terrorist is the proximate cause of the arson. The terrorist in this example happens to have a high level of intentionality regarding the harm to the building. However, as noted earlier, proximate causation does not depend on a high level of intentionality on the part of the secondary actor. For example, the secondary actor may have negligently started a forest fire. If the primary actor sees the fire and plows a fire break to stop its progress, the fact that the primary actor faced a very limited set of choices will make the secondary actor the proximate cause of the damage done by the primary actor's foreseeable conduct.

If the primary actor lacks information or volition regarding the harm he or she commits, the secondary actor remains the proximate cause of the harm committed by the primary actor. The lack of blameworthiness on the part of the primary actor leads to the conclusion that the secondary actor should be viewed as a proximate cause of the harm.

In *Trail Smelter*, Canada's liability could be thought of as derivative or direct. Conceiving of Canada's liability as derivative or direct could make a difference in the scope of its liability. The next section of this chapter will analogize transboundary harm to the derivative and direct liability doctrines discussed earlier.

EXPLORING CANADA'S DERIVATIVE LIABILITY IN THE *TRAIL SMELTER* ARBITRATION

Focusing first on derivative liability in the civil context, agency principles would lead to the conclusion that Consolidated's actions in operating the smelter should not be attributed to Canada as the secondary actor. Consolidated was not Canada's agent since it did not act on Canada's behalf.[7] Because it was not employed to perform services in the affairs of Canada, neither was Consolidated Canada's servant. Therefore, agency law would not attribute the conduct of Consolidated to Canada, even though, like the customers, suppliers, and employees of Consolidated, Canada may have benefited from Consolidated's conduct.

Analogizing to criminal law theories of derivative liability is more complex. Under those theories, Canada's liability could be based on an omission, that is, permitting its territory to be used to injure parties in the United States. This indeed, was the famous language employed by the Tribunal: "no State has the right to use or permit the use of its territory in such a manner as to cause injury by fumes in or to the territory of another or the properties or persons therein."[8] An omission will implicate a secondary actor as an accomplice only if there is an underlying duty. However, accomplice liability requires more than just an omission in the presence

[7] Restatement (Third) of Agency, 1.01 (Tentative Draft No. 2, 2001).

[8] *Trail Smelter* (1941), *supra* note 1, at 716.

of a duty. Determining whether Canada was an accomplice to Consolidated's pollution requires an analysis of intentionality. Canada did have the purpose of promoting the smelting of ore at Trail. It may even have intended to promote the process that released the sulfur gas. However, Canada presumably did not have the purpose to promote the harm caused in the United States by the release of the gas. But, like the accomplice encouraging machine gun fire into the air discussed earlier, an accomplice who intends to promote the conduct of the principal, may be liable even without intending that the conduct cause the harmful result. This is so if the accomplice's level of intentionality would be sufficient to impose liability on it as a principal.[9] In *Trail Smelter,* Canada may have had the purpose of promoting the industrial activity that resulted in the release of sulfur gas and may have been at least negligent with regard to whether harm would occur in the United States.

The intentionality issue would be substantially more complex in a case where the government did not know of the hazardous activity. Such a situation is especially likely where the hazard is of low probability but high harm, for example, a very low risk that an industrial plant may explode with environmentally catastrophic consequences because a hazardous process is being used without adequate safety measures. In such a case, the government may not know that the hazardous process is being used or, alternatively, that safety measures are inadequate. Under these circumstances, the State would lack the intent to promote the conduct of the principal required for accomplice liability.

By contrast, in *Trail Smelter* Canada knew that Consolidated was conducting the smelting activity which emitted the sulfur gas. Indeed, Consolidated had a history of conflict and litigation over the smoke damage inflicted on its Canadian neighbors.[10] The central question for Canada's derivative liability is whether Canada knew that Consolidated's conduct would cause transboundary harm. If a state did not intend to promote the polluting activities of a firm within its borders, however, the level of intention regarding the conduct of the primary actor required for accomplice liability would be absent.

In summary, derivative liability requires a high level of intentionality in either the civil or criminal settings. On the facts of *Trail Smelter* this level of intentionality may be present. However, in many cases of transboundary environmental harm the level of intentionality of the polluting state may not meet this standard. If we conclude that derivative liability is the basis for Canada's liability under the *Trail Smelter* principle, the scope of liability defined by that principle would be quite narrow.

9 Model Penal Code 2.06(4)(1962).

10 *See* Allum in this volume; *see also* John D. Wirth, Smelter Smoke in North America 13–14 (2000).

EXPLORING CANADA'S DIRECT LIABILITY IN THE *TRAIL SMELTER* ARBITRATION

As explained earlier, the alternative basis for liability of a secondary actor is direct liability. In the context of transboundary environmental harms, we might characterize the actor as the polluting state, rather than the polluting industry in the state. Such a characterization would lead to the imposition of direct liability on the polluting state. If the polluting state is directly liable, the standard of liability could be much broader than that for derivative liability.

One basis for direct liability is voluntary assumption of an obligation. As noted earlier, in the agency context a principal can become directly liable to third parties by entering into contracts presented by an agent or ratifying an unauthorized contract. Similarly in the international environmental context, a state may voluntarily assume an obligation to other states, for example, by treaty. In the arbitration agreement that gave rise to the *Trail Smelter* decisions, Canada voluntarily assumed liability for any transboundary harm Consolidated had caused in the United States.[11] If Canada's liability was defined solely on the basis of voluntarily assumed direct liability, then *Trail Smelter* is limited to its unique facts, with little to offer the ongoing conversation about state liability for transboundary harms. However, voluntary assumption is not the only basis for direct liability.

Direct liability can be based on the violation of an obligation by the secondary actor to the victim. In the civil context a secondary actor can incur direct liability by negligently selecting or supervising an agent who causes harm. In the criminal context, a secondary actor who is a cause in fact and proximate cause of the prohibited harm will similarly be held directly liable. In the *Trail Smelter* situation analogies can be drawn to both civil and criminal theories of direct liability. The civil analogy would view Canada as having an obligation to the United States to appropriately select and supervise those who smelt ore within its territory. If Canada negligently selected or supervised a smelter and harm was caused to the United States, it would be liable.

The direct liability criminal analogy is more complex as it involves establishing that Canada was a cause in fact, and a proximate cause of the harm to the United States. To be a cause in fact of the harm to the United States, Canada would need to be a but-for cause of the harm. Since Canada could have prohibited Consolidated from operating the smelter altogether and thereby prevented the harm to the United States, Canada might satisfy a but-for test. Proximate causation requires more extended analysis.

A secondary actor can be viewed as the proximate cause of the primary actor's behavior if the primary actor is not blameworthy. A primary actor that lacks

[11] Convention on Damages Resulting From Operation of Smelter at Trail, British Columbia, June 5, 1935, U.S.-Canada, 49 Stat. 3245. See Annex to this volume.

information or with impaired volition would be considered not blameworthy. In the package delivery examples discussed earlier, the intentional or negligent shipper is viewed as the proximate cause of the harm to the recipient of the package because the package delivery service was uninformed about the content of the package. The package delivery service is not blameworthy because it did not know nor should it have known of the harmful nature of the contents of the package. A primary actor also may not be blameworthy if its volition is impaired. In the examples provided earlier involving primary actors causing harm when responding to terrorist threats or forest fire emergencies, the primary actors are not blameworthy since their choices were constrained by the conduct of the secondary actors. In such situations the secondary actor is considered a proximate cause of the harm committed by the primary actor.

In *Trail Smelter*, Consolidated was obviously not uninformed about its own activities. Nor was Consolidated forced in some way by Canada to emit sulfur fumes. However, international law still could view Canada as the proximate cause of Consolidated's emissions if, for purposes of international law, Canada is perceived as having a higher level of blameworthiness than Consolidated. Austen Parrish, in his contribution to this volume, has observed that the *Trail Smelter* principle is about the accommodation of the sovereignty of neighboring states.[12] In reaching such an accommodation, the conduct of private actors could be viewed as inconsequential. The vast powers of a sovereign under international law may make it relatively more blameworthy than a private actor, at least for the purposes of international law. This assessment of relative blame as between the sovereign and the private actor could be a basis for treating the sovereign as a proximate cause of harm committed by the private actor.

Treating the sovereign as a proximate cause of transboundary harm done by private actors leaves open the question of the degree of intentionality required to impose direct liability. The important point is that the answer to this question is not constrained by the intentionality requirements of derivative liability. Once the secondary actor is deemed a proximate cause, the level of culpability required for liability is the same for the secondary actor as it is for any other person who causes harm. The fact that the secondary actor is secondary becomes irrelevant. The question is merely one of the level of culpability required for parties who cause harm. Nothing in the nature of the two-party scenario constrains the question of culpability.

One attempt to address the issue of the intentionality of states regarding transboundary harm is contained in The Draft Articles on Prevention of Transboundary Harm from Hazardous Activities.[13] The Draft Articles require states to "take all

[12] *See* Parrish in this volume.

[13] Draft Articles on Prevention of Transboundary Harm from Hazardous Activities, Report of the International Law Commission on the Work of Its Fifty-third Session, UN GAOR, 56th Sess., Supp. No. 10, Chpt. V.E.1, UN Doc. A/56/10 (2001). For attribution of internationally wrongful

appropriate measures to prevent significant transboundary harm or at any event to minimize the risk thereof."[14] The commentary to the Draft Articles states that due diligence is the standard to be applied under the articles.[15] The due diligence standard would presumably apply both to the state's diligence in discovering the existence of an activity and to the state's diligence in discovering the likelihood of harm from a known activity. Using this low level of intentionality for states is consistent with the direct liability approach to state liability outlined earlier.

It is interesting to note that the Draft Articles adopt this low level of intentionality for a limited range of transboundary harms that cause significant physical harms. "It was agreed by the Commission that in order to bring the topic within a manageable scope, it should exclude transboundary harm which may be caused by State policies in monetary, socio-economic or similar fields."[16] In this aspect, the Draft Articles mirror the liability principles articulated and applied in the *Trail Smelter* arbitration. The Tribunal rejected the U.S. claims of invisible injury,[17] limiting liability to physical harms "when the case is of serious consequence and the injury is established by clear and convincing evidence."[18]

This result may be one of the major disadvantages of using the *Trail Smelter* arbitration as a baseline. In a postindustrial, information age, it is unclear why a line should be drawn between physical consequences and other consequences. Nonphysical consequences in a state's financial markets, on its reputation with travelers and business people as a safe place to travel and do business, or to its access to telecommunications could be equally devastating. Perhaps a better course would be to set different appropriate levels of intentionality for different kinds of harms, rather than categorically exclude nonphysical harms.

CONCLUSION

In conclusion, liability of a secondary actor in a two-party scenario may be based on a derivative or a direct approach. A high level of intentionality is usually a precondition to derivative liability of a secondary actor. Direct liability of a

acts by organs of states and others closely connected to states, *see* Draft Articles on Responsibility of States for Internationally Wrongful Acts, in Report of the International Law Commission on the Work of Its Fifty-third Session, UN GAOR, 56th Sess., Supp. No. 10, Chpt. IV.E.1, UN Doc. A/56/10 (2001).

[14] Draft Articles on Prevention of Transboundary Harm from Hazardous Activities, *supra* note 13, at Article 3.

[15] *See* Commentary to the Draft Articles on Prevention of Transboundary Harm from Hazardous Activities, *supra* note 13, *republished in* JAMES CRAWFORD, THE INTERNATIONAL LAW COMMISSION'S ARTICLES ON STATE RESPONSIBILITY: INTRODUCTION, TEXT AND COMMENTARIES (2002).

[16] *Id.* at Commentary to Article 1.

[17] *See Trail Smelter* (1939), *supra* note 1, at 194; WIRTH, *supra* note 10 at 39; *see also* Miller in this volume.

[18] *See Trail Smelter* (1941), *supra* note 1, at 716.

secondary actor may be based on lower levels of intentionality. In *Trail Smelter*, Canada voluntarily assumed liability, so choosing between a derivative approach and a direct approach to its liability was unimportant. The Draft Articles on Prevention of Transboundary Harm from Hazardous Activities adopt a low level of intentionality for a limited set of transboundary harms. This is consistent with a direct liability approach.

10 Transboundary Pollution, Unilateralism, and the Limits of Extraterritorial Jurisdiction: The Second *Trail Smelter* Dispute

Neil Craik

INTRODUCTION

On December 11, 2003, the U.S. Environmental Protection Agency (EPA) issued a unilateral administrative order pursuant to the Comprehensive Environmental Response, Compensation and Liability Act (CERCLA),[1] regarding contamination of the Upper Columbia River in Washington State. What is remarkable about this order is that it is directed against a Canadian company, Teck Cominco Metals, Ltd. (TCML), and concerns historical contamination arising from TCML's smelting and refinery operation located in Trail, British Columbia. The facility in question is, of course, the famed Trail smelter. By seeking to impose liability against a company operating outside of U.S. territory, this order marks an unprecedented extension of the EPA's jurisdiction under CERCLA, prompting a formal diplomatic response from the Canadian government.[2]

A second transboundary pollution dispute concerning the Trail smelter presents an intriguing opportunity to consider the continuing relevance of the *Trail Smelter* arbitration.[3] The intrigue arises not only because of the legally historic significance of the facility itself, but because the approach taken by the EPA in the second dispute appears to be a radical departure from the traditional approach respecting transboundary environmental harm that had its genesis in the original *Trail Smelter* arbitration.

[1] 42 U.S.C. §§ 9601 *et seq.* [hereinafter "CERCLA"].

[2] Canadian Embassy, Washington, D.C., *Diplomatic Note No. 0001, dated January* 8, 2004 (on file with author) [hererinafter "Diplomatic Note, January 8, 2004"].

[3] *Trail Smelter Arbitral Decision*, 33 American Journal of International Law 182 (1939) [hereinafter "*Trail Smelter* (1939)"]; *Trail Smelter Arbitral Decision*, 35 American Journal of International Law 684 (1941) [hereinafter "*Trail Smelter* (1941)"]. *See* Annex to this volume.

At the heart of the *Trail Smelter* arbitration and the subsequent development of the *harm principle* in international environmental law is a recognition that the duty to prevent harm is mediated by an opposing right of a state to exploit its own natural resources in accordance with its own environmental policies.[4] This structure, because it involves reconciling competing claims of sovereignty, has resulted in a highly contextual and process-oriented regime of transboundary harm. The approach associated with the *Trail Smelter* arbitration stands in stark contrast to the approach taken by the EPA in issuing its order against TCML. By seeking to impose retroactive liability on TCML for historical environmental harm under CERCLA, the EPA does not concern itself with the rights of the Canadian or British Columbian government to regulate domestic environmental matters. Where the *Trail Smelter* approach is procedural and cooperative, the EPA's imposition of an administrative order against TCML is unilateral and highly provocative. Yet, it also must be acknowledged that the extraterritorial application of one of the U.S. federal government's most far-reaching environmental laws has the potential to be an effective mechanism for imposing liability against transboundary polluters in accordance with the *polluter pays principle*,[5] something the *Trail Smelter* approach and the *harm principle* in general have been criticized for failing to do.[6]

The turn to unilateralism raises the question of whether such an approach is legally justified in light of international rules regarding the extraterritorial application of domestic law. The jurisdictional limits placed by international law on a state in the exercise of its prescriptive powers indicate that the EPA does not have an unfettered right to dictate a liability regime for an activity that occurs outside of its territory. But those limits are far from clear, requiring an assessment of the competing interests of the U.S. and Canadian authorities in the activity being regulated. In this regard, a comparison between the past and present *Trail Smelter* disputes reveals both contrast and confluence: contrast in the opposing approaches to transboundary pollution in each case; confluence in that international law operates in both instances to require the parties to balance their opposing interests in a reasonable and cooperative fashion. In short, while the EPA may seek to address transboundary pollution in a unilateral fashion, international law may draw it back towards the more conciliatory and cooperative approach that lies at the heart of the *Trail Smelter* arbitration.

4 *See* Stockholm Declaration, United Nations Conference on the Human Environment, Stockholm Declaration, June 16, 1972, Principle 21, UN Doc. A/conf.48/14, *reprinted in* 11 INTERNATIONAL LEGAL MATERIALS 1416 (1972) [hereinafter "Stockholm Declaration"].

5 The "polluter pays principle" requires those responsible for pollution to bear the costs of prevention and remediation. *See* United Nations Conference on Environment and Development, Rio Declaration on Environment and Development, June 14, 1992, Principle 16, UN Doc. A/conf.151/5/Rev.1, *reprinted in* 31 INTERNATIONAL LEGAL MATERIALS 874 (1992).

6 *See, e.g.*, John Knox, *The Myth and Reality of Transboundary Environmental Impact Assessment*, 96 AMERICAN JOURNAL OF INTERNATIONAL LAW 291 (2002).

CONTRAST: THE TWO *TRAIL SMELTER* DISPUTES

The *Trail Smelter* Arbitration of 1938 and 1941

It is largely accepted that the question of liability was not formally before the *Trail Smelter* Tribunal, having been conceded by Canada in the *compromis*.[7] However, it was necessary for the panel to articulate a theory of liability in order to determine whether the remedy of cessation could be imposed against Canada, as Canada could only be enjoined from engaging in activities that were contrary to its legal obligations. Thus, although Canada may have agreed to be liable for damages arising from the operation of the Trail smelter, such as they were, it did not concede that the injuries suffered warranted an injunction preventing further emissions without qualification.

The Tribunal's approach was to frame the issue of liability in light of the general duty of states "to protect other States against injurious acts of individuals from within its jurisdiction."[8] The key for the panel was determining what constituted a sufficiently injurious act in order to engage this obligation. In the absence of international jurisprudence on this issue, the Tribunal turned to U.S. Supreme Court decisions respecting interstate rights in the American federalist system,[9] which, in the Tribunal's view, conformed to the general rules of international law.[10] In the cases cited, the U.S. Supreme Court appears well aware that determining the respective rights of quasi-sovereigns differed from a private law context.[11] Part of the concern is constitutional; that the Court should only interfere with the rights of the state to regulate its internal affairs where the interference with another state's rights is of a "serious magnitude."[12] In addition, because the plaintiff state regulates similar activities within its own borders, there must be "clear and convincing evidence" that the threat complained of emanates from the defendant's activities alone.[13] But once causation of damages in the affected state is established, this line of cases holds that the affected state is entitled to an injunction preventing

7 *See* John E. Read, *The Trail Smelter Dispute*, 1 CANADIAN YEARBOOK OF INTERNATIONAL LAW 213 (1963); Karin Mickelson, *Rereading Trail Smelter*, 31 CANADIAN YEARBOOK OF INTERNATIONAL LAW 219 (1993). Both republished in this volume.

8 *Trail Smelter* (1941), *supra* note 3, at 713 (quoting CLYDE EAGLETON, RESPONSIBILITY OF STATES IN INTERNATIONAL LAW 80 (1928)).

9 State of Missouri v. State of Illinois, 200 U.S. 496 (1906); Kansas v. Colorado, 185 U.S. 125 (1907); New York v. New Jersey, 256 U.S. 296 (1921); New Jersey v. New York, 283 U.S. 473 (1931); State of Georgia v. Tennessee Copper Company and Ducktown Sulphur, Copper and Iron Company, Limited, 206 U.S. 230 (1907).

10 *Trail Smelter* (1941), *supra* note 3, at 713.

11 *See* New York v. New Jersey, 256 U.S. 296, 309 (1921). *See also* Georgia v. Tennessee Copper Company, 206 U.S. 230, 238 (1907).

12 State of Missouri v. State of Illinois, 200 U.S. 496, 521 (1906) (cited in *Trail Smelter* (1941), *supra* note 3, at 714–715).

13 *Id.* at 521–22.

further damages. On this basis, the *Trail Smelter* Tribunal put forward its famous conclusion that,

> No State has the right to use or permit the use of its territory in such a manner as to cause injury by fumes in or to the territory of another or the properties or persons therein, when the case is of serious consequence and the injury is established by clear and convincing evidence.[14]

Despite the Tribunal's decision to hold a state strictly liable for damages arising from activities by a private corporation operating within the state's jurisdiction,[15] the subsequent development of international rules respecting liability for transboundary harm has been more equivocal in its treatment of the opposing rights of the source state and the affected state.[16] Principle 21 of the Stockholm Declaration explicitly recognizes that the right of a state not to be subject to transboundary harm is balanced against the right of economic development.[17] As such, the threshold requirement of the principle announced in *Trail Smelter* acknowledges that states must withstand a tolerable level of transboundary harm. Moreover, the duty to prevent harm is now understood as an obligation of due diligence, not strict liability, requiring an affected state to show that the source state acted unreasonably and that the harm in question was foreseeable.[18]

The difficulty that arises is that determinations of "significance" and "reasonableness" depend on the particular circumstances of each case, such as the nature of the harm in question, the risk it poses, the location of the harm in relation to natural features and human activity, as well as the particular capabilities of the state in question.[19] The result is a rule that is unpredictable to apply given the absence of quantifiable substantive standards. As well, the rule is often underlain by a concern that a precedent could negatively affect future economic activity, what Karin Mickelson refers to as "reciprocity of risk."[20] In essence, states have an incentive to maintain a level of ambiguity in respect of due diligence obligations as a shield against their own transgressions.[21]

In the face of competing claims based on sovereign rights, the *harm principle* has tended to turn toward process, requiring states to carry out prior environmental impact assessments, notify and consult with states potentially affected by those

[14] *Trail Smelter* (1941), *supra* note 3, at 716.

[15] *See* Anderson in this volume.

[16] *See* Jacobson in this volume.

[17] Stockholm Declaration, *supra* note 4, Principle 21.

[18] Knox, *supra* note 6, at 293–294.

[19] *See, e.g.*, United Nations Convention on the Law of the Sea, 1982, *reprinted in* 21 INTERNATIONAL LEGAL MATERIALS 1261 (1983), (entered into force 16 November 1994), Art 194. *See also* Pierre Dupuy, *Due Diligence in the International Law of Liability*, *in* LEGAL ASPECTS OF TRANSFRONTIER POLLUTION 369, 375–376 (OECD, 1977).

[20] Mickelson, *supra* note 7, at 228–9.

[21] *See also* Jutta Brunnée, *Of Sense and Sensibility: Reflections on International Liability Regimes as Tools for Environmental Protection*, 53 INTERNATIONAL & COMPARATIVE LAW QUARTERLY 351, 354 (2004).

activities, but not explicitly requiring states to refrain from potentially harmful activities.[22] To do otherwise (i.e., to require the consent of the affected state) would be to grant to the affected state a veto over any potentially harmful activity in the state of origin and impose an unacceptable limit on the polluting state's sovereignty. The apotheosis of this procedural approach is found in the International Law Commission's *Draft Articles on Prevention of Transboundary Harm From Hazardous Activities*, adopted in 2001,[23] where consultations between states are required to be based on "an equitable balance of interests."[24] This approach draws on the concept of "equitable utilization" adopted in shared watercourse regimes.[25] It includes a set of factors to be taken into account, including recognition that environmental factors must be balanced against economic and social considerations.[26] This is similar to the path taken by the *Trail Smelter* arbitration. In this respect, the Tribunal already recognized that the proceduralization of the *harm principle* and the resort to equity are inevitable products of the dyadic structure of transboundary harm.

The Contemporary Dispute over the *Trail Smelter*

In addition to the air emissions that were the subject of the *Trail Smelter* arbitration, a further by-product of the smelting process at the Trail smelter is slag, a sandlike material that contains metals such as lead and zinc. In the case of the Trail smelter, large amounts of slag have been discharged by TCML into the Columbia River and carried downstream into the United States.[27] About twenty-four kilometers (fifteen miles) south of the U.S. border, the Upper Columbia River becomes a reservoir, Lake Roosevelt. In 1999, the Coleville Indian tribe petitioned the EPA to conduct an assessment of hazardous substance contamination in the Upper Columbia River south of the U.S.-Canadian border.[28] The assessment was conducted between 1999 and 2003 and disclosed the presence of hazardous substances consistent with slag releases.[29]

[22] *Report of the International Law Commission, Fifty-third Session*, UN GAOR, 56th Sess., Supp. No. 10., art. 9(3), UN Doc. A/56/10 (2001)[hereinafter "Draft Articles on Transboundary Harm"] (noting that where consultations between states fail to produce an agreed-upon solution, the state of origin may still proceed with the activity, although it must, as a matter of good faith take the affected state's interests into account). *See* Drumbl in this volume.

[23] *Id.*

[24] *Id.* at art. 9(2).

[25] UN Convention on the Law of the Non-navigational Uses of International Watercourses, *reprinted in* 36 International Legal Materials 719 (1997) (not in force).

[26] *Id.* at art. 10.

[27] Matthew Preusch, *Pollution Dispute in Northwest Straddles Border*, [N.Y. Times], March 20, 2004. *See* John D. Wirth, Smelter Smoke in North America (2000) (There was some interest, though summarily dismissed, in the smelter's water-borne pollution in the original *Trail Smelter* dispute.).

[28] The assessment was requested pursuant to section 105 of CERCLA. *See* CERCLA, *supra* note 1.

[29] Under the CERCLA regime, a site subject to an assessment is given a Hazardous Ranking System score, which determines the site's eligibility for the National Priorities List (NPL) and Superfund

After the completion of the site assessment, the EPA and Teck Cominco America Inc. (TCAI), the American affiliate of TCML, entered into discussions regarding voluntary measures to be taken to address the contamination. Although TCAI was prepared to expend considerable funds to address the contamination, the EPA wanted TCAI and TCML to enter into a consent order under CERCLA, which TCML was not prepared to do.[30] Consequently, the EPA issued a unilateral administrative order (the Order) to TCML under section 106 of CERCLA.[31] The Order requires TCML, a Canadian firm, to conduct a "Remedial Investigation/Feasibility Study," which includes the further investigation of the contamination and a determination of appropriate remedial action. This is a preliminary step towards effecting a cleanup of the site.

Liability for the cleanup of contaminated sites under CERCLA is expansive.[32] Among those who can be held responsible for cleanup costs is any person who arranges for the disposal of a hazardous substance at a facility where there is a release or threatened release into the environment.[33] There is no set threshold for determining whether a substance is "hazardous."[34] Liability attaches regardless of the quantity or concentration of the disposal.[35] Liability under CERCLA is retroactive and it is imposed regardless of fault.[36] Where a person is determined to be responsible for a release, they can be held liable for the entire cleanup regardless of the contribution of others.[37] The EPA or others, such as state and local governments, can undertake to clean up the site themselves and may seek to recover those costs from a responsible person,[38] or, as is the case with TCML, the EPA can order a responsible person to undertake or finance a response action where there is an imminent or substantial endangerment to the public health or welfare of the environment.[39]

From the standpoint of the EPA, one can see why the imposition of an administrative order under CERCLA would be an attractive option to address

cleanup funds. The site subject to the TCML order received a score high enough to make it eligible for inclusion on the NPL.

30 The structure of the proposal to allow TCAI to enter the consent order, with TCML being bound under a "tolling agreement" is described in the EPA Order. *See* U.S. EPA, "Unilateral Administrative Order for Remedial Investigation/Feasibility Study issued to Teck Cominco Metals, Ltd. December 11, 2003, *available at* http://yosemite.epa.gov/ R10/CLEANUP.NSF/ [hereinafter "EPA Order"].

31 *Id.*

32 *See* Gerald George, *Over the Line – Transboundary Application of CERCLA*, 34 ENVIRONMENTAL LAW REPORTS 10275, 10275 (2004). Even the United States courts, while upholding CERCLA's liability scheme, have commented on its potential for unfairness. *See* U.S. v. Alcan Aluminum Co., 990 F.2d 711, 716–7 (2nd Cir. 1993).

33 CERCLA, *supra* note 1 at § 9607(a)(3),(4).

34 *Id.* at § 9601(14).

35 *Id.*

36 *See* New York v. Shore Reality Corp., 759 F.2d 1032, 1039–42 (2d Cir. 1985); United States v. Shell Oil Co., 605 F. Supp. 1064, 1072–73 (D. Colo. 1985).

37 U.S.EPA v. Monsanto Co., 858 F.2d 160 (4th Cir. 1988); United States v. Chem Dyne Corp., 572 F.Supp. 802, 811 (S.D. Ohio 1983).

38 CERCLA, *supra* note 1, at § 9607(a)(4)(A)–(D).

39 *Id.* at § 9604(a).

contamination from the Trail smelter. The EPA does not have to show that the damage in question was foreseeable or that TCML breached a due diligence standard, as appears to be required for establishing state liability under international environmental law. Nor does the EPA have to show that the disposal of hazardous substances may result in a "substantial" or "significant" impact on the environment. The question of proving causation is also simplified under CERCLA. The EPA only has to show that TCML was responsible for disposing hazardous substances and that there is a release from the facility. Here the facility in question is not the Trail smelter itself (situated in Canada), but the affected area of the Upper Columbia River in the vicinity of Lake Roosevelt (and well within the territory of the United States).[40] The theory of liability being pursued is that TCML arranged for the disposal of its effluent, which constitutes a "hazardous substance," in a "facility" (the Upper Columbia River), and that the presence of the hazardous substances in the Upper Columbia River or their potential migration constitutes an actual or threatened "release."[41]

The imposition of liability under CERCLA against TCML for discharges of effluent occurring in Canada offers a strikingly different approach to liability than that found in the *Trail Smelter* arbitration. For example, the requirements announced in the *Trail Smelter* arbitration that a transboundary discharge must have "serious consequences" and be demonstrated by "clear and convincing" evidence, if liability is to attach, are absent. As discussed earlier, under the imposition of CERCLA liability, the discharge can be benign, and there is no requirement to demonstrate a causal link between the discharge and the need for a cleanup. These issues are very much contested by TCML, which maintains that the release of slag into the Columbia River has no discernable adverse environmental impacts and that they are not solely responsible for the presence of hazardous substances in the Upper Columbia River. A further benefit of proceeding under CERCLA is that it provides a clear set of remedies in the case of nonperformance and enforcement can be pursued, in the first instance, through the U.S. domestic courts. The prevailing standard of liability, as between states, namely *due diligence*, also is not reflected in the application of CERCLA to transboundary pollution. There is a jarring disjuncture between the requirement under public international law to show unreasonable behavior, on the one hand, and the ability under CERCLA to impose liability for acts that were lawful and accorded with industry standards at the time of discharge, on the other. Finally, the equivocal and cooperative posture of the *harm principle* is substituted by an approach that is unilateral and simply fails to account for the social and economic impact that the imposition of liability may have in the state of origin.

In summary, the EPA's extension of CERCLA liability to include liability for activities occurring outside U.S. territory would appear from an environmental

[40] The definition of "facility" includes, "any site or area where a hazardous substance has been deposited, stored, disposed of, or placed, or otherwise come to be located."*Id.* at § 9601(9)(b).

[41] EPA Order, *supra* note 30, at 5.

perspective to provide for a more definitive and efficient approach to transboundary pollution than that which has emerged in international environmental law from the *Trail Smelter* precedent. Structurally, the approach is more in keeping with the *polluter pays principle* and recognizes (implicitly, at least) that there may be little incentive for a source state to impose or enforce a strong pollution control regime where there is little domestic environmental harm. However, from a broader perspective, a unilateral approach, because it fails to recognize the sovereign right of the source state to determine the balance between environmental protection and economic development, is likely to result in a conflictual and unstable regime for transboundary pollution. In this case, the potential for conflict is not speculative, as the Canadian government has already voiced its concern through formal diplomatic channels.[42]

Although the EPA itself has not sought to enforce the order, private citizens have brought an action under CERCLA's citizen's suit provision to seek a court order requiring compliance.[43] As of this writing, the matter remains before the federal courts in the United States.[44] Even in the event that the invocation of the citizen's suit provision is successful, there remains a question as to whether a judgment compelling TCML to comply with the order would be enforced by the Canadian courts.[45] Significantly, there are some indications that the EPA's unilateral approach is not an isolated response to the new *Trail Smelter* dispute, but is part of a broader intention to apply CERCLA to transboundary contexts.[46]

CONFLUENCE: EXTRATERRITORIAL JURISDICTION

The EPA is not, however, unrestricted in its ability to apply CERCLA to activities originating in Canada, as international law limits the extent of a state's jurisdiction to prescribe laws. The principal jurisdictional basis for a state to prescribe laws is territorial sovereignty.[47] That is, a state has, presumptively at least, jurisdiction

42 Diplomatic Note, January 8, 2004, *supra* note 2.

43 CERCLA, *supra* note 1, at § 9659.

44 Pakootas and Michel v. Teck Cominco Metals, Ltd., U.S. District Court for the Eastern District of Washington, Case No. CV-04-256-AAM, Complaint For Injunctive and Declaratory Relief and for Civil Penalties, dated July 21, 2004. TCML brought a motion to dismiss the claim on the basis that the claim failed to disclose a cause of action because of a lack of personal and subject matter jurisdiction. The District Court dismissed the motion. Pakootas v. Teck Cominco Metal, Ltd. (2004) U.S. Dist. LEXIS 23041. The decision to dismiss the motion was appealed to the Federal Court of Appeal.

45 For a discussion of the circumstances under which Canadian courts will enforce foreign judgments, *see* J.-G. Castel and J. Walker, Canadian Conflict of Law § 14.5 (5th ed. 2004). In the event that TCML has sufficient U.S. assets enforcement in Canada may not be required.

46 *See* George, *supra* note 32, at 10275 (citing Environmental Law Institute, Strengthening U.S-Mexico Transboundary Environmental Enforcement: Legal Strategies For Preventing The Use of the Border as a Shield Against Liability (2002), *available at* http://www.eli.org).

47 For a description of the different principles upon which jurisdiction may be asserted, *see* Malcolm Shaw, International Law 527 (5th ed. 2003). *See also* Peter Malanczuk, Akehurst's Modern Introduction to International Law 109 (7th ed. 1997).

to legislate in respect of activities within its territory. In addition to the *territorial principle*, international law recognizes, *inter alia*, the possibility that a state may prescribe laws in respect to conduct outside of its territorial jurisdiction that has impacts within its jurisdiction; an approach, referred to as the *effects doctrine*.[48] The theoretical foundation of the effects doctrine is itself based on the principle of territoriality as the occurrence of the effects themselves within a state's territory is a sufficient basis for the assertion of jurisdiction.[49]

One difficulty with the effects doctrine is that, because the activity in question concerns conduct occurring in another state, there is an inherent possibility for conflict between states exercising concurrent jurisdiction. That is, the activity in question, here the discharge of hazardous substances, may be regulated by the source state on the basis that the activity is carried out in its territory, and by the affected state on the basis of the *effects doctrine*. The United States has adopted legislation with extraterritorial reach in connection with economic matters since the 1940s, and has sought to justify these extensions of jurisdiction with reference to the effects doctrine.[50] However, the exercise of extraterritorial jurisdiction has been resisted by other countries, and has been tempered from time to time by the American courts.[51]

Rosalyn Higgins, in rejecting an absolutist approach either in favor of or opposed to the exercise of extraterritorial jurisdiction, suggests that resolution of jurisdictional disputes of this nature should not be made with blind invocation of state sovereignty for its own sake but, rather, should be addressed more functionally with reference to the common values of the international community.[52] On a broad level, I take Higgins to mean that there should be a level of harmony between exercises of jurisdiction by individual states and their commitment to substantive rules and principles of international law. There is support for this approach in the interpretive canon that "an act of congress ought never to be construed to violate the law of nations if any other construction remains."[53] This rule is a recognition that rational states do not seek to undermine their own international commitments through legislative enactments. On this basis, the extraterritorial application of environmental statutes should account for the existing body of

48 *See* U.S. v. Alcoa, 148 F.2d 416 (2nd Cir. 1945); Timberlane Lumber Co. v. Bank of America, 549 F.2d 597 (9th Cir. 1976) [hereinafter "Timberlane"].

49 *See* MALANCZUK, *supra* note 47, at 116.

50 For examples of U.S. extraterritorial legislation, *see* Cuban Liberty and Democratic Solidarity (Libertad) Act, 22 U.S.C. §6021; Iran and Libya Sanctions Act, 50 U.S.C. §1701. *See* A. V. Lowe, *U.S. Extraterritorial Jurisdiction: the Helms-Burton and D'Amato Act*, 46 INTERNATIONAL & COMPARATIVE LAW QUARTERLY 378 (1997).

51 ROSALYN HIGGINS, PROBLEMS AND PROCESS: INTERNATIONAL LAW AND HOW WE USE IT 76 (1995) (citing Timberlane Lumber Co. v. Bank of America, 549 F.2d 597 (9th Cir. 1976); Mannington Mills v. Congoleum Corp., 595 F. 2d 1287 (3rd Cir. 1979)).

52 *Id.* at 77.

53 Murray v. Schooner Charming Betsy (1804) 2 Cranch 64, 118 (U.S.S.C). *See* F. Hoffman-La Roche Ltd. v. Empagran S.A., 124 S.Ct. 2359, 2366 (2004). In Canada, *see* Ordon Estate v. Grail [1998] 3 SCR 437.

international regulation of transboundary pollution. A recent decision of the U.S. Supreme Court on extraterritorial application of domestic statutes indicates that the effects in question must be intended and must produce some "substantial effect in the United States," if jurisdiction is to be successfully asserted under this doctrine.[54] Thus, it would appear that, in the United States at least, where the effects are direct, substantial, and foreseeable, jurisdiction to prescribe could be claimed.[55]

The concern over the possibility of conflict has led to the development of the *principle of comity* as a way of mediating these conflicts. *Comity* finds its principal expression, in the U.S. context, in Section 403 of the *Restatement (Third) of Foreign Relations Law of the United States*.[56] The rights of states to prescribe law in respect to "conduct outside of its territory that has or is intended to have substantial effect within its territory,"[57] is recognized, subject to the limitation that that the exercise of jurisdiction in respect of persons or activities having connections with other states is reasonable.[58] The determination of reasonableness is conceived as a balancing of interests and, to that end, the *Restatement* sets out a list of factors that should be considered in this determination, including the extent to which the extraterritorial regulation is consistent with the traditions of the international system; the extent to which another state regulates, or may have an interest in regulating, the activity; and the likelihood of conflict with the regulation of another state.[59] On this last point, the U.S. Supreme Court in the *Hartford Insurance* case suggests that a conflict only exists in the event of the impossibility of dual compliance, and in the absence of such a conflict, there is no need to consider *comity*.[60] A view more sympathetic to the competing interests of both states recognizes that conflict between foreign laws is a continuum, requiring consideration of the extent of the conflict, as part of the broader consideration of each state's interests under the *principle of comity*.[61]

Applying the rules of extraterritorial jurisdiction to the second *Trail Smelter* dispute, an approach emerges that has many parallels to the original. First, before

54 Hartford Fire Insurance Co. v. California, 113 S. Ct. 2891, 2909 (1993) [hereinafter "Hartford Insurance"].

55 The phrase "direct, substantial, and reasonably foreseeable effect" is derived from United States antitrust legislation that seeks to assert extraterritorial jurisdiction, but this language has been incorporated into more general requirements for exercising extraterritorial jurisdiction. *See* AMERICAN LAW INSTITUTE, RESTATEMENT (THIRD) OF FOREIGN RELATIONS LAW OF THE UNITED STATES § 403(2)(a) (1987)[hereinafter "Restatement"].

56 *Id.*

57 *Id.* at § 402.

58 *Id.* at § 403.

59 *Id.* at § 403(2).

60 Hartford Fire Insurance Co. v. California, 113 S. Ct. 2891, 2911 (1993); *but see Id.* at 2920–2922 (Scalia J., dissenting).

61 *See* Sosa v. Alvarez-Machain, 124 S.Ct. 2739, 2782 (2004). In Canada, *see* Libman v. R. [1985] 2 SCR 178. *See also* Andeas Lowenfeld, *Conflict, Balancing of Interests, and the Exercise of Jurisdiction to Prescribe: Reflections on the Insurance Antitrust Case*, 89 AMERICAN JOURNAL OF INTERNATIONAL LAW 42 (1995).

prescriptive jurisdiction can be exercised, the effects of the activity subject to regulation must be "substantial, direct and foreseeable."[62] This requirement imposes a threshold similar to the requirement for "serious" or "significant" harm contained in the *harm principle*. In both cases, the rule recognizes that states must accept a minimal level of interference from activities outside their borders. The "foreseeability" requirement suggests that the effects from the activity regulated should not be accidental or otherwise unintentional.

Second, both the *harm principle* and the rules for extraterritorial jurisdiction indicate that a state's right to act will be predicated on a balancing of interests. The approach in section 403 of the Restatement has clear parallels to the "equitable balancing" approach to transboundary harm taken in the ILC *Draft Articles on Transboundary Harm*,[63] and under the rules respecting the non-navigational uses of international watercourses.[64] Without wanting to overstate this point, both approaches turn to contextual considerations oriented towards ensuring that the interests of other states affected by assertions of jurisdiction or potentially harmful activities are accounted for. Ultimately, the application of extraterritorial jurisdiction is qualified by the same sensitivity to opposing sovereign rights that dictated the form of the *harm principle* in the *Trail Smelter* arbitration.

Applying the rules of extraterritorial jurisdiction to the extension of CERCLA liability for discharges occurring outside the United States, a number of factors militate against imposition of the Order, or at the very least, suggest careful consideration of those issues not captured by the statutory requirements under CERCLA. First, the threshold requirement for "substantial" effects does not mesh with the approach under CERCLA, which allows for the imposition of liability without a quantity or concentration requirement. It is far from clear on the facts set out in the Order whether the slag discharges would meet such a "substantial effects" threshold. Second, the retroactive nature of CERCLA liability and the absence of the need to show a causal link between the cleanup ordered and the release are at odds with the requirement that the effects that form the basis of an extraterritorial application of a domestic law be direct and foreseeable. In the case of historical discharges, it is certainly open to question whether activities that were considered both benign and legal at the time of their release into the environment could lead to foreseeable environmental effects requiring extensive cleanup today or in the future. Again, the point here is not that the effects sought to be addressed by the EPA are necessarily unforeseeable, but that under the CERCLA regime foreseeability of effects is not considered.

A further consideration that must be accounted for in determining jurisdiction is the extent to which the activity is already subject to regulation by Canadian

62 Restatement, *supra* note 55.

63 *Id.* at §403(2); Draft Articles on Transboundary Harm, *supra* note 22, at art. 10.

64 UN Convention on the Law of the Non-navigational Uses of International Watercourses, *supra* note 25, at art. 5.

and British Columbian authorities. Although the possibility of a direct conflict, in the sense of an impossibility of dual compliance, is unlikely, it is important to bear in mind that the allowable discharge levels have economic consequences. An extraterritorial attempt to regulate transboundary discharges through the imposition of a strict liability regime will likely interfere with economic and social policies of the state of origin. In this regard, the source state's right to economic development must be balanced against the affected state's right to be free from environmental harm. The concern for conflict appears to be at the heart of the Canadian government's objection, whose diplomatic note indicated that the EPA's action "may set an unfortunate precedent by causing transboundary environmental liability cases to be initiated in both Canada and the United States."[65]

Finally, the extraterritorial application of CERCLA conflicts with the rules and principles of international environmental law accepted by the United States and Canada. The undesirability of unilateral approaches to transboundary environmental issues forms the basis of the *duty to cooperate*, which requires that states approach conflicts over shared resources in a spirit of good-neighborliness and good faith.[66] Principle 12 of the *Rio Declaration* specifically notes that "[u]nilateral actions to deal with environmental challenges outside the jurisdiction of the importing country should be avoided."[67] Unilateral approaches have been rejected in the context of trade-related environmental measures by GATT and WTO dispute settlement panels.[68] More specific to the dispute in question, the international rules respecting shared watercourses also require cooperation,[69] and very clearly recognize that the use of shared watercourses is subject to balancing the opposing rights of each state, an approach that necessarily rejects a wholly unilateral approach.

CONCLUSION

One could rightly question whether the rules for exercising transboundary prescriptive jurisdiction result in some of the same difficulties confronted in the *Trail Smelter* arbitration. The balancing of interests requirement under Article

[65] Diplomatic Note, January 8, 2004, *supra* note 2.

[66] *See* Lac Lanoux Arbitration (France v. Spain), 1957, 24 I.L.R. 101. *See also* Alan Boyle, *The Principle of Co-operation: the Environment*, *in* THE UNITED NATIONS AND THE PRINCIPLES OF INTERNATIONAL LAW 120 (Lowe and Warwick eds., 1994).

[67] Rio Declaration, *supra* note 5, Principle 12.

[68] *See* United States – Restrictions on Imports of Tuna, 30 INTERNATIONAL LEGAL MATERIALS 1594 (1991); United States – Import Prohibition of Certain Shrimp and Shrimp Products, Report of the Panel, WT/DS58/R (15 May 1998); United States – Import Prohibition of Certain Shrimp and Shrimp Products, Appellate Body, WT/DS58/AB/R, reprinted in 38 INTERNATIONAL LEGAL MATERIALS 118 (1999).

[69] UN Convention on the Law of the Non-navigational Uses of International Watercourses, *supra* note 25, at art. 8. *See also* STEPHEN MCCAFFREY, THE LAW OF INTERNATIONAL WATERCOURSES: NON-NAVIGATIONAL USES398–404 (2001).

403 of the *Restatement*, much like that which is implicit in the *harm principle*, has been the subject of trenchant criticism because its case-by-case approach and ambiguous, often incommensurate, consideration of a range of associated factors makes for an unprincipled and unpredictable approach.[70] The result is to invest (American) courts with a high degree of discretion.

It is possible that, through judicial interactions such as these, more generalized rules can emerge – a sort of transnational law of transboundary pollution that more accurately reflects the blurred national/international and public/private distinctions that characterize transboundary pollution problems.[71] However, these interactions themselves are likely to be bilateral as courts in source jurisdictions will be called upon to enforce foreign judgments, and through that process will have an opportunity to engage their judicial counterparts in a dialogue on the extent of extraterritorial jurisdiction.[72] In a related fashion, extraterritorial exercises of jurisdiction can overcome some of the structural inertia that has plagued the imposition of liability for transboundary pollution *via* the *harm principle*. States may have little incentive to agree to abstract rules regarding transboundary pollution that may hinder their future activities, but on a case-by-case basis, they may be more inclined to cooperate.

Even if the imposition of CERCLA liability for historical discharges of slag by the Trail smelter appears to be misguided, the EPA's turn to extraterritoriality has the potential to be a positive development for addressing transboundary environmental harm. However, much will depend on the willingness of the courts to take the *principle of comity* seriously. In this regard, courts and legislatures may be wise to consider the principal lesson from the *Trail Smelter* arbitration, which emphasized cooperation over conflict.

[70] *See* Larry Kramer, *Extraterritorial Application of American Law After the Insurance Antitrust Case: A Reply to Professors Lowenfeld and Trimble*, 89 American Journal of International Law 750, 755 (1995).

[71] For a broader discussion of the implications of judicial interaction for transnational law, *see* Harold Koh, *Transnational Public Law Litigation*, 100 Yale Law Journal 2347 (1991); Harold Koh, *Transnational Legal Process*, 75 Nebraska Law Review 181 (1996).

[72] *See* Hestermeyer in this volume.

PART TWO

Trail Smelter and Contemporary Transboundary Harm – The Environment

11 *Trail Smelter* in Contemporary International Environmental Law: Its Relevance in the Nuclear Energy Context

Günther Handl

INTRODUCTION

Although the *Trail Smelter* arbitration[1] has often and justifiably been referred to as the *fons et origo* of international law on transboundary environmental harm[2] and has over the years attracted considerable attention in the literature,[3] significant differences of opinion, as well as basic misperceptions continue to exist among commentators and experts with respect of the arbitration's true international legal import. Critical views range from outright dismissals of *Trail Smelter* as of "limited precedential value,"[4] to the belief that, while *Trail Smelter* recognizes

[1] *Trail Smelter Arbitral Decision*, 33 AMERICAN JOURNAL OF INTERNATIONAL LAW 182 (1939) [hereinafter "*Trail Smelter* (1939)"]; *Trail Smelter Arbitral Decision*, 35 AMERICAN JOURNAL OF INTERNATIONAL LAW 684 (1941) [hereinafter "*Trail Smelter* (1941)"]. *See* Annex to this volume.

[2] *See especially* Alfred P. Rubin, *Pollution by Analogy: The Trail Smelter Arbitration*, 50 OREGON LAW REVIEW 259 (1971), republished in this volume; Günther Handl, *Territorial Sovereignty and the Problem of Transnational Pollution*, 69 AMERICAN JOURNAL OF INTERNATIONAL LAW 50, 60 (1975); Robert Quentin-Baxter, Second Report on International Liability for Injurious Consequences Arising out of Acts not Prohibited by International Law, UN Doc. A/CN.4/346 and Add.1 & 2, reprinted in 2(1) YEARBOOK OF THE INTERNATIONAL LAW COMMISSION 103, 108–112 (1981).

[3] *See, e.g.*, John Read, *The Trail Smelter Dispute*, 1 CANADIAN YEARBOOK OF INTERNATIONAL LAW 213 (1963), republished in this volume; Rubin, *supra* note 2; Frederic L. Kirgis Jr., *Technological Challenge to the Shared Environment: United States Practice*, 66 AMERICAN JOURNAL OF INTERNATIONAL LAW 290, 291–94 (1972); Handl, *supra* note 2; Kevin J. Madders, *Trail Smelter Arbitration, in* 4 ENCYCLOPEDIA OF PUBLIC INTERNATIONAL LAW 900 (Rudolf Bernhardt, ed., 1992); Quentin-Baxter, *supra* note 2, at 108–112; Karin Mickelson, *Rereading Trail Smelter*, 31 CANADIAN YEARBOOK OF INTERNATIONAL LAW 219 (1993), republished in this volume; *see generally* Draft Articles on the Prevention of Transboundary Harm from Hazardous Activities, together with Commentaries, Report of the International Law Commission on the Work of Its Fifty-Third Session, UN GAOR, 56th Sess., Supp. No. 10, at V.E.1, UN Doc. A/56/10 (2001) (hereinafter "Draft Articles on Prevention").

[4] Some commentators point to the "unique circumstances" surrounding both the decision to submit to arbitration and the Tribunal's decision itself, as factors that severely curtail the significance of the case. *See, e.g.*, Mickelson, *supra* note 3, at 223. Others deny altogether the decision's precedential value. *See, e.g.*, Patrick Kelly, *The Twilight of Customary International Law*, 40 VIRGINIA JOURNAL OF INTERNATIONAL LAW 449, 472 (2000).

international liability for transboundary damage, it does not impose an obligation to prevent such damage.[5] Similarly, the fact that the Tribunal took a nuanced approach in assessing Canada's liability – setting out the grounds for both a due diligence-based and strict liability argument[6] – appears often to have gone unnoticed.[7] This collection of essays, which seeks to assess for the first time the overall legacy of *Trail Smelter* and in this process to redress some of the underappreciation of the arbitration's scope and complexity, is thus particularly welcome and relevant.

This paper will not seek to track all the various strands of the *Trail Smelter* arbitration that have had a discernible impact on the evolution of international environmental law.[8] Rather, its objective is necessarily more modest. Given the assigned topic – the relevance and applicability of *Trail Smelter* in the context of nuclear energy – this chapter will focus on *Trail Smelter's* most relevant implication for nuclear power-related activities: a state's obligation to prevent or minimize the risk of transboundary harm.[9]

SOME GENERAL OBSERVATIONS ON THE *TRAIL SMELTER* ARBITRATION

The *Trail Smelter Dictum* as a Principle of Customary International Law

As is well known, the Tribunal concluded that

> under principles of international law, as well as the law of the United States, no State has the right to use or permit the use of its territory in such a manner as to cause injury by fumes on or to the territory of another or the properties or persons therein, when the case is of serious consequence and the injury is established by clear and convincing evidence.[10]

5 *See infra* text accompanying note 48.

6 *See* Günther Handl, *Liability as an Obligation Established by a Primary Rule of International Law*, 16 Netherlands Yearbook of International Law 49, 61 (1985).

7 *But see* Quentin-Baxter, *supra* note 2, at paras. 28, 38–40; Johan G. Lammers, Pollution of International Watercourses 524–25 (1984).

8 The term as used here refers loosely to what Birnie and Boyle call "the entire corpus of international law, public and private, relevant to environmental issues and problems." Patricia W. Birnie & Alan E. Boyle, International Law and the Environment 1–2 (2nd ed. 2002).

9 Other parts of the *Trail Smelter* decision have served as catalysts for the emergence of normative standards of special significance to nuclear power activities. One example is the procedural safeguard implications of *Trail Smelter*, specifically the source state's "obligation to consult" with the risk-exposed state(s), which may well have metamorphosed into an obligation to accept joint or multilateral on-site visits/inspections of the nuclear facility concerned. *See* Günther Handl, Grenzüberschreitendes nukleares Risiko und völkerrechtlicher Schutzanspruch 81–87 (1992).

10 *Trail Smelter* (1941), *supra* note 1, at 716.

With minor variations,[11] this *dictum* is reflected in general international practice,[12] incorporated either directly or indirectly by reference in such "foundational" international environmental legal instruments and documents as Principle 21 of the Stockholm Declaration on the Human Environment[13] and Principle 2 of the Rio Declaration on Environment and Development,[14] and has been a cornerstone in the International Law Commission's recent environment-related work.[15] It has also been endorsed as "part of the corpus of international law on the environment" by the International Court of Justice (ICJ) in its 1996 Advisory Opinion on the *Legality of the Threat or Use of Nuclear Weapons*[16] and its 1997 judgment in the *Gabčikovo* case.[17]

Despite this practice and the specific imprimatur by the ICJ and other tribunals,[18] doubts continue to be voiced today – not merely as to the precise legal implications of the *Trail Smelter* formula but also as to the very status of the *Trail Smelter* obligation as part of present customary international law. Critics assert that the existence of a customary legal obligation to "avoid significant harm," while admittedly attracting wide and uniform verbal support, is not adequately reflected in actual state practice,[19] or in any event, is not borne out by any systematic empirical study. In a similar vein, they dismiss the ICJ's express endorsement of

[11] *See infra* § 2B.

[12] For the reasons explained *infra*, it will be unnecessary to set out the evidence in support of this conclusion. One might add, however, a reference to Birnie and Boyle, for whom the existence of a customary legal norm to this effect is simply "beyond serious doubt." BIRNIE & BOYLE, *supra* note 8, at 109.

[13] Report of the Stockholm Conference, U.N. Doc. A/CONF.48/14, at 7, *reprinted in* 11 INTERNATIONAL LEGAL MATERIALS 1416, 1420 (1972).

[14] Report of the United Nations Conference on Environment and Development, Annex I, U.N. Doc. A/CONF.151/26 (Vol. I), *reprinted in* 31 INTERNATIONAL LEGAL MATERIALS 874 (1992). At the World Summit on Sustainable Development, states reaffirmed their commitment to the Rio Declaration. *See* Johannesburg Declaration on Sustainable Development para. 8, *in* Report of the World Summit on Sustainable Development, U.N. Doc. A/CONF.199/20, 2 (2002).

[15] *See* Draft Articles on the Non-Navigational Uses of International Watercourses, art. 7, *Report of the International Law Commission on the Work of its Forty-Sixth Sess.*, U.N. GAOR, 52nd Sess., Supp. No.10, U.N. Doc. A/49/10 (1994) [hereinafter Draft Articles on Non-Navigational Uses]; *see also* Draft Articles on Prevention, art 3, *supra* note 3, at 372; Draft Articles on Transboundary Aquifers and Aquifer Systems, art 7, *in* Chusei Yamada, Third Report on Shared Natural Resources: Transboundary Groundwaters, U.N. Doc. A/CN.4/551, at 11 (2005).

[16] Legality of the Threat or Use of Nuclear Weapons, Advisory Opinion, 1996 I.C.J. 226, para.29 (July 8).

[17] *See* Gabčikovo-Nagymaros Project (Hun./Slov.), 1997 I.C.J. 7, at 41, para.53 (Sept. 25).

[18] *See, e.g.*, the recent decision in the Iron Rhine *("Ijzeren Rijn")* Railway case (Belgium v. The Netherlands), Award of the Arbitral Tribunal, May 24, 2005, para.59.

[19] *See, e.g.*, Daniel Bodansky, *Customary (and Not So Customary) International Environmental Law*, 3 INDIANA JOURNAL OF GLOBAL LEGAL STUDIES 105, 116 (1995) (critically referring to such norms as "declarative law").

states' customary legal obligation to ensure that activities within their jurisdiction or control "respect" the environment of other states or of areas beyond national control as normatively too ambiguous to be meaningful.[20]

There is no gainsaying that at times states act in a manner clearly incompatible with the obligation laid down in the *Trail Smelter* arbitration.[21] However, those who seize upon these instances of "noncompliance" as evidence of a fatal lack of state practice appear blinded to the possibility that, in at least in some parts of the world, there might be routine compliance with the no-significant-harm obligation.[22] More important, the criticism of inadequate state practice misses a larger point. The legal status of the basic obligation to avoid significant transboundary harm is directly related to the sovereign equality of states as the most axiomatic premise of the international legal order. This no-significant-harm obligation seeks to reconcile one state's sovereign right to use its territory and resources with another state's defensive invocation of the very same sovereignty-based right. The validity of this obligation thus does not depend on confirmation through the usual inductive process of proving customary international law. "The proposition that no State may disregard the rights and interests of its neighbours is an offspring of conceptual reasoning. It [cannot] . . . be invalidated by evidence pointing to manifold departures from the conduct required."[23] This is not to suggest, of course, that in general it might be possible or permissible to infer precise conduct-specific international legal rules from abstract fundamental principles, such as sovereign equality, rather than inductively from the practice of states.[24] However, the no-significant-harm principle is sufficiently determinate and has further been circumscribed in state practice to provide pertinent guidance as to

[20] *See* John Knox, *The Myth and Reality of Transboundary Environmental Impact Assessment*, 96 AMERICAN JOURNAL OF INTERNATIONAL LAW 291, 295 (2002).

[21] These instances include occasional spats among countries that traditionally have enjoyed good neighborly relations. Along these lines, *Trail Smelter* has recently re-emerged as a transboundary environmental irritant between Canada and the United States. *See, e.g.*, Matthew Preusch, *Pollution Dispute in the Northwest Straddles the Border*, NEW YORK TIMES, March 20, 2004, A8. *See* Craik in this volume.

[22] Given the degree of existing and growing environmental interdependence among states, compliance with "*Trail Smelter* obligation" seems to be a routine occurrence that does not attract special attention. This seems to be true of much of the transboundary environmental interferences in Western Europe and North America. Conversely, instances of noncompliance might be notorious precisely because they represent exceptional situations. Thus, irrespective of whether individual instances of noncompliance represent pathological or normal behavior of states, opponents are at the very least guilty of the same evidentiary sin of which they accuse proponents of the customary law status.

[23] Christian Tomuschat, *Obligations Arising for States without or against Their Will*, *in* 241 RECUEIL DES COURS 195, 295 (1993). *See also* ULRICH FASTENRATH, LÜCKEN IM VÖLKERRECHT 203–06 (1991).

[24] *See, e.g.*, Delimitation of the Maritime Boundary in the Gulf of Maine Area (Can. v. U.S.), 1984 I.C.J. 299, para.111 (Oct. 12).

the basic allocation of states' corollary rights and obligations in a situation of interdependent natural resource uses.[25]

The Utility of the *Trail Smelter* Threshold

Some commentators take issue with the Tribunal's formula, which pegs the wrongfulness of state conduct to transboundary environmental effects "of a serious consequence." They criticize this threshold as either inherently ambiguous, and thus incapable of being applied except in the rare instances where states have agreed to submit the issue to a third-party decision maker, or as not corresponding to how transboundary environmental interferences are actually handled in state practice.[26] However, these doubts regarding the threshold's guiding function appear overdrawn.

Without revisiting the debate over how to define the customary legal threshold of prohibited transboundary interference,[27] one might note that there is a general consensus that states are enjoined only from causing transboundary harm that is "serious," "significant," "substantial," "appreciable," or similarly qualified.[28] These qualifying words are not free of ambiguity. However, they all convey the idea that the harm concerned must be more than *de minimis*, trivial, or simply "detectable,"[29] and as such reflect a factual judgment about "a real detrimental effect on matters such as . . . human health, industry, property, environment or agriculture in other States."[30] States are also subject to far-reaching procedural obligations – both treaty-based and customary[31] – that facilitate clarification of whether transboundary effects cross the threshold of "significant harm" regardless of the precise circumstances of any individual case. Indeed, certain types of transboundary effects involving, for example, radiological, toxic, or highly dangerous

25 Some generalists, however, continue to question the present-day existence of environment-related specific customary international rules. *See, e.g.*, ANTONIO CASSESE, INTERNATIONAL LAW 490 (2nd ed. 2005).

26 For example, John Knox asserts that "none of these variations [of the *Trail Smelter* threshold of prohibited transboundary harm when correctly understood as implying a due diligence obligation of prevention] necessarily comports to state practice." Knox, *supra* note at 20, at 294.

27 *See instead* Günther Handl, *National Uses of Transboundary Air Resources: The International Entitlement Issue Reconsidered*, 26 NATURAL RESOURCES JOURNAL 405, 412–27 (1986); ANDRÉ NOLLKAEMPER, THE LEGAL REGIME FOR TRANSBOUNDARY WATER POLLUTION: BETWEEN DISCRETION AND CONSTRAINT (1993); STEPHEN C. MCCAFFREY, THE LAW OF INTERNATIONAL WATERCOURSES: NON-NAVIGATIONAL USES 365–80 (2001).

28 *See* Handl, *supra* note 27, at 412.

29 *See* Commentary on the Draft Articles on Prevention, art. 2, *supra* note 3, at 388.

30 *Id.*

31 These procedural obligations include the traditional state-to-state duties to inform, consult, and so on, and also encompass internationally guaranteed individual rights, like that of access to environmental information, which contribute to greater transparency regarding human-induced environmental changes, including transboundary environmental effects.

substances otherwise affecting public health, endangering lives, or producing serious irreversible conditions, are likely to be *a priori* deemed significantly harmful.[32] Thus, in any given case of transboundary environmental interference the task of defining "significant harm" is probably much less challenging than critics make it out to be.

That said, some genuine uncertainty may arise over the consequences of a plausible *prima facie* demonstration of such transboundary effects. *Trail Smelter* suggests that any such effects would constitute internationally prohibited conduct. However, this straightforward linkage between "significant harm" and legal "prohibition" has not been universally accepted. A competing conceptualization of the threshold for internationally prohibited transboundary environmental effects traces its origin to legal principles that evolved in relation to non-navigational uses of international watercourses.[33] According to this latter view, "significant harm" is only one factor to consider in balancing the interests of concerned states. This "dual element" (significant harm *plus* inequitable use) analysis might relax the relative strictness of a duty simply pegged to "significant harm." From an environmental policy perspective, a persuasive case can and has been made against this "dual element" characterization of the legal threshold of prohibited transboundary environmental effects.[34] Moreover, Birnie and Boyle conclude that "neither the international case law nor treaty definitions of harm . . . support thresholds determined by equitable balancing."[35] However, the picture that emerges from an analysis of state practice, including recent work of the ILC, is somewhat more ambiguous.

It is true that, following the instructions of the *compromis*,[36] the Tribunal basically applied a balancing of interests approach to the *Trail Smelter* dispute. However, in its *dictum* on Canada's obligation to abate transboundary air pollution, the Tribunal conspicuously omitted any mention of "inequitable use," and imposed no prerequisites to Canada's international responsibility other than the finding of significant harm. Some instances of state practice relating to transboundary air pollution[37] similarly suggest international acceptance of a factually determined

32 *See* Convention on the Non-Navigational Uses of International Watercourses, *opened for signature* May 21, 1997, 36 INTERNATIONAL LEGAL MATERIALS 700, arts. 21(2) (harm to human health and safety) and 23 (alien species) (1997).

33 *See* Helsinki Rules on the Uses of Waters of International Rivers, art. X, International Law Association (ILA), Report of the Fifty-Second Conference, Aug. 14–20, 1966, 497–99. *See also* MCCAFFREY, *supra* note 27, at 349.

34 *See* Handl, *supra* note 27, at 417–21; BIRNIE & BOYLE, *supra* note 8, at 124; NOLLKAEMPER, *supra* note 27, at 66–69. *But see* Patricia Wouters, *The Legal Response to International Water Conflicts: The UN Watercourses Convention and Beyond*, 42 GERMAN YEARBOOK OF INTERNATIONAL LAW 293, 322–23 (1999).

35 BIRNIE & BOYLE, *supra* note 8, at 123.

36 *See* Convention on Damages Resulting From Operation of Smelter at Trail, British Columbia, June 5, 1935, U.S.-Canada, art. IV, 49 Stat. 3245. *See* Annex to this volume.

37 For details, *see* Handl, *supra* note 27, at 422–23.

threshold of "significant harm," as the hallmark of prohibited transboundary harm. On the other hand, it is also true that, with regard to international watercourses, the notion of "equitable" or "reasonable use" continues to affect the threshold of impermissible transboundary environmental effects. Thus, the commentary to Article 7 (the obligation not to cause significant harm) of the ILC Draft Articles on the Law of the Non-Navigational Uses of International Watercourses, clearly envisaged circumstances in which "prohibited harm" is not equated with significant transboundary harm but instead is conditioned on such harm representing an inequitable balance of interests as between the states concerned.[38] The United Nations Convention retains this approach.[39]

One might be justified in questioning the appropriateness, in this day and age, of ranking "significant harm" as merely one element – albeit an important one – in a reasonable or equitable use test for the management of international watercourses.[40] At any rate, and whatever one's view of the matter in relation to nonnavigational uses of international watercourses, there is substantial agreement among commentators that, in general, the threshold for prohibited harm ought not be conditioned on equitable balancing.[41] This view has found expression in the work of the ILC on the Prevention of Transboundary Damage from Hazardous Activities.[42] Unlike the approach opted for in draft articles on International Watercourses,[43] the Draft Articles on Prevention do not subordinate the obligation to prevent significant transboundary harm to an equitable balancing of interests. Thus, in discussing the relationship between Article 3 (on prevention) and what was then Art. 12 (on equitable balancing of interests among States concerned), the Commission's Special Rapporteur expressly denied that the latter was meant in any way to dilute the obligation of prevention.[44]

38 "[T]he fact that an activity involves significant harm, would not in and of itself necessarily constitute a basis for barring it. In certain circumstances 'equitable and reasonable utilization'... may still involve significant harm to another watercourse State." *See* Draft Articles on Non-Navigational Uses, *supra* note 15.

39 *See* Convention on the Non-Navigational Uses of International Watercourses, art. 7, *supra* note 32; *see also* McCaffrey, *supra* note 27, at 369–70; *contra* Birnie & Boyle, *supra* note 8, at 124.

40 *See, e.g.*, Nollkaemper, *supra* note 27, at 69.

41 *See* Johan G. Lammers, *"Balancing the Equities" in International Environmental Law*, *in* The Future of the International Law of the Environment 153, 163 (Rene Jean Dupuy ed., 1985) ("The principle of equitable utilization or apportionment cannot be deemed to apply to instances of transfrontier pollution.")

42 Draft Articles on Prevention, *supra* note 3.

43 The Commission, however, appears to follow the 1997 Watercourses Convention and to opt again for a dual element approach in defining prohibited utilization of a transboundary aquifer or aquifer systems. *See* Yamada, *supra* note 15, proposed arts.5 & 7.

44 *See* Pemmaraju S. Rao, Third Report on International Liability for Injurious Consequences Arising out of Acts Not Prohibited by International Law, (Prevention of Transboundary Harm for Hazardous Activities), para.21, U.N. Doc. A/CN4.510 (2000). This clarification was prompted by expressions of concern in the General Assembly's Sixth Committee regarding the possibility that the Draft Articles on Prevention might be subordinated to equitable balancing.

Finally, there is general agreement that the no-significant-harm obligation is one of due diligence.[45] A state is accordingly only obliged to adopt reasonable restraints on transboundary injurious activities or agencies. This fact introduces a degree of uncertainty regarding – not the definition – but the actual operation of the threshold. For what is "reasonable" under the circumstances will, of course, vary from case to case. However, the proliferation of pertinent environmental and technical standards, such as best available technology, best environmental practice, and so on,[46] which may provide a *de facto*, if not *de jure*, authoritative point of reference, reduces – sometimes dramatically – the uncertainty over what constitutes due diligence aimed at preventing or abating transboundary harmful effects. Any remaining unpredictability, it ought to be remembered, bears on whether there has been a breach of a state's obligation entailing responsibility for the transboundary harm occasioned. It does not in any way affect that *Trail Smelter*–based threshold of what constitutes prohibited state conduct.

TRAIL SMELTER AND THE MANAGEMENT OF SIGNIFICANT TRANSBOUNDARY NUCLEAR RISKS

It was not that long ago that experts drawing principally on *Trail Smelter* still ventured the opinion that positive international law did not address transboundary environmental problems preventively.[47] This view is, of course, no longer tenable. Today, it is generally accepted that international law goes beyond requiring states to redress transboundary harmful effects *ex post facto*, and instead obliges states to take adequate measures to control and regulate in advance sources of potential significant transboundary harm.[48] Moreover, the view that *Trail Smelter* itself could not accommodate an obligation of prevention has been similarly jettisoned. In this vein, the ILC's General Commentary on the Draft Articles on Prevention simply notes that "the well-established principle of prevention was highlighted in the arbitral award in the *Trail Smelter* case . . . "[49]

Indeed, prevention presents itself as an essential aspect of the *Trail Smelter* obligation not to cause significant transboundary environmental harm. In this sense, the Tribunal's formula clearly covers situations in which transboundary

[45] *See, e.g.*, Draft Articles on Prevention, *supra* note 3, at 391–96.

[46] For a discussion of the normative implications of "best available technology" in the nuclear power context, *see* HANDL, *supra* note 9, at 51–55; ASTRID EPINEY & MARTIN SCHEYLI, STRUKTURPRINZIPIEN DES UMWELTVÖLKERRECHTS 141–44 (1998).

[47] *See* Alexandre. C. Kiss, *Un case de pollution internationale: L'Affaire des Boues Rouges*, JOURNAL DU DROIT INTERNATIONAL 207, 235 (1975); *see also* Rüdiger Wolfrum, *Die grenzüberschreitende Luftverschmutzung im Schnittpunkt von nationalem Recht und Völkerrecht*, 99 DEUTSCHES VERWALTUNGSBLATT 493, 495 (1984).

[48] *See, e.g.*, the General Commentary to the Draft Articles on Prevention, *supra* note 3, at 377–79.

[49] *Id.* at 378. *See also* BIRNIE & BOYLE, *supra* note 8, at 109.; *but see* ULRICH BEYERLIN, UMWELTVÖLKERRECHT 7 (2000).

harm is merely threatened, provided the risk thereof is significant – meaning a high probability of nontrivial consequences. By contrast, the Tribunal's decision does not directly cast a light on the status of threatened transboundary harms when the possible consequences are extremely serious but are of low probability. Writing at the time of the 1972 Stockholm Conference on the Human Environment, Frederic Kirgis first raised this issue of *Trail Smelter*'s normative reach to low probability/serious consequence transboundary risks.[50] Building on the risk of harm as a composite of probability and consequence, Kirgis proposed modifying the *Trail Smelter* standard so that a "potentially greater harm call[s] for abstention from conduct under a proportionately lesser showing that harm will occur."[51] This modified-burden-of-proof approach attracted instant support in the literature.[52] Its general acceptance in the practice of states, however, is less clear.

Given the nature of nuclear transboundary risks, which more often than not represent low probability/high consequence events, it is not surprising that much of the relevant international practice concerns nuclear activities of states.[53] International legal instruments on the management of transboundary nuclear risk provide only limited support for the present-day relevance of the *Trail Smelter* modified-burden-of-proof approach. For example, Article 6 of the Nuclear Safety Convention[54] calls on contracting states to upgrade the safety of existing nuclear installations and, if such upgrading cannot be achieved, to shut down the nuclear installation as soon as possible. Although this (due diligence) obligation to upgrade installation safety is couched in mandatory terms, the call for the installation's

[50] *See* Kirgis, *supra* note 3, at 294 (asking rhetorically "whether a disinterested decision-maker thirty years after *Trail Smelter*, in a world awakened to the existence of environmental deterioration, would find the clear and convincing standard literally applicable when there are plausible consequences magnified far beyond those considered in that case.")

[51] *Id.*

[52] *See* Luzius Wildhaber, *Die Öldestillieranlage Sennwald und das Völkerrecht der grenzüberschreitenden Luftverschmutzungen*, 21 ANNUAIRE SUISSE DE DROIT INTERNATIONAL 97, 119 (1975); Albrecht Randelzhofer & Bruno Simma, *Das Kernkraftwerk an der Grenze, in* FESTSCHRIFT FÜR FRIEDRICH BERBER 389, 416–17 (Dieter Blumenwitz & Albrecht Randelzhofer eds., 1973); and Günther Handl, *An International Legal Perspective on the Conduct of Abnormally Dangerous Activities in Frontier Areas: The Case of Nuclear Power Plant Siting*, 7 ECOLOGY LAW QUARTERLY 1 (1978).

[53] The precautionary principle represents another significant change in the standard for assessing transboundary risks. It directs states not to postpone cost-effective measures to prevent environmental degradation despite lack of full scientific certainty, in the face of possible serious or irreversible damage. *See* UNCED Report, *supra* note 14. The precautionary principle loosens the traditional – the Tribunal's "clear and convincing" – evidence standard and shifts the burden of proof to the risk-creating state. Its nature and purpose thus differ from the *Trail Smelter* – based modified-standard-of proof approach. In practice, however, the two often merge into a single line of argument about the international legal permissibility of state conduct that carries a significant risk of transboundary harm. *See generally* ARIE TROUWBORST, EVOLUTION AND STATUS OF THE PRECAUTIONARY PRINCIPLE IN INTERNATIONAL LAW (2002). *See also* Bratspies in this volume.

[54] Convention on Nuclear Safety, September 20, 1994, 1963 U.N.T.S. 293, *reprinted in* 33 INTERNATIONAL LEGAL MATERIALS 1514 (1994).

shutdown employs merely hortatory language. Moreover, in timing a shutdown, the source state is entitled to take into account "the whole energy context and possible alternatives as well as the social, environmental and economic impact." Similarly, Chapter 22 of Agenda 21, on "Safe and environmentally sound management of radioactive wastes"[55] echoes *Trail Smelter's* modified-burden-of-proof in its call upon states to abstain from storing or disposing of radioactive wastes near the marine environment, unless they determine that such practice does not pose an "unacceptable risk to people and the marine environment. . . ." This portion of Agenda 21 is also couched in hortatory, rather than mandatory, language and Agenda 21 is, of course, formally a nonbinding instrument.

The bulk of directly relevant practice consists of diplomatic protests by risk-exposed states challenging the source state's right to create significant transboundary nuclear risks. Most frequently, these risks involve siting nuclear power plants[56] or waste facilities in border areas,[57] or threats to internationally shared resources, such as international watercourses or coastal seas.[58] Although these protests may vary in detail and legal sophistication, ranging from simple appeals invoking the principle of good neighborliness[59] to implied military threats in case of nonabandonment of the transboundary risk-creating project,[60] they evidently are inspired by the same international legal rationale, the modified-burden-of-proof approach. However, it is not clear that these instances demonstrate an acceptance of the risk-exposed state's underlying legal claim. In situations in which source states ceased or adjusted their transboundary risk-creating conduct, their reasons for doing so have remained ambiguous, or at least cannot be readily characterized

55 UNCED, Agenda 21, Chapter 22, *Safe and environmentally sound management of radioactive wastes*, para.22.5(c), *supra* note 14.

56 *See, e.g.*, Eric Frey, *Austria Demands EU Act on Czech Reactor*, FINANCIAL TIMES, October 13, 2000, p.2; *see also* the references *supra* note 9.

57 For details on relevant state practice, *see* HANDL, *supra* note 9 at 42–46.

58 *See, e.g.*, David Buchen and Quinten Peel, *Finns Fear Russia's Pollution Timebomb*, FINANCIAL TIMES, November 25, 1997, at 2. Vessels carrying radioactive materials pose an analogous set of risks when they pass through the territorial sea or exclusive economic zone of coastal states. Concerned coastal states have espoused a modified-burden-of-proof approach in attacking the legality of this risk-creating activity. Thus, Caribbean countries consider transshipments of nuclear waste legally unacceptable given the extraordinary risk of harm, albeit of a low probability, associated with a transport accident in light of "the economic importance and ecological fragility of the Caribbean Sea and the well-being of the millions of people who depend on this unique resource for their very existence." Canute James, *Nuclear Waste Raises the Heat on its Way through Caribbean*, FINANCIAL TIMES, April 30, 1999, at 3.

59 *See* the discussion of Austria's diplomatic protest against the Swiss (Rüthi) power plant project, *in* Handl, *supra* note 52, at 29.

60 One of the more dramatic statements in this respect, although not an official United States Government position, is the "sense of the Congress" reflected in Sec.101 of the Cuban Liberty and Democratic Solidarity (*Liberdad*) Act of 1996, 22 USCA § 6031, according to which Cuba's "completion and operation of any nuclear power facility . . . will be considered an act of aggression which will be met with an appropriate response in order to maintain the security of the national borders of the United States and the health and safety of the American people."

as a recognition of the basic legal position advanced by the risk-exposed states.[61] A further factor contributing to the difficulty in assessing the effectiveness of claims advancing the modified-risk-of-harm argument is the defensive invocation by source states of the *nemo tenetur ad impossible* principle.[62] For example, Western governments' attempts to secure the closure of high-risk, Soviet-design nuclear reactors in Eastern Europe, including the lone operating unit at Chernobyl,[63] on the grounds of, *inter alia*, their unacceptable transboundary risk potential, have been resisted by the source state(s) as socially and economically infeasible.[64] Although some of the targeted facilities were ultimately shut down or given indispensable safety-upgrades, these reductions of transboundary nuclear risks were accomplished only upon the granting of financial or economic assistance,[65] making these accommodations suspect as precedent for the *Trail Smelter*–based risk approach.

A somewhat inconclusive picture also emerges from an analysis of the handful of international judicial or arbitral decisions reviewing transboundary nuclear risks. First, these cases, as framed by the parties' arguments, do not expressly raise the issue of the low probability/catastrophic consequence risk that we are concerned with here. Instead, the modified-burden-of-proof standard is only implicitly in play. Second, these proceedings resulted in only a few, relatively meager pronouncements on the legality of transboundary nuclear risk-creation. With these caveats in mind, the first case to be examined, one of the earliest instances in which the modified evidentiary standard can be said to have been relied on, is the *Nuclear Tests* cases.[66] Australia and New Zealand went before the ICJ requesting interim measures of protection to stop France from conducting atmospheric nuclear tests, which, the applicants alleged, carried significant transboundary risks of harm.[67] To the extent that the Court or its individual members addressed any aspects of the merits of applicants' claim, the transboundary risk-based argument did not fare well. For example, in his dissenting opinion Judge Ignacio-Pinto, although sympathetic to Australia's request, flatly denied that there existed "legal means in the present state of the law which would authorize a State to come before the Court asking it to prohibit another State

[61] Switzerland's cancellation of the Rüthi nuclear power plant project on the border with Austria could have been a direct result of Austria's diplomatic protest against the planned project, of a downward revision of energy demand projections rendering Rüthi unnecessary, or a combination of factors. *See* Handl, *supra* note 52, at 29–30.

[62] The degree of due diligence to be exercised by a state is inter alia a function of that state's means. *See, e.g.,.* British Claims in the Spanish Zone of Morocco, 2 R.I.A.A. 615, 644 (1924).

[63] *See, e.g., Ukraine to Keep Chernobyl in Operation*, New York Times, October 22, 1993, at A8.

[64] *See, e.g.*, World Nuclear Association, *Early Soviet Reactors and EU Accession*, January 2005, at www.world-nuclear.org/info/printable_information_papers/inf44print.htm.

[65] *See, e.g.*, Tom Warner, *Ukraine Set to Receive EBRD N-Plant Finance*, Financial Times, November 8, 2001, p. 6.

[66] *See* Nuclear Tests, (Austl. v. Fr.) I.C.J. Pleadings, Vol. I at 185.

[67] Nuclear Tests (Austl. v. Fr.), *Interim Measures* 1973 I.C.J. 99 (June 22).

from carrying out on its own territory such activities, which involve risks to its neighbours."[68]

Similar arguments about transboundary nuclear risks were again at the core of New Zealand's request in 1995 for a reexamination of the international legality of French nuclear testing in the South Pacific.[69] None of New Zealand's customary international law-based claims drew specifically on a characterization of French underground testing as posing a significant risk of a low probability/high consequence transboundary harm. Still, the allegation of a transboundary risk with a catastrophic impact potential was very much at the heart of New Zealand's application.[70] New Zealand specifically argued that, consistent with the precautionary principle, a state wishing to engage in transboundary risk-bearing conduct had the burden of proof of showing in advance that its conduct would not cause harm.[71] The ICJ, however, dismissed the request as not being within the limits of the original – the 1973–1974 – dispute and thus did not even touch on the merits of New Zealand's claim. Nevertheless, some dissenting judges expressed support for New Zealand's view that, legally speaking, the creation of a significant risk of radioactive contamination of the marine environment was equivalent to the actual introduction of nuclear materials into the marine environment which is prohibited under present general international law.[72]

More recently, in *The MOX Plant case*,[73] Ireland filed a request with the International Tribunal of the Law of the Sea for provisional measures of protection.[74] Ireland's aim was to secure its claimed right not to be polluted or exposed to the risk of transboundary radioactive pollution allegedly associated with the United Kingdom's commissioning of a mixed-oxide nuclear fuel plant at Sellafield on the Irish Sea.[75] The nature of the risk underlying this claim is not free of ambiguity, as

68 *Id.* at 131 (Dissenting Opinion of Judge Ignacio-Pinto).

69 Request for an Examination of the Situation in Accordance with Paragraph 63 of the Court's Judgment of 20 December 1974 in the Nuclear Tests (N.Z. v. Fr,) Case, 1995 I.C.J. 288 (Sept. 22).

70 The risk of potentially catastrophic transboundary effects was described in terms of a possible destabilization of the geological structure of the Mururoa and Fangataufa atolls – the French underground nuclear test sites – resulting in "a sudden spill-out of part of the radioactive 'stockpile' into the sea and the formation of a tidal wave" that was "a genuine threat to coasts as far away as New Zealand and Australia." *See id.* at 367 (Dissenting Opinion of Judge Koroma), (discussing New Zealand's allegations).

71 *Id.* at 290–91, para.6; and 298–99, paras.33–35.

72 *See id.* at 371 and 374. *See also* Dissenting Opinions of Judges Weeramantry, *id.* at 345–47; and Palmer, *id.* at 406–09.

73 The MOX Plant Case (Ir.v. U.K.) 2001 I.T.L.O.S. No. 10, Provisional Measures, (Dec. 3), *reprinted in* 41 INTERNATIONAL LEGAL MATERIALS 405 (2002).

74 In the second phase of the MOX Plant proceedings, the Irish Government's renewed request for provisional measures of protection was couched in terms that do not readily fit the modified-burden-of-proof mold. *See* Permanent Court of Arbitration, The MOX Plant Case (Ir. v. U. K.), Order No 3, June 24, 2003 (Suspension of Proceedings on the Merits, Request for Further Provisional Measures) *reprinted in* 42 INTERNATIONAL LEGAL MATERIALS 1187 (2003).

75 *See* The MOX Plant Case, Request for Provisional Measures and Statement of Case of Ireland, November 9, 2001, para.107.

Ireland complained of potential harm from both ongoing radioactive pollution, and from accidental or intentional (because of terrorist acts) releases into the marine environment. The Tribunal refused to accede to Ireland's request, apparently because it concluded that the requirement of urgency was not satisfied.[76] In rejecting Ireland's request, it did not answer the question of whether evidence of a risk of increased radioactivity in the Irish Sea – just as evidence of an actual increase – might trigger international legal protective measures. However, there is some indication that individual members of the Tribunal might have been inclined to accept such an argument.[77]

Because neither the evidence drawn from relevant state practice, nor international case law permits unambiguous conclusions regarding the extent to which the modified-burden-of-proof approach is part of present-day international law, the ILC's Draft Articles on the Prevention of Transboundary Harm from Hazardous Activities[78] acquire special importance. At first sight, they seem to fully confirm the persuasiveness of the *Trail Smelter*–based-modified-burden-of proof approach. Article 1 defines the articles' scope of application as extending to activities that involve "a risk of significant transboundary harm." Article 2(a) – consistent with the modified-burden-of-proof approach – defines "risk of causing significant transboundary harm" as including both risks with a "high probability of causing significant transboundary harm and a low probability of causing disastrous transboundary harm."[79] Article 3, the Draft Articles' key substantive provision then calls on the state of origin to "take all appropriate measures to prevent significant transboundary harm or at any rate to minimize the risk thereof."[80] However, as the Commission's specification of the core concept that underpins states' fundamental legal entitlement to be free from transboundary risks of harm, this formulation is rather disappointing: At the very least, it is potentially misleading and apt to raise questions about its consistency with international public policy.

Because both Article 3 obligations relate to future events or realizations of risk,[81] the distinction between "significant transboundary harm" and a "risk thereof" carrying differing obligations to "prevent" and to "minimize," is problematic. For this differentiation, between certain harm (to be prevented) and less-than-certain harm (to be minimized), is based on the probability of harm alone, rather than on the composite of probability and consequence of the harm. In this normative scheme, a "mere" risk of significant transboundary harm does not attract an obligation of prevention. The Commission's commentary does acknowledge the

76 *See* The MOX Plant Case, Provisional Measures, *supra* note 73, at para. 81.

77 *See id.* Separate Opinion of Judge Treves, paras. 7–8.

78 *See* Draft Articles on Prevention, *supra* note 3, at 370.

79 *Id.* at 371.

80 *Id.* at 372.

81 Article 1 and the Commission's commentary thereto make it clear that the Draft Articles are not intended to cover situations involving continuously or routinely occurring transboundary harm (such as "creeping pollution"). Rather, the Draft Articles cover exclusively activities that involve a "risk of causing significant harm and, as the commentary further notes, "the element of 'risk' is by definition concerned with future possibilities." *Id.* at 385, paras.13–14.

due diligence implications of transboundary risks associated with ultrahazardous activities. But in describing a variety of factors bearing on the proportionately higher standard of care, it speaks in essence about "measures to minimize risk."[82] Not once does it expressly acknowledge that, at the extreme end of the spectrum of threatened consequences, the heightened standard of due diligence might force the conclusion that the activity itself would be impermissible.[83] In the end, uncertain future harm – indeed, any harm that is "less than certain" if one follows the Commission's logic[84] – no matter how potentially catastrophic its nature and scope, does not *eo ipso* attract a legal obligation to end the risk-bearing activity. The Commission's efforts at softening this impression are unconvincing.[85]

The Draft Articles might be premised on the assumption that these Article 3 obligations, in conjunction with the Article 4 obligation to cooperate in good faith, and the attendant procedural obligations,[86] would ensure that states arrive at a proper balance of their respective rights to engage in and to be protected against transboundary risk-carrying activities. And they may well further be based on the assumption that such an understanding might include a ban if the activity concerned entailed an irremediable risk of significant transboundary harm. But these assumptions are not spelled out.

Although the Commission's preference for providing maximum leeway for states seeking an accommodation may be understandable, its failure to unambiguously specify the outer limits of permissible risk-creation is a different matter. The Commission's decision to err on the side of protecting states' freedom to engage in transboundary risk-bearing conduct does not readily conform to international public policy as reflected in the principle of sustainable development or, indeed, the precautionary principle. It is also not in accord with the *Trail Smelter*–based modified-burden-of-proof approach, or the substantial doctrinal

82 *See id.* at 394.

83 The argument that an acceptable mitigation of transboundary risk would always be achievable through a careful manipulation of risk factors, and thus rendering unnecessary the prohibition of the activity altogether, is unpersuasive. For example, where location is a critical factor – as, e.g., in the Caribbean nations' complaints, *supra* note 58 – change of location, the sought-after risk-reduction strategy, in essence implies a demand for the termination of the transboundary risk-carrying activity.

84 Since the Draft Articles apply to "future possibilities," Article 3 obligations must be located on a sliding scale of probability. Where along this scale the obligation to minimize uncertain harms would arise remains undefined.

85 For example, the Commission's commentary emphasizes prevention as "the primary duty of the State of origin" and explains that only where prevention is "not fully possible [the source State] should exercise its best efforts to minimize the risk thereof." *See* Commentary to the Draft Articles on Prevention, art. 3, *supra* note 3 at 390. By the same token, the Article 4 obligation that states cooperate in preventing harm or minimizing the risk thereof, cannot hide the fact that the Commission was not willing to accept as a general normative notion that a mere risk of harm might trigger the prohibition of the transboundary risk-carrying conduct.

86 *Id.* at arts. 7–9, covering risk assessment, notification, and consultations on preventive measures, respectively.

support the latter has attracted whose quintessential point it is that the international legal impermissibility of an activity may be first and foremost a function, not of the degree of certainty of the harm threatened, but of its magnitude. Finally, the Draft Articles give short shrift to the many diplomatic claims by risk-exposed states – whether formally or informally accepted or rejected by the source state(s) concerned – in which prevention of the risk-carrying activity itself, not just adjustments of its modalities, was viewed as the only way to eliminate the transboundary risk of harm.[87]

CONCLUSION

Notwithstanding these somewhat disappointing findings regarding the ILC Draft Articles, it is evident that the *Trail Smelter* decision, in particular the *Trail Smelter*–based modified-burden-of-proof rationale has had a huge impact on the structure and the contents of the Commission's Draft. Its influence can be traced all the way back to the musings of the Commission's first Special Rapporteur on the topic of prevention.[88] Indeed, international environmental practice in general, as well as practice specific to managing transboundary nuclear risks and other hazardous activities, have been fundamentally shaped by the Tribunal's *dictum*. The arbitration marks "no-significant harm" as the dividing line between internationally permissible and impermissible conduct. It evinces a surprisingly modern position by not subjecting this threshold to any balancing of interests and leaves no doubt that the obligation of prevention is one of due diligence. These particular facets of the Tribunal's decision are part and parcel of present-day international environmental law. Although *Trail Smelter* poses some conceptual challenges, it would thus be difficult to deny the decision's usefulness – notwithstanding its age – as a pivotal concept of present-day customary international law applicable to the environment.

87 For example, in relation to nuclear facilities in border areas, complaining states have frequently sought the termination or prevention of the risk-carrying activity, rather than a mere mitigation of the transboundary risk. For details see HANDL, *supra* note 9, at 23–28, 43–45. Further, in the *Request for an Examination* case, *supra* note 69, at 291, New Zealand's principal objective was to obtain a declaratory judgment that France's proposed nuclear underground tests would violate New Zealand's international legal rights, in other words to put a stop to the risk-creating conduct. Only secondarily did New Zealand seek to force France to undertake an environmental impact assessment "according to internationally accepted standards," a remedy that might minimize the risk involved.

88 *See supra* note 2.

12 Through the Looking Glass: Sustainable Development and Other Emerging Concepts of International Environmental Law in the *Gabčikovo-Nagymaros* Case and the *Trail Smelter* Arbitration

James F. Jacobson

INTRODUCTION: LESSONS FROM THE PAST REFLECTING INTO THE FUTURE

The *Gabčikov-Nagymaros* case (*Danube Dam* case)[1] provides an insightful glimpse of emerging concepts in international environmental law. These concepts may not have been decisive for the holding of the decision, but they had an underlying and indirect influence on the outcome. These underlying concepts provide an important framework for analyzing difficult international environmental law cases and will yet play a significant role in determining the outcome of future international environmental law disputes. The same concepts made an earlier appearance in the historic *Trail Smelter* arbitration.[2] I intend to open the *Trail Smelter* arbitration as a window onto the contemporary nature and status of the following principles, which also found some expression in the *Danube Dam* case: (1) sustainable development; (2) environmental impact assessment; and (3) continued environmental monitoring.

THE *DANUBE DAM* CASE

On September 16, 1977, the Hungarian People's Republic and the Czechoslovak People's Republic concluded a treaty (Treaty) by which they agreed to build a series of locks and dams along a stretch of the Danube River between Gabčikovo in Czechoslovak territory and Nagymaros in Hungarian

[1] Gabčikovo-Nagymaros Project, 1997 I.C.J. 7 (Sept. 25), *reprinted in* 37 International Legal Materials 162 (1998) [hereinafter "*Danube Dam* case"].

[2] *Trail Smelter Arbitral Decision*, 33 American Journal of International Law 182 (1939) [hereinafter "*Trail Smelter* (1939)"]; *Trail Smelter Arbitral Decision*, 35 American journal of International Law 684 (1941) [hereinafter "*Trail Smelter* (1941)"]. See Annex to this volume.

territory.[3] The Treaty provided for the building of a series of dams and weirs along the Danube, in both Hungary and Slovakia.[4] The objects of the Treaty were the production of hydroelectricity, the improvement of navigation on the relevant portion of the Danube, and flood control.[5]

The project was to be a joint effort between the two countries in several ways. Article 5 of the Treaty required that the cost and ownership of the project be borne in equal measure.[6] A joint contractual plan, which was to be created and executed in connection with the Treaty, became the means whereby the joint provisions of the Treaty were to be implemented.[7]

Articles 15, 19, and 20 of the Treaty allowed for special consideration of environmental concerns. Article 15 required the parties, by means specified in the joint contractual plan, to ensure that the quality of water in the Danube was not impaired by the project.[8] Article 19 also required the parties, through means specified in the joint contractual plan, to comply with obligations to protect the environment in connection with the project.[9] Article 20 provided a means for protecting fishing interests in the Danube in connection with the project.[10] Thus, the joint contractual plan was key to the fulfillment of the Treaty, particularly as it related to the environmental provisions of the Treaty.[11]

Two Protocols were issued in connection with the project. The first, issued in October 1983, called for a reduction in the work and an extension of the time frame for completion of the project.[12] The second, issued in February 1989, called for the acceleration of work on the project.[13] However, during that same period, concern and criticism in Hungary about the project became intense.[14] Concerns existed as to the economic benefit of the project, but more significant were those related to environmental harms that might result from the project.[15] In light of those concerns, Hungary, on May 13, 1989, suspended work on its portion of the project at Nagymaros in order to conduct studies with respect to possible environmental impacts. On October 27, 1989, Hungary abandoned all work at Nagymaros.[16]

In 1991, Slovakia responded to Hungary's abandonment of the project by formulating and beginning construction of an alternative project known as Variant C.[17] Variant C would reroute the Danube through Slovakia so as to bypass the

[3] *Danube Dam* case, *supra* note 1, at 174. All references to the Treaty and the two protocols issued in connection with it are in reliance on the ICJ opinion. For an English-language translation of the Treaty, *see* http://www.gabcikovo.gov.sk/doc/it1977en.treaty.html.

[4] *Id.*

[5] *Id.*

[6] *Id.* at 175.

[7] *Id.*

[8] *Id.* at 176.

[9] *Id.*

[10] *Id.*

[11] *Id.* at 177.

[12] *Id.*

[13] *Id.*

[14] *Id.* at 180.

[15] *Id.*

[16] *Id.* at 177.

[17] *Id.* at 177.

Nagymaros section.[18] On May 19, 1992, Hungary announced its termination of the Treaty, effective May 25, 1992.[19] A series of negotiations and attempts to come to a resolution of the dispute came to no avail. Both parties then agreed to submit the dispute to the ICJ for decision.[20]

The Special Agreement, submitted to the ICJ pursuant to Article 40 of its Statute, set forth three main issues for decision by the ICJ. The first was "[w]hether the Republic of Hungary was entitled to suspend and subsequently abandon, in 1989, the works on the Nagymaros Project and on the part of the Gabčikovo Project for which the Treaty attributed responsibility to the Republic of Hungary."[21]

Hungary argued that a state of ecological necessity existed that justified its termination of its obligations under the Treaty.[22] Hungary maintained as its key concerns that the completion of the project would seriously impair the supply of groundwater in the area and, through decreased riverbank filtration, the drinking water to two-thirds of Budapest.[23]

Slovakia responded by asserting that ecological problems could arise but would be of such an extent that they could be remedied.[24] In Slovakia's view, Hungary's characterization of the environmental harm of the project and its operation was overly pessimistic and "of an extreme kind."[25]

In order to determine whether a "state of necessity" existed in the case justifying Hungary's deviation from the terms of the Treaty, the ICJ turned for guidance to the "state of necessity" provision of the Draft Articles on the International Responsibility of States as set forth by the International Law Commission,[26] concluding that a state's ecological interest in all or some of its territory constitutes an essential interest under the terms of the Draft Articles.[27] This interpretation of a "state of necessity" found further support in the ICJ's advisory opinion regarding the *Threat or Use of Nuclear Weapons*, in which the Court stated:

> The environment is not an abstraction but represents the living space, the quality of life and the very health of human beings, including generations unborn. The general obligation of States to ensure that activities within their jurisdiction and control respect the environment of other States or of areas beyond national control is now part of the corpus of law relating to the environment.[28]

[18] *Id.*

[19] *Id.*

[20] Statute of the International Court of Justice, June 26, 1945, art. 40, para. 1, 59 Stat. 1055, 33 U.N.T.S. 993.

[21] *Danube Dam* case, *supra* note 1, at 179.

[22] *Id.* at 182.

[23] *Id.*

[24] *Id.* at 183.

[25] *Id.*

[26] Article 33 of the Draft Articles on the International Responsibility of States. *Id.* at 184 (quoting Draft Articles on the International Responsibility of States, Yearbook of the International Law Commission, 1980, Vol. II, Part 2, p. 34).

[27] *Id.*

[28] *Id.* at 184–185 (citing Legality of the Threat or Use of Nuclear Weapons, Advisory Opinion, 1996 I.C.J. 226, 241–242 (July 8)).

Having found that environmental matters can form the basis of a "state of necessity," the Court nonetheless found no imminent environmental peril in the case at hand, concluding that a significant degree of uncertainty still existed regarding any environmental harm or its imminence, that is, its certainty or inevitability.[29]

The second question was whether Slovakia was entitled to proceed, in November 1991, to the "provisional solution" (Variant C), and to put it into operation beginning in October 1992.[30] The ICJ concluded that the implementation of Variant C did not constitute an effective countermeasure because it was disproportionate and violated the provisions of the Treaty, which required the joint construction and utilization of the project. With these terms of the Treaty in mind, the Court concluded that Variant C also constituted an internationally wrongful act.[31]

Finally, the ICJ was asked to consider the legal effects of Hungary's May 19, 1992 termination of the Treaty.[32] Given that both Hungary and Slovakia committed internationally wrongful acts, the Court concluded that both were under an obligation to pay compensation and entitled to obtain compensation.[33]

THE EVOLUTION OF SUSTAINABLE DEVELOPMENT AND RELATED CONCEPTS

Particular focus should be given to the ICJ's conclusion that the parties were under a legal obligation to consider, in their subsequent negotiations, the ways in which the multiple objectives of the Treaty could best be served, while keeping in mind that all the Treaty objectives should be fulfilled.[34] This mandate would require the parties to give consideration to the project's impact on and its implications for the environment, which was a key objective of the Treaty,[35] without compromising the development goals central to the entire project. With respect to the environmental risks of the Treaty project, the ICJ stated that current standards must be taken into consideration.[36] This was not only allowed by the wording of Articles 15 and 19 of the Treaty, the Court explained, but concern for the environment was prescribed by the Treaty, to the extent that Articles 15 and 19 imposed a continuing obligation on the parties to maintain the quality of the water of the Danube and to protect nature.[37] The reconciliation of these distinct and potentially contradictory obligations is a project not unlike that which is at the core of the general international environmental law concept of sustainable development.[38]

29 *Id.* at 185.
30 *Id.* at 187.
31 *Id.* at 190.
32 *Id.*
33 *Id.*
34 *Id.* at 200.
35 *Id.*
36 *Id.*
37 *Id.*
38 *Id.* at 201.

Sustainable Development in International Law

The majority opinion of the Court defined sustainable development very simply as an attempt to reconcile the need for economic development with the protection of the environment.[39] The majority opinion refers to sustainable development as a concept of international law,[40] declining to bestow on it the status of a principle of international law, in contrast to Judge Weeramantry in his separate opinion.[41] Although genuine debate exists as to the status of the doctrine in international law, sustainable development's stature has been growing, particularly in the last thirty years.[42]

One of the most fundamental documents dealing with sustainable development is the Stockholm Declaration.[43] Two principles, in particular, offer an adequate overview and are reflective of the dual-right perspective inherent to sustainable development. Principle 2 of the Stockholm Declaration states that "the natural resources of the earth . . . must be safeguarded for the benefit of present and future generations through careful planning or management, as appropriate."[44] Principle 8 of the Stockholm Declaration articulates the right to development in this manner, "[e]conomic and social development is essential for ensuring a favorable living and working environment for man and for creating conditions on earth that are necessary for the improvement of the quality of life."[45] These two principles highlight the rights to a healthy environment and to development as anthropocentric, while reflecting the inherent need for harmony between both.

The United Nations Convention on Environment and Development, and, more particularly, the Rio Declaration, builds upon the commitment to sustainable development set forth in the Stockholm Declaration.[46] Principle 1 of the Rio Declaration affirms that sustainable development is an anthropocentric principle.[47] Principle 3 of the Rio Declaration states that development is a fundamental right, with Principle 5 of the Rio Declaration characterizing the main thrust of that right as the eradication of poverty and the improvement of the world's standard of living.[48] Principle 4 of the Rio Declaration emphasizes the integral link between

39 *Id.*

40 *Id.*

41 *Id.* at 204, 207 (separate opinion of Judge Weeramantry). Judge Weeramantry's separate opinion expands significantly the definition of sustainable development. He defines sustainable development as the reconciliation of two fundamental human rights, the right to development and the right to environmental protection in order to sustain human life.

42 Sumudu Atapattu, *Sustainable Development, Myth or Reality? A Survey of Sustainable Development Under International Law and Sri Lankan Law*, 14 Georgetown International Environmental Law Review 265, 283–284 (2001).

43 Report of the Stockholm Conference, U.N.Doc. A/CONF. 48/14, *reprinted in* 11 International Legal Materials 1416 (1972).

44 *Id.* Principle 2.

45 *Id.* Principle 8.

46 Report of the United Nations Conference on Environment and Development, Annex I, U.N. Doc. A/CONF.151/26 (Vol. I) *reprinted in* 31 International Legal Materials 874 (1992).

47 *Id.* Principle 1.

48 *Id.* Principles 3 and 5.

environmental protection and the development process, with Principle 7 of the Rio Declaration placing responsibility on all states, but particularly developed states, to conserve, protect, and restore the health and integrity of the Earth's ecosystem.[49]

The Rio Declaration also lays out more specifically the obligations of states in harmonizing these two interests. Principle 8 of the Rio Declaration requires states to reduce and eliminate unsustainable patterns of production and consumption.[50] Principle 9 speaks to the need for exchange of scientific understanding and development of new technologies in order to reduce the impact of development upon environmental concerns.[51] Principles 8 and 9 work in concert with one another, and neither can be truly fulfilled without the other. Principle 12 of the Rio Declaration offers a key warning that "trade policy measures for environmental purposes [should] . . . not constitute a means of arbitrary or unjustifiable discrimination or a disguised restriction on international trade."[52]

The concept of sustainable development has not only garnered support among developed countries, it also has been seen as essential to developing countries as they seek economic and social growth. The heads of state for the Central African countries, recognizing the vast resource and potential wealth of their shared tropical forests, affirmed their commitment to develop their forest resources through conservation, rational use, and sustainable development.[53] In doing so, these African states committed to pursuing greater involvement of the rural populations and economic operators in the sustainable management and conservation of forest ecosystems.[54] Also, the African Union called on its members to take all necessary actions to give support to the Union's initiatives regarding sustainable development.[55]

An examination of the key European Union treaties reveals not only a commitment to sustainable development but also an insightful view as to how reconciliation between development and environmental protection can take place. Article 2 of the Treaty on the European Union establishes, as one of the Union's objectives, the promotion of "economic and social progress and a high level of employment and to achieve balanced and sustainable development."[56] Yet, the European Communities recognized that development objectives

49 *Id.* Principles 4 and 7.

50 *Id.* Principle 8.

51 *Id.* Principle 9.

52 *Id.* Principle 12.

53 Summit of Central African Heads of State on the Conservation and Sustainable Management of Tropical Forests: The Yaounde Declaration, March 17, 1999, *reprinted in* 38 International Legal Materials 783 (1999).

54 *Id.*

55 African Union: The Durban Declaration in Tribute to the Organization of African Unity on the Occasion of the Launching of the African Union, Jul. 10, 2002, *reprinted in* 41 International Legal Materials 1029 (2002).

56 Treaty on the European Union, Feb. 7, 1992, Article 2, O.J. (C 191) 1 (1992) [hereinafter EU Treaty].

should not trump legitimate concerns about protecting the environment and its resources. Article 174 (ex Article 130r) outlines Community objectives of preserving, protecting, and improving the quality of the environment while prudently and rationally utilizing natural resources.[57] That same article further states that the Community's policy on the environment should aim at a high level of protection.[58]

Some may argue that Article 174 (ex Article 130r) represents a retrenchment from objectives focused on progressive economic development under growing political pressure over environmental concerns. However, none of the objectives outlined above have been recanted or stricken. If the objectives of economic and social progress and a high level of environmental protection are both to be met, then reconciliation of the objectives must be the focus.

It is clear that "movement toward sustainable development will be feasible only if based on reliable cost/benefit projections and feedback regarding the true environmental impact of change."[59] That information is, in part, acquired through environmental impact assessment and continued environmental monitoring.

The Necessity of an Environmental Impact Assessment

Basic Framework of an Environmental Impact Assessment

Without the requisite information regarding the environmental impact of a particular development project, reconciling development and environmental protection, as well as making wise and far-sighted decisions, becomes impossible. Thus, an environmental impact assessment is essential to the reconciliation that is at the heart of sustainable development.

No uniform criteria exist for conducting an environmental impact assessment,[60] but Erica Preiss has identified the following as components of an international obligation to conduct an environmental impact assessment: the definition of the proposed activity;[61] the identification of the impacts that will be studied, including the magnitude, extent, significance, with special attention to be given certain areas and certain harms;[62] the completion of a base line study that identifies mitigating measures and compares alternatives;[63] and, finally, governmental agency review and public participation are solicited, all culminating in a full documentation of the environmental impact assessment.[64]

57 Treaty Establishing the European Community, Nov. 10, 1997, Article 174, O.J. (C 340) 3 (1997) [hereinafter EC Treaty].

58 *Id.*

59 *Id.*

60 Erika L. Preiss, *The International Obligation to Conduct An Environmental Impact Assessment: The ICJ* Case Concerning the Gabčikovo-Nagymaros Project, 7 New York University Environmental Law Journal 307, 308 (1999).

61 *Id.*

62 *Id.*

63 *Id.*

64 *Id.*

International Support for an Environmental Impact Assessment

The Rio Declaration again provides a touchstone for the rising practice of conducting an environmental impact assessment. Principle 17 of the Rio Declaration states, "Environmental impact assessment, as a national instrument, shall be undertaken for proposed activities that are likely to have a significant adverse impact on the environment and are subject to a decision of a competent national authority."[65] That principle underscores the important role environmental impact assessment plays in providing a practical way of reconciling the interests of development and environment. Over 70% of the world's nations have implemented some form of "informal or mandatory environmental impact assessment requirements."[66]

The state plays a vital role in ensuring that environmental impact assessments take place. The Convention on Environmental Impact Assessment in a Transboundary Context (EIA Convention) announces a clear role for states regarding the enforcement of the obligation to conduct impact assessments. Article 2(3) of the EIA Convention imposes the obligation on the state undertaking one of the listed activities to ensure that an environmental impact assessment is undertaken.[67] Article 4 of the EIA Convention describes, by reference to Appendix II, the contents of the environmental impact assessment, and requires the submission of the assessment to the competent authority in the generating state.[68] Article 3(7) of the EIA Convention allows the party states, when they cannot agree whether there is likely to be a significant adverse transboundary impact, to submit the question to an inquiry commission or to agree upon an alternate method of settling the question.[69] Furthermore, the state's duty to prevent transboundary pollution not only underscores "the global interdependence of ecological relationships [but] can obligate a state to ensure that proper mechanisms are in place to prevent transboundary pollution in accordance with the doctrine of state responsibility."[70]

The United Nations Convention on the Protection and Use of Transboundary Watercourses and International Lakes does not mention environmental impact assessments specifically, but it does affirm obligations that are the equivalent of an impact assessment.[71] Article 3 of the Watercourses Convention requires "measures for the preservation, control, and reduction of water pollution shall be taken, where possible, at source."[72] Article 5 also requires

65 Rio Declaration, *supra* note 46, Principle 17.

66 Kevin R. Gray, *International Environmental Impact Assessment: Potential for a Multilateral Environmental Agreement*, 11 COLORADO JOURNAL OF INTERNATIONAL ENVIRONMENTAL LAW & POLICY 83, 89 (2000).

67 Convention on Environmental Impact Assessment in a Transboundary Context, Feb. 25, 1991, art. 2(3), 30 INTERNATIONAL LEGAL MATERIALS 802 (1991) [hereinafter "EIA Convention"].

68 *Id.* at art. 4.

69 *Id.* at art. 3(7).

70 Gray, *supra* note 66, at 100.

71 Convention on the Protection and Use of Transboundary Watercourses and International Lakes, Mar. 17, 1992, 31 INTERNATIONAL LEGAL MATERIALS 1312 (1992).

72 *Id.* at art. 3.

the parties to the convention to cooperate in research and development of effective means of preventing, controlling, and reducing transboundary impact.[73]

Perhaps one of the most important advances with respect to an international obligation to perform an environmental impact assessment is the United Nations Convention on Environmental Impact Assessment in a Transboundary Context.[74] That convention obligates state parties to give notice when a proposed activity is likely to cause a significant adverse transboundary impact to any party that may be affected by that activity.[75] Then, the affected party may respond that it intends to participate in the environmental impact assessment procedure.[76] On receipt of the response, the state proposing the activity must provide all relevant information regarding the environmental impact assessment procedure and the possible significant, adverse transboundary impact.[77] Significantly, the convention requires the parties to ensure that "due account is taken of the environmental impact assessment" in the final decision regarding the proposed activity.[78]

The Commission on Sustainable Development (CSD) has, as one of its major tasks, the assessment and evaluation, at the international level, of the implementation of Agenda 21.[79] "Agenda 21 calls for the assessment of impacts upon the environment and the monitoring of those effects and changes."[80] Thus, the CSD plays an important information gathering role, a role that was augmented in 1993 by a change in the rules to allow NGOs to make written presentations to the CSD.[81] Although some frustration may be exhibited toward the relative ineffectiveness of the CSD, patience has the potential of proving virtuous with respect to compliance in the future.[82] The perfection of information gathering regarding environmental impact will lay a much-needed foundation on which effective compliance mechanisms can be built.

Environmental Monitoring to Ensure Future Sustainability

Judge Weeramantry, in his separate opinion in the *Danube Dam* case, broadened the scope of an environmental impact assessment by stating that it is "not merely an assessment prior to the commencement of a project, but a continuing assessment and evaluation as long as the project is in operation."[83] This expansion of the impact assessment obligation over the life of a project makes good sense. Why go to great effort to determine the potential impact at the outset of the project without then monitoring the project and ascertaining more information regarding

[73] *Id.* at art. 5.

[74] EIA Convention, *supra* note 67.

[75] *Id.*

[76] *Id.*

[77] *Id.*

[78] *Id.*

[79] Günther Handl, *Controlling Implementation of and Compliance with International Environmental Commitments: The Rocky Road from Rio*, 5 Colorado Journal of International Environmental Law & Policy 305, 318 (1994).

[80] Gray, *supra* note 66, at 91.

[81] Handl, *supra* note 79.

[82] *See id.* at 313–315.

[83] *Danube Dam* case, *supra* note 1, at 214.

the actual impact? It would be difficult, if not impossible, to ascertain all of the environmental danger a project would pose from the very beginning.[84] Yet, it must be acknowledged that certain development projects pose a greater need for continued monitoring. Judge Weeramantry added that "the greater the size and the scope of the project, the greater is the need for a continuous monitoring of its effects."[85]

The Convention on Environmental Impact Assessment also makes provision for continued monitoring of development projects.[86] Any party concerned may request a post-project analysis, which includes surveillance of the activity and monitoring for compliance, for an activity for which an environmental impact assessment has been undertaken.[87]

THE *TRAIL SMELTER* ARBITRATION AS A REFLECTION OF THE FUTURE

Whereas sustainable development and related concepts hovered in the background to inform and shape the ICJ's resolution of the *Danube Dam* case, the *Trail Smelter* arbitration, in pioneering fashion, more explicitly engaged these concepts of international environmental law.

Sustainable Development in the *Trail Smelter* Arbitration

The *Trail Smelter* Tribunal's concern for sustainable development represents a reasoned means for reconciling the seemingly competing interests of the smelter's continued operation and the farmers' right to a healthy environment in which to live and cultivate their crops.

At the time of the arbitration, the Trail smelter had become one of the most modern and best-equipped smelting plants in North America.[88] Furthermore, the Trail smelter comprised one of the largest employers in British Columbia, and the Canadian government received, on average, $1 million in tax revenue from it annually. "Canadian officials were highly conscious of [Trail smelter's] significance in the local and national economy."[89] Thus, any action to completely shut down the smelter would have had adverse effects on the Canadian economy, including leaving hundreds of people in the local area unemployed.

On the opposite side, the damage to the agricultural areas in northern Washington was sufficiently evident that the Tribunal had little difficulty in finding that such damage had occurred.[90]

84 *Id.*
85 *Id.*
86 EIA Convention, *supra* note 67.
87 *Id.*
88 *Trail Smelter* (1939), *supra* note 2, at 190.
89 *Id.* at 222.
90 *Id.* at 208.

The contemporary manifestation of the concept of sustainable development was foreshadowed by the *Trail Smelter* Tribunal's attempt to reconcile the Canadian interest in economic development and the U.S. interest in the protection of its environment. Application of Principle 8 of the Stockholm Declaration and Principle 5 of the Rio Declaration would have required that the continued operation of the Trail smelter be made contingent on the prevention of a decline in the standard of living among the local residents. Also, Principle 7 of the Rio Declaration would place upon Canada the obligation to conserve and protect the environment, not only within its own state boarders but also in the transboundary context as it affected the United States. Elimination of unsustainable patterns of production and consumption would also be required under Principle 8 of the Rio Declaration. Whereas the exact line at which the smelter's activities would drop below a sustainable level may be difficult to draw, it is unlikely that the concept of sustainable development would have tolerated the release of 10,000 tons of sulfur into the air per month.

The *Trail Smelter* Tribunal's attempt to reconcile these interests was mandated by the Convention creating the Tribunal, which required the Tribunal to reach a conclusion "just to all parties."[91] One commentator has described the Tribunal's charge as a matter of balancing the interests of the agricultural community and the industrial community and "endeavor[ing] to adjust the conflicting interests by some 'just solution' which would allow the continuance of the operation of the Trail smelter but under such restrictions and limitations as would, as far as foreseeable, prevent damage in the United States."[92] This desire to reconcile developmental and environmental interests is evidenced by the standard articulated by the Tribunal in assessing responsibility.[93] If the Tribunal required the smelter to refrain from all activities producing an effect on a foreign territory, then the smelter would be forced to cease its operations completely.[94] The Tribunal's compromise, requiring actual and significant damage, allowed the smelter to continue its operations as long as it kept them below an injurious level.[95]

Environmental Impact Assessment and Continued Monitoring in *Trail Smelter*

There can be little doubt that Canada would be required to produce an environmental impact assessment if the smelter were to be built today,[96] and that requirement would have been reinforced both domestically and internationally,

91 Convention on Damages Resulting From Operation of Smelter at Trail, British Columbia, June 5, 1935, U.S.-Canada, art. IV, 49 Stat. 3245. *See* Annex to this volume.

92 Alfred P. Rubin, *Pollution by Analogy: The Trail Smelter Arbitration*, 50 Oregon Law Review 259, 271 (1971). Republished in this volume.

93 *Id.* at 266.

94 *Id.*

95 *Id.*

96 *See* Canadian Environmental Assessment Act (unofficial version) available at http://laws.justice.gc.ca/en/C-15.2/text.html.

particularly from the United States. Perhaps the most significant precedent arising out of the *Trail Smelter* decision is the obligation the Tribunal imposed for a continued environmental monitoring regime.[97] Yet, it is important to remember that such a regime need not be a separate enforcement mechanism, occurring after the harm, as it was in *Trail Smelter*. Continued monitoring can be part and parcel of the duty to conduct an environmental impact assessment. This interconnectedness between impact assessment and continued monitoring in the international context is reinforced by the Convention on Environmental Impact Assessment, which allows a party to request continued monitoring on a project for which an environmental impact assessment was undertaken.

The continued monitoring regime imposed by the *Trail Smelter* Tribunal also presents a useful analogy with contemporary international law obligations. The monitoring regime was to be executed in two parts.[98] A trial period was established for the growing seasons of 1938–1940 and the winter seasons of 1938–1939 and 1939–1940.[99] At the conclusion of the trial period, the Tribunal would decide the type of permanent regime to be put into place.[100] The temporary regime provided for two consultants who would have authority to require regular reports from the smelter as to its methods of operation and could direct the smelter's operations based on the results of those reports.[101] For times of increased crop and tree sensitivity to the sulfur dioxide emissions, limits were placed on those emissions of not more than 100 tons per day.[102]

The final regime put in place by the Tribunal in 1941 contained several elements.[103] First, the Tribunal specified two meteorological instruments to be used to measure wind direction, velocity, and turbulence.[104] Three automatic sulfur dioxide recorders were to be positioned at various locations in northern Washington to measure the concentrations.[105] The smelter was not permitted to place the stacks at a height any lower than their present height.[106] Maximum permissible emissions were set based on the time of day and the type of wind conditions at that time.[107] Other more specific restrictions became applicable depending on the weather conditions, the season of the year, or the visual observations of trained observers.[108] The Tribunal also provided, significantly, for the suspension of the regime should technological advances allow for sufficient reduction in the emissions so as not to pose any undue risk on the U.S. interest.[109]

Today, a continued monitoring regime such as this would be available, according to the Convention on Environmental Impact Assessment, as a matter of party request, and would attach, at least according to Judge Weeramantry's separate opinion in the *Danube Dam* case, as an implication of a treaty relationship.

97 *See* Bratspies and Miller in this volume.
98 *Trail Smelter* (1939), *supra* note 2, at 211.
99 *Id.*
100 *Id.*
101 *Id.* at 210.
102 *Id.* at 211.
103 *Trail Smelter* (1941), *supra* note 2.
104 *Id.* at 727
105 *Id.*
106 *Id.* at 728.
107 *Id.*
108 *Id.* at 728–730.

CONCLUSION

As we look to the future, more opportunities will arise for application to international environmental disputes of the concepts of sustainable development, environmental impact assessment, and continued monitoring. When they come, those decisions will owe much to the *Danube Dam* case, and perhaps even more to the *Trail Smelter* arbitration.

13 *Trail Smelter's* (Semi)Precautionary Legacy

Rebecca M. Bratspies

INTRODUCTION

Although almost every discussion of state responsibility begins with its talismanic invocation,[1] time has not been kind to the *Trail Smelter* arbitration. Its primary contributions to international law have been the statement that: "no State has the right to use or permit the use of its territory in such a manner as to cause [environmental] injury . . . in or to the territory of another,"[2] and its requirement that Canada pay the United States compensation for damages. Although these *Trail Smelter* principles have become customary international environmental law, the arbitration itself is often viewed as a quaint remnant of a bygone world. As Mark Drumbl succinctly explains, many scholars view *Trail Smelter's* marginalization as inevitable in light of international law's evolution from a state-to-state realm to one of multilateral, consensus-based actions.[3] Others have suggested that the arbitration's impact is blunted by the fact that harm was not contested before the Tribunal.[4] This unique combination of characteristics leads many to conclude that *Trail Smelter* has little relevance for resolving the thorny transboundary environmental challenges that beset our ever-globalizing world.

I think the case has much to teach modern international environmental law, but for somewhat unconventional reasons. In the context of global warming, Russell Miller points out that the arbitration offers some procedural lessons as well as its famous *Trail Smelter* principles.[5] This chapter explores one of the

[1] Alfred Rubin, *Pollution by Analogy: The Trail Smelter Arbitration*, 50 OREGON LAW REVIEW 259 (1971), ("Every discussion of the international law of pollution starts, and must end, with a mention of the *Trail Smelter Arbitration* . . ."). Republished in this volume.

[2] *See Trail Smelter Arbitral Decision*, 33 AMERICAN JOURNAL OF INTERNATIONAL LAW 182 (1939) [hereinafter "*Trail Smelter* (1939)"]; *Trail Smelter Arbitral Decision*, 35 AMERICAN JOURNAL OF INTERNATIONAL LAW 684 (1941) [hereinafter "*Trail Smelter* (1941)"]. *See* Annex to this volume.

[3] *See* Drumbl in this volume.

[4] *See generally*, Rubin, *supra* note 1; Karin Mickelson, *Rereading Trail Smelter*, 31 CANADIAN YEARBOOK OF INTERNATIONAL LAW 219 (1993). Republished in this volume. *See also* Ellis in this volume.

[5] *See* Miller in this volume.

arbitration's least considered facets – the decisional process itself. Hampered by a lack of scientific evidence, the *Trail Smelter* Tribunal crafted an adaptive decisional structure in order to fulfill its charge to be "just to all parties"[6] while resolving a conflict over pollution flowing across the Canadian border and causing harm in Washington State. The Tribunal's innovative and far-reaching solution, which I am calling (semi)precautionary, was to craft an interim regime from the available but incomplete information, with a clear understanding that the interim period would be used to develop more information. This newly developed information was then used to create a permanent regime designed to minimize harms while permitting the smelter to continue operations.

This structure – using preliminary measures to prevent harm while information sufficient to create a permanent regime fair to all parties is developed – is the *Trail Smelter* arbitration's (semi)precautionary legacy. Regardless of the critiques of the arbitration's holdings or its normative relevance, this (semi)precautionary legacy resonates profoundly in modern international environmental law. For example, this early case presaged much of the contemporary debate about appropriate regulation, like that surrounding genetically modified organisms (GMOs). Viewing this controversy through *Trail Smelter's* (semi)precautionary lens might reveal an appropriate middle ground between the competing claims for regulatory legitimacy made by advocates and opponents of the precautionary principle.

THE PRECAUTIONARY PRINCIPLE

We live in a global society full of risks. In a world of scarce resources and many risks, regulators must make difficult choices to prioritize a regulatory agenda. Uncertainties surrounding potential environmental harms have frequently been used to push protection of the environment lower on the priority list. As a result, environmentally harmful activities have historically been regulated, if at all, on a *post-hoc* basis – with regulatory action coming only after serious or irreversible environmental degradation had occurred.[7] This is the great weakness in the traditional "damage and compensation" regulatory tools embodied by most environmental laws – they are unable to prevent catastrophic or irreversible harms. In case after case (DDT, lead, and asbestos come readily to mind) regulatory action took place only after disaster had struck. Scientific uncertainties ensured that the delay between first knowledge of harm and ultimate regulatory action was lengthy, with

[6] *Trail Smelter* (1939), *supra* note 2, at 184 (expressing that the Tribunal was guided by the direction that it "reach a solution just to all parties concerned."); *Trail Smelter* (1941), *supra* note 2, at 713 (same).

[7] For a discussion of how the precautionary principle shifts scientific inquiry in an attempt to avoid rather than mitigate harm, *see* Carolyn Raffensperger and Peter L. deFur, *Implementing the Precautionary Principle: Rigorous Science and Solid Ethics*, 5 HUMAN AND ECOLOGICAL RISK ASSESSMENT 933 (1999).

serious harms continuing during the interim. With powerful economic interests benefiting from activities that threaten the environment, lack of scientific certainty has often provided regulators reluctant to jeopardize the economic *status quo* with an excuse, or even a mandate for doing nothing.

Much of the international community views the precautionary principle[8] as an antidote to this kind of regulatory failure. To that end, virtually every recent international environmental agreement invokes this principle. The 1981 World Charter for Nature first gave the principle international recognition,[9] and the precautionary principle was most famously enshrined as Principle 15 of the Rio Declaration:

> In order to protect the environment, the precautionary approach shall be widely applied by States according to their capabilities. Where there are threats of serious or irreversible damage, lack of full scientific certainty shall not be used as a reason for postponing cost-effective measures to prevent environmental degradation.[10]

Since the Rio Declaration in 1992, many international agreements have incorporated the precautionary principle, including *inter alia*: the 1995 Straddling Fish Stocks Agreement;[11] the Bamaka Convention;[12] the Framework Convention on Climate Change;[13] the Convention on Biological Diversity;[14] and the POPs Treaty.[15] The Biosafety Protocol to the Convention on Biodiversity unambiguously confirmed the principle's centrality in environmental decision making.[16]

8 *See* James E. Hickey Jr. and Vern R. Walker, *Refining the Precautionary Principle in International Environmental Law*, 14 Virginia EnvironmentaL Law Journal 423 (1995); Gregory D. Fullem, *Comment, The Precautionary Principle: Environmental Protection in the Face of Scientific Uncertainty*, 31 Willamette Law Review 495 (1995).

9 World Charter for Nature, G.A. Res. 7, U.N. GAOR, 37th Sess., Supp. No. 51, at 21, U.N. Doc. A/37/L.4 and Add.1 (1982) (when "potential adverse effects are not fully understood, the activities should not proceed.").

10 The Rio Declaration was issued at the 1992 United Nations Conference on Environment and Development. *See* Rio Declaration on Environment and Development, (June 14, 1992) U.N. Doc. A/Conf. 151/26 Annex I, princ. 15, *reprinted in* 31 International Legal Materials 874, 879.

11 Agreement for the Implementation of the Provisions of the United Nations Convention on the Law of the Sea of 10 December 1982 Relating to the Conservation and Management of Straddling Fish Stocks and Highly Migratory Fish Stocks, *reprinted in* 34 International Legal Materials 1542 (1995).

12 Bamako Convention on the Ban of the Import Into Africa and the Control of Transboundary Movement and Management of Hazardous Wastes Within Africa, Jan 30, 1991, Art. 4(3)(f), *reprinted in* 30 International Legal Materials 773, 781.

13 United Nations Framework Convention on Climate Change, May 9, 1992, Art 3.3, *reprinted in* 31 International Legal Materials 849.

14 United Nations Convention on Biological Diversity, June 5, 1992, *reprinted in* 31 International Legal Materials 818, 822 (1992).

15 Stockholm Convention on Implementing International Action on Certain Persistent Organic Pollutants, May 22, 2001, Arts. 1, 8.9, *reprinted in* 40 International Legal Materials 532 (2001).

16 Cartagena Protocol on Biosafety to the Convention on Biological Diversity, Jan. 29, 2000, Art. 10, *reprinted in* 39 International Legal Materials 1027 (2000).

The European Union regards the precautionary principle as a "full-fledged and general principle of international law,"[17] and in his dissent in the *Nuclear Test Case*, Judge Weeramantry described the precautionary principle as part of customary international law.[18] The recently published United Nations Corporate Norms explicitly direct corporations to observe the precautionary principle.[19]

The precautionary principle has gained content and dimension as it evolved from its first iteration in the World Charter for Nature. As a result, various international agreements contain slightly different formulations of the precautionary principle. Some, like the Climate Change Treaty, require that positive steps be taken to resolve uncertainty,[20] whereas others simply prevent a lack of full scientific certainty from being a justification for not taking action.[21] Common to all formulations is the recognition that certainty regarding an environmental harm should not be a prerequisite for taking action to avert it.[22] The open-ended nature of this central principle has enabled new horizons of cooperative environmental protection.[23] The primary tool for achieving this end is an environmental impact assessment, which places a burden on those who would undertake an activity to determine its effects and to consider less harmful alternatives.[24]

Widespread adoption of the precautionary principle has predictably led to a backlash. Some scholars characterize the precautionary principle as too indeterminate and no more meaningful than saying "take care"[25] or "better safe than sorry."[26] These critics decry the precautionary principle as imposing unnecessary costs to address remote and improbable harms. The basis for this critique is

[17] *See* Commission of the European Communities, Communication from the Commission on the Precautionary Principle, COM (00)1 at 11, (Feb. 2, 2000) *available at* http://europa.eu.int/comm/dgs/health_consumer/library/pub/pub07_en.pdf.

[18] International Court of Justice: Advisory Opinion on the Legality of the Threat or Use of Nuclear Weapons, 1996 I.C.J. 226, 443 (Oct. 15), *reprinted in* 35 International Legal Materials 809 (1996) (dissenting opinion of Judge Weeramantry).

[19] Norms on the Responsibilities of Transnational Corporations and Other Business Enterprises with Regard to Human Rights, U.N. ESCOR 55th Sess., Agenda Item 4, E/CN.4/Sub.2/2003/12/Rev.2 (2003).

[20] Framework Convention on Climate Change, *supra* note 13, ("The Parties should take precautionary measures to anticipate, prevent or minimize the causes of climate change and mitigate its adverse effects"); *see also* Bamako Treaty, *supra* note 12 (calling on parties "to adopt and implement the preventive, precautionary approach to pollution problems").

[21] Convention on Biological Diversity, *supra* note 14, ("*Noting also* that where there is a threat of significant reduction or loss of biological diversity, lack of full scientific certainty should not be used as a reason for postponing measures to avoid or minimize such a threat.")

[22] *See* Ellen Hey, *The Precautionary Concept in Environmental Policy and Law: Institutionalizing Caution*, 4 Georgetown International Environmental Law Review 303, 304 (1992).

[23] *See* Hickey and Walker, *supra* note 8; *but see* Frank Cross, *Paradoxical Perils of the Precautionary Principle*, 53 Washington & Lee Law Review 851 (1996).

[24] *See* Jacobsen in this volume.

[25] *See, e.g.*, Christopher D. Stone, *Is There a Precautionary Principle?*, 31 Environmental Law Reports 10,790 (2001).

[26] Cross, *supra* note 23.

obvious. Precautionary regulation usually involves restrictions on human actions that, by definition, cannot be justified by unambiguous scientific evidence, yet may impose substantial costs.

This backlash has been lead by scholars and officials in the United States.[27] Rather than embracing the precautionary principle embodied in many international environmental agreements, the United States has developed an elaborate regulatory system based on quantitative risk assessment. This system rarely permits regulation without scientific evidence of a "significant risk," and typically strikes a responsive rather than precautionary stance. Prompted by this focus on quantitative risk assessment, U.S. officials have characterized the precautionary principle as "a mythical concept, like a unicorn."[28] Indeed, many opponents of the precautionary principle view it as undermining the quantitative risk analysis that supports sophisticated environmental regimes.[29] Because these regimes begin from the assumption that economic expansion and technological innovation increase overall social welfare, its advocates perceive the precautionary principle as an unwelcome and technically unsound deviation from science-based regulation.

The critical difference between the precautionary principle and quantitative risk assessment is the balance struck between the risks posed by underregulation or by overregulation. Both under- and overregulation pose risks for society: underregulation by exposing the public and the environment to harm; and overregulation by imposing higher monetary costs on consumers and creating barriers to innovation. Quantitative risk assessment generally strikes a balance that emphasizes minimizing overregulation, even at the expense of accepting greater harms from underregulation. The precautionary principle reverses this balance. Somewhat surprisingly, certain structural aspects of the *Trail Smelter* arbitration speaks almost directly to this debate and offers an alternative path, a (semi)precautionary hybrid between risk assessment and precautionary approaches.[30]

TRAIL SMELTER'S (SEMI)PRECAUTIONARY LEGACY

Commentators often refer to *Trail Smelter* as articulating a "do no harm" principle.[31] What constitutes "harm" is obviously open to a great deal of interpretation.

[27] *See, e.g.*, Cass R. Sunstein, *Beyond the Precautionary Principle*, 151 University of Pennsylvania Law Review 1003 (2003); Jonathan B. Wiener, *Whose Precaution After All? A Comment on the Comparison and Evolution of Risk Regulatory Systems*, 13 Duke Journal of Comparative & International Law 207 (2003).

[28] *See*, John D. Graham, *The Role of Precaution in Risk Assessment and Management: An American's View*, Address Before The European Commission, (Jan. 11–12, 2002), *available at* www.whitehouse.gov/omb/inforeg/eu_speech.html.

[29] Cross, *supra* note 23.

[30] For a skeptical read of this proposition, *see* Handl in this volume.

[31] Ved P. Nanda & George (Rock) Pring, International Environmental Law for the 21st Century 227 (2003); *see also* Parrish in this volume.

Confronting this question, the *Trail Smelter* Tribunal concluded that there must be evidence that a "serious harm" has, or is likely to occur before the affirmative "do no harm" obligation is implicated. Accordingly, the *Trail Smelter* duty seems to hinge on whether there exists a threatened or actual harm "serious" enough to trigger the duty to take preventive or ameliorative measures.

Discussions of *Trail Smelter* tend to omit an important part of the backstory – the extensive scientific research that went into determining whether the smelter was causing harm. In retrospect, we tend to view the *Trail Smelter* arbitration as a dispute in which unambiguous harms had already occurred, but the idea that air emissions could cause distant harm was much less settled in those days.[32] As a result, the smelter's, and indeed Canada's, first line of defense had been that no harm had occurred in Washington State; or that no harm attributable to the smelter had occurred. The United States and Canada jointly conducted extensive investigations into the effects of smelter smoke on the land in question. The question of precautionary measures was referred to the Tribunal only after there had been a showing of harm, but the allegations of harm triggered an obligation to investigate. This approach recognized that sometimes the need for early detection and action is greater than the need for definition and quantification.

This prearbitration inquiry was a scientific one, and was jointly funded by the states in question. In other words, the *Trail Smelter* arbitration suggests that those whose behavior is alleged to be creating a transboundary harm have an obligation, in the face of such an allegation, to participate in the evidence gathering, and bear an obligation to come forward with information and funds to develop the necessary scientific information upon which an assessment of harm can be made. It was this commitment to information gathering by both sides of the dispute that made resolution possible. This principle remains valid even though the Tribunal underestimated the damage caused by sulfur dioxide to crops and timber, and entirely ignored its effects on human health.[33]

One of the questions ultimately referred to the Tribunal was "whether the Trail Smelter should be required to refrain from causing damage in the State of Washington and, if so, to what extent?"[34] While the Tribunal considered this question, it instituted a temporary regime aimed primarily at pursuing a "more adequate and intensive study ... of meteorological conditions in the valley."[35] Interim measures and further research were the Tribunal's response to uncertainty. To that end, the Tribunal ordered a detailed series of scientific investigations,[36] and appointed "Technical Consultants" to head the research effort. This research later played a critical role in the Tribunal's decisions about risk management

[32] *See generally*, JOHN D. WIRTH, SMELTER SMOKE IN NORTH AMERICA (2000).

[33] *Id.* at 84, 99–102. The Tribunal also ignored the environmental degradation from other pollutants, including lead emissions and slag pollution of the Columbia River.

[34] *Trail Smelter* (1939), *supra* note 2 at 209.

[35] *Id.*

[36] *Id.* at 184.

under conditions of uncertainty. The Tribunal thus created a dynamic regulatory regime that evolved along with the scientific information.

GMOs as an Example of the Precautionary/Quantitative Divide

Trail Smelter's backstory and the current debate about GMOs share a striking parallel – once again producer nations argue that no harms have been attributed to their conduct and that therefore no duty to prevent transboundary harm is applicable. There is real disagreement over both the scientific reasonableness of fears about GMOs and over the degree of proof necessary before preventive measures should be required. What the *Trail Smelter* arbitration can offer this discussion is a burden-sharing process for developing an answer these questions.

The debate surrounding GMOs pits the United States, as their primary producer and advocate, against the European Union[37] and much of the developing world. The stakes in this debate are high – GMOs might generate vast profits and improved crops but unintended side-effects could jeopardize food security and irreparably damage the earth's ecosystems. The safety or danger posed by GMOs is hotly contested, with both advocates and opponents forced to extrapolate from incomplete scientific data. The most that can be said is that GMOs pose a possible but unknown risk that is of unknown but potentially serious dimensions.[38] Truly accurate statements about safety or risk are not possible without significantly more research. For this reason, GMOs have been an international flashpoint for at least a decade, leaving erstwhile trade partners instead trading WTO actions and overheated rhetoric.

At its heart, the GMO debate can be reduced to a single choice: precautionary regulation versus quantitative risk assessment. The precautionary principle calls for action to avert risks of serious or irreversible harm to the environment or human health *in the absence* of scientific certainty about the harm. Quantitative risk assessment, by contrast, builds regulation on a foundation of extensive evidence about harm and causation. Where the naked precautionary principle is vulnerable to the critique that it finesses the factual record to justify precautionary regulatory action,[39] unadulterated quantitative analysis is hard pressed to respond proactively to possible or likely harms.[40]

37 *See* Treaty Establishing the European Community, Art. 174, 1997 Official Journal (C 340) 3 (Nov. 10, 1007) (adopting the precautionary principle in the context of environmental decisionmaking).

38 The Commission on European Communities plainly had GMOs in mind when it described its position on the precautionary principle in 2000. *See* Commission of the European Communities, *supra* note 17.

39 For advocacy for this position, *see* Gail Charnley & E. Donald Elliott, *Risk Versus Precaution: Environmental Law and Public Health Protection*, 32 Environmental. Law Reports 10,363, 10,364–66 (2002).

40 *See* Rena I. Steinzor, *"You Just Don't Understand" – The Right and Left Conversation*, 32 Environmental. Law Reports 11,109–11,112–13 (2002) (explaining differing conceptions of precaution and risk assessment).

Some of the precautionary principle's most vocal critics are those who favor advancing the role of biotechnology in food production. This camp generally dismisses the precautionary principle as "antiscientific," with a tendency to ban first and ask questions later.[41] These advocates of biotechnology argue that GMOs are the "substantial equivalent" of their more conventional counterparts and, therefore, under the logic of quantitative risk assessment, need no special regulation. Opponents of biotechnology find this reliance on "substantial equivalence" misguided because it relies wholly on chemical similarity, with no biological, toxicological, or immunological data to back up the assumption of safety.[42] They also contend that "substantial equivalence" is used to exempt GMOs from rigorous scientific assessments of their potential to cause harm, and thus acts as a barrier to further research.[43] Where biotechnology's advocates see the precautionary principle as overly indeterminate, biotechnology's critics view "substantial equivalence" as equally vague.

International Decision Making, Precaution, and GMOs

The official United States position with regard to GMOs is one of "substantial equivalence."[44] With well over 90% of GMOs produced in three countries with significant commodity export industries (the United States, Canada, and Argentina)[45] the GMO debate is no longer solely about domestic law. For example, the EU refuses to import many GMOs grown in the United States because these products have not been approved under its more precautionary regulatory standards.[46] Because GMOs and nonmodified commodities are routinely commingled in the United States, this amounts to a *de facto* ban on many American commodity exports. With billions of dollars of trade at stake, the United States

[41] *See, e.g.*, Ronald Bailey, *EU Fear-Mongers' Lethal Harvest*, LA TIMES, August 18, 2002, at M3.

[42] *See, e.g*, Erik Millstone, Eric Brunner, & Sue Mayer, *Beyond Substantial Equivalence*, 401 NATURE 525 (Oct. 7, 1999).

[43] *See* Royal Society of Canada, The Canadian Academy of Sciences and Humanities, Expert Panel on the Future of Food Biotechnology, Aub. 29, 2001, Chapter 7, available at http://www.rsc.ca/index.php?page_id=119.

[44] Coordinated Framework for the Regulation of Biotechnology, 51 Fed. Reg. 23,302, 23,305 (June 26, 1986).

[45] The United States accounts for most of the genetically modified crops planted in the world and each year a larger percentage of the American harvest is comprised of GM plantings. In 2004, GM soybeans accounted for 85% of the soybean acreage planted in the United States; GM cotton for 76% of the cotton, and GM corn for 45% of the corn. For these and other data pertaining to GM crops, *see* PEW INITIATIVE ON FOOD & BIOTECHNOLOGY, AUG. 2004 FACTSHEET, GENETICALLY MODIFIED CROPS IN THE UNITED STATES (Aug. 2004).

[46] European Parliament & Council Directive 2001/18/EC, 2001 OFFICIAL JOURNAL (L 106) Art 4.1, *available at* http://www.biosafety.be/PDF/2001_18.pdf. This Directive explicitly employs the precautionary principle to "ensure that all appropriate measures are taken in order to avoid adverse effects on human health and the environment which might arise from the deliberate release or the placing on the market of GMOs."

and its GMO-producing allies argue that the EU and likeminded GMO-skeptics must demonstrate a clear risk of harm before they can legitimately regulate GMOs in the name of safety. GMO opponents, by contrast, want the producer nations to demonstrate basic safety before introducing these products into international commerce.

At first blush, the *Trail Smelter* arbitration may seem to have little to offer this situation. The arbitration's central "do no harm" holding only applies when "the case is of serious consequence and the injury is established by clear and convincing evidence,"[47] and there is very little evidence either way about GMOs. What the *Trail Smelter* arbitration does offer, however, is a clear structural model for how best to engage in a process of investigation and decision making in the face of this uncertainty.

Two principal international agreements govern the sale of GMOs in international commerce: the WTO Agreement on Sanitary and Phytosanitary Measures, (SPS Agreement)[48] and the Cartagena Protocol on Biosafety to the Convention on Biological Diversity (Biosafety Protocol).[49] These agreements reflect the international divergence of opinion over whether quantitative risk assessment or precautionary regulation is the appropriate decisional matrix. As a result, the agreements offer very different paths for resolving the question of whether sale of GMOs should trigger *Trail Smelter*–like obligations to avoid harm. Both agreements can be improved, and their methods made more consistent, if they are read through *Trail Smelter's* (semi)precautionary lens.

The SPS Agreement permits sanitary or phytosanitary measures only to the extent "necessary to protect human, animal or plant life or health, as demonstrated by scientific principles."[50] WTO Appellate Body decisions have interpreted this language to mean that a state adopting an SPS regulation must demonstrate "a probability, not just a possibility" of harm,[51] and must do so through a scientific risk assessment.[52] The SPS Agreement puts the burden of information collection squarely on a state seeking to regulate and requires that the state demonstrate sufficient risk to justify the decision. A primary factor in this analysis must be the effect the restrictive measure will have on trade, and the availability of less restrictive alternatives.[53] The SPS Agreement thus seeks to avoid the errors of overregulation, even at the risk of incurring harms because of underregulation.

47 *Trail Smelter* (1941), *supra* note 2, at 716.

48 WTO Agreement on the Application of Sanitary and Phytosanitary Measures, Apr. 15, 1994, Marrakesh Agreement Establishing the World Trade Organization, Annex 1A (1994) [hereinafter SPS Agreement], http://www.wto.org/english/docs_e/legal_e/legal_e.htm#import.

49 Biosafety Protocol, *supra* note 16, art. 10.

50 SPS Agreement, *supra* note 47, at art. 2.1.

51 WTO Report of the Appellate Body, Japan – Measures Affecting the Importation of Apples, WTO Doc WT/DS245/R (Nov. 26, 2003) para. 168, 202.

52 SPS Agreement, *supra* note 47.

53 *Id.* at arts. 5.3, 5.5.

By contrast, the Biosafety Protocol defines its scope as applying to the transboundary movement, transit, handling, and use of all GMOs that *may have* adverse effects. The Protocol specifically states that: "lack of scientific certainty due to insufficient relevant scientific information about the extent of harm shall not prevent a State from taking a decision in order to avoid or minimize such potential adverse effects."[54] In other words, the Protocol employs the precautionary principle to allow an importing country to ban or restrict a GMO on a precautionary basis. And the Protocol permits an importing state to require that the exporting state conduct, and pay for, a risk assessment before lifting those restrictions. The Biosafety Protocol thus errs on the side of overregulation rather than underregulation.

In the atmosphere of uncertainty surrounding GMOs, the Biosafety Protocol's bias toward overregulation conflicts head on with the SPS Agreement's bias toward underregulation. Moreover, the SPS Agreement and the Biosafety Protocol use alternative conceptions of the scientific threshold necessary for determining when regulatory action is appropriate – with the SPS Agreement requiring scientific evidence of harm and the Biosafety Protocol requiring similar evidence demonstrating safety. The two agreements then impose very different burdens of proof and cost when evaluating the potential for harm. Neither approach is particularly satisfactory. The SPS Agreement is likely to permit harmful activities to continue until and unless a clear scientific case for regulation can be made, whereas the Biosafety Protocol is likely to inhibit useful activity unnecessarily.

Trail Smelter can be read as offering a third way – a (semi)precautionary model capable of harmonizing these two agreements in a fashion that avoids their separate pitfalls while capitalizing on their strengths. This (semi)precautionary model may be summarized as follows: (1) determination of harm must involve a scientific inquiry; (2) the investigation should not be funded solely by the complaining party; and (3) while permanent preventive and precautionary measures must be based on an adequate showing of possible harm, interim precautionary measures are an appropriate response to uncertainty. It was by following this model that the Tribunal was able to resolve the *Trail Smelter* dispute in a fashion that, at the time seemed "just to all" parties. Those same points could be the basis of a (semi)precautionary resolution of the international GMO controversy.

Regulating GMOs through a (Semi)Precautionary Lens

The SPS Agreement, applies to all "sanitary and phytosanitary measures which may directly or indirectly affect international trade."[55] Its primary goal is to minimize impacts on free trade. To that end, the SPS Agreement requires that states employ the least trade-restrictive means available to achieve health and safety

[54] Biosafety Protocol, *supra* note 16.

[55] SPS Agreement, *supra* note 47. art. 1.1.

goals.[56] Since the formation of the WTO, the Appellate Body has issued a series of decisions clarifying the scope of these SPS obligations.

In the *EC – Hormones* case,[57] and again in the *Japan – Agricultural Products* case,[58] the Appellate Body has clearly stated that Article 5.1 of the SPS Agreement requires that an SPS measure be based on a risk assessment, and that Article 2.2 requires sufficient scientific evidence to justify an SPS measure. Much like the *Trail Smelter* Tribunal, the Appellate Body insisted that decisions be grounded in solid science. However, the Appellate Body also acknowledges that, in determining what will constitute "sufficient scientific evidence" it is important to bear in mind that "responsible, representative governments commonly act from perspectives of prudence and precaution where risks of irreversible, e.g. life-terminating, damage to human health are concerned."[59]

Indeed, despite repeatedly stating that the precautionary principle "has not been written into the SPS Agreement"[60] the Appellate Body has been careful to point out that the precautionary principle "finds reflection" in various provisions of the SPS Agreement."[61] In particular, the SPS Agreement explicitly provides that even when the necessary scientific evidence is not present, a state may nevertheless adopt provisional SPS measures if the conditions of Article 5.7 have been satisfied. Article 5.7 provides in relevant part:

> [i]n cases where relevant scientific evidence is insufficient, a Member may provisionally adopt sanitary or phytosanitary measures on the basis of available pertinent information.... In such circumstances, Members shall seek to obtain the additional information necessary for a more objective assessment of risk and review the sanitary or phytosanitary measure accordingly within a reasonable period of time.[62]

So far, it sounds very much like the *Trail Smelter* arbitration. Indeed, a robust interpretation of Article 5.7, would produce an approach to uncertainty akin to that envisioned in the Biosafety Protocol, and embodied by the *Trail Smelter* arbitration. Such an interpretation would permit provisional measures when there is insufficient scientific information to assess risk, and also when a risk assessment produces conflicting, inconclusive or uncertain results.

[56] *Id.* at art. 5.6.

[57] WTO, Report of the Appellate Body, EC Measures Concerning Meat and Meat Products (Hormones) [hereafter EC – Hormones], WT/DS26/AB/R, (Feb. 13 1998) para. 175; WTO, Report of the Appellate Body, Japan – Measures Affecting Agricultural Products, [hereafter Japan – Agricultural Products] WTO Doc WT/DS76/AB/R (Feb. 22, 1999), para. 79.

[58] Japan – Agricultural Products, *supra* note 56, paras.75–78; Australia – Measures Affecting Importation of Salmon, [hereafter Australia – Salmon] WT/DS18/AB/R (Oct. 20 1998), para. 125 ("the 'risk' evaluated in a risk assessment must be an ascertainable risk; theoretical uncertainty is 'not the kind of risk which, under Article 5.1, is to be assessed.' ") citing EC – Hormones, *supra* note 57, para. 186.

[59] EC – Hormones, *supra* note 56, at para. 124.

[60] *Id.*; *see also* Japan – Agricultural Products, *supra* note 56, at para. 81.

[61] *Id.*

[62] SPS Agreement, *supra* note 48, art. 5.7.

Unfortunately, the Appellate Body has signaled its unwillingness to permit robust implementation of this provision – it has repeatedly rejected measures asserted to protect environmental safety under the SPS Agreement and has limited states' ability to impose provisional measures by reading the Article 5.1 "sufficient scientific evidence" requirement narrowly.[63] The precautionary principle was a central element of the *EC – Hormones* dispute, with the question of whether "the precautionary principle was relevant in the interpretation of the SPS Agreement" clearly raised by the parties.[64] The Appellate Body sidestepped this question,[65] but made it clear that "in the event a sanitary measure is not based on a risk assessment as required in Articles 5.1 and 5.2, this measure can be presumed, more generally, not to be based on scientific principles or to be maintained without sufficient scientific evidence."[66] Moreover, the Appellate Body is on record as asserting that unknowns and uncertainty do not justify a departure from the otherwise applicable requirements of assessing the likelihood of harm before imposing restrictions.[67]

Although these decisions represent a trend away from *Trail Smelter's* (semi)precautionary interim measures, they should not be read as entirely foreclosing a more balanced approach. For one thing, these decisions involved permanent rather than provisional measures, and, indeed, the European Community explicitly refrained from invoking Article 5.7 in the *EC – Hormones* case. Moreover, in the *European – Asbestos* case, the Appellate Body issued its first opinion approving a putative restraint on trade as a legitimate health and safety measure. The precautionary principle was clearly the animating force behind the French asbestos ban upheld in that case. So, there may still be room in the SPS Agreement for adoption of *Trail Smelter*–like (semi)precautionary regimes in the face of significant uncertainties.

Existing interpretations of the SPS Agreement also fall short of the *Trail Smelter* arbitration's (semi)precautionary model on another front. The Appellate Body has strongly indicated that a state imposing an SPS measure bears the sole burden of coming forward with sufficient scientific evidence to justify the restriction.[68] The *Trail Smelter* arbitration's joint action requirement offers a better path here – the resource disparity between states would otherwise place legitimate SPS measures beyond the reach of less-technologically advanced countries.

By contrast, the Biosafety Protocol is a much narrower document, addressing only the question of international transport of GMOs.[69] Like the SPS Agreement, the Biosafety Protocol purports to balance the competing values of free trade, modern technology, sound science, and environmental protection. It does

[63] Japan – Agricultural Products, *supra* note 56, para. 76; Australia – Salmon, *supra*, note 57, para. 123; EC – Hormones, *supra* note 56, at para. 193.

[64] *Id.* at para. 96 (c).

[65] *Id.* at, para 123.

[66] Australia – Salmon, *supra* note 57, para. 137, 138.

[67] *Id.* at para. 130.

[68] Japan – Agricultural Products, *supra* note 56, at para. 137.

[69] The Protocol actually employs the term "living modified organism" or LMO.

so, however, from a strikingly different perspective. Where the SPS Agreement is concerned primarily with trade, the Biosafety Protocol's core obligations are to protect biodiversity and human health. To that end, the Biosafety Protocol embraces the precautionary principle[70] and gives a high level of deference to states imposing restrictions on import of GMOs.[71]

The combined effect of these two Biosafety Protocol provisions may come perilously close to crossing another important *Trail Smelter* baseline. The Biosafety Protocol seems to permit a state to unilaterally impose restrictions with no corresponding duty to look for ways to minimize economic harms. Because the *Trail Smelter* Tribunal viewed its charge as creating a solution that would be "just to all parties," the Tribunal rejected a flat out ban on the Trail smelter's activities, and, instead, established a regime designed to allow the smelter to continue operation in a fashion that would prevent damage in the United States. This principle of balance remains critical. There is still room to interpret the Biosafety Protocol, which only came into force on September 11, 2003,[72] through the lens of the *Trail Smelter* arbitration to create shared the burdens of production and proof, rather than merely shifted burdens.

CONCLUSION

In May 2003, the United States, Canada, and Argentina initiated a WTO action challenging to the European Union's GMO policy as a violation of the SPS Agreement.[73] The outcome of the dispute may elucidate the relationship between the SPS Agreement and the Biosafety Protocol. Where the Biosafety Protocol would employ an abundance of caution in light of a possible risk and thus likely permit the European approach, the SPS Agreement embodies a reluctance to permit regulation absent solid quantitative evidence.

There are two possible responses to these differing mandates – either there will be a search for common ground and a harmonization of objectives, or one approach will simply "win" outright. The *Trail Smelter* arbitration offers a model for harmonization. Such a model is desperately needed because the power dynamics between the two agreements otherwise make a win/lose situation likely. Given that the SPS Agreement can draw on the WTO's effective dispute resolution and sanctions mechanism, while any proceedings under the Biosafety Protocol are limited by the Convention on Biodiversity's weak compliance mechanism,

70 Biosafety Protocol, *supra* note 16, at art. 1.

71 Biosafety Protocol, *supra* note 16, at art. 10.

72 The United States is not a party to the Biosafety Protocol.

73 WTO, Request for the Establishment of a Panel by the United States, European Communities – Measures Affecting the Approval and Marketing of Biotech Products, WTO Doc WT/DS291/23 (Aug. 8, 2003) (contending that the EU's de facto moratorium on approving and marketing GMOs violates the various WTO agreements including the SPS Agreement).

violations of the SPS Agreement are likely to trigger serious sanctions while violations of the Biosafety Protocol are unlikely to carry significant consequences. This disparity means that, absent a harmonizing approach, the SPS Agreement is likely to "win" and the Biosafety Protocol to "lose" – creating a bias in favor of the values embodied in the SPS Agreement at the expense of those in the Biosafety Protocol. Moreover, a WTO ruling that compliance with the Biosafety Protocol is not a defense to an SPS Agreement challenge would create a clear hierarchy, with trade values at the top, and all other values, including environmental values, below.

There is a real concern that GMO-producing nations will leverage the WTO's raw power to establish an international system in which promoting enhanced international trade takes precedence over all nontrade values, including environmental protection. This concern is not unique to the context of GMOs – economic and environmental priorities frequently overlap or conflict. Thus, answering the GMO question has implications for how international law will respond to the myriad uncertain harms that lurk in the modern world.

Focusing on an end result that is "just to all parties," a *Trail Smelter*–inspired (semi)precautionary approach would consider trade alongside biodiversity rather than privileging one over the other. After all, an environmental regime built around an obligation to wait for harm to occur before taking preventive steps would be self-defeating, particularly when the harms at stake would be irreversible should they come to pass.

More than anything, there is a compelling need to develop the information necessary to assess the risks posed by GMOs. A central element of a (semi)precautionary approach would be the adoption of an interim regime, imposed with a clear understanding that the interim period would be used to develop more information. This research process would be jointly funded by states producing GMOs and states seeking to regulate these novel products. The information generated from this research could then serve as the basis for a permanent regime designed to minimize harms while permitting the development of this new technology. Although such a regime might be more flexible than initially contemplated by the Biosafety Protocol, it would be far more protective of the environment than would naked trade rules, while still permitting trade.

14 Surprising Parallels between *Trail Smelter* and the Global Climate Change Regime

Russell A. Miller

INTRODUCTION

The substantive principles of international law for which the *Trail Smelter* arbitration[1] has come to serve as shorthand are widely viewed as the foundation stones on which all international environmental law is constructed.[2] Such expansive, almost mythological claims for the *Trail Smelter* case have, in due course, invited severe criticism.[3] Still, the *Trail Smelter* principles persist in international instruments, casebooks, and scholarly footnotes, as the "*locus classicus*" of international environmental law.[4] So it is not at all surprising that one need only scratch the surface to find trace elements of these *Trail Smelter* principles in the global climate change regime,[5] where they make their most conspicuous appearance in

[1] *See Trail Smelter Arbitral Decision*, 33 American Journal of International Law 182 (1939) [hereinafter "*Trail Smelter* (1939)"]; *Trail Smelter Arbitral Decision*, 35 American Journal of International Law 684 (1941) [hereinafter "*Trail Smelter* (1941)"]. *See* Annex to this volume. For an introduction to the case, *see* Allum, Read, McCaffrey, and Rubin in this volume.

[2] Alexandre Kiss and Dinah Shelton, International Environmental Law 107 (1991).

[3] *See, e.g.*, Samuel Bleicher, *An Overview of International Environmental Regulation*, 2 Ecology Law Quarterly 1 (1972); Alfred P. Rubin, *Pollution by Analogy: The Trail Smelter Arbitration*, 50 Oregon Law Review 259 (1971); Günther Handl, *Territorial Sovereignty and the Problem of Transnational Pollution*, 69 American Journal of International Law 50 (1975); Karin Mickelson, *Rereading* Trail Smelter, 31 Canadian Yearbook of International Law 219 (1993). Rubin and Mickelson republished in this volume. *See also* Jaye Ellis and John Knox in this volume.

[4] Handl, *supra* note 3, at 60.

[5] *See* Andrew L. Strauss, *The Legal Option: Suing the United States in International Forums for Global Warming Emissions*, 33 Environmental Law Reports 10185 (2003); Michael Weisslitz, *Rethinking the Equitable Principle of Common But Differentiated Responsibility*, 13 Colorado Journal of International Environmental Law & Policy 473, 499 (2002); Isabel Rauch, *Developing A German and an International Emissions Trading System*, 11 Fordham Environmental Law Journal 307, 429 (2000); J. Chris Larson, *Note, Racing the Rising Tide: Legal Options for the Marshall Islands*, 21 Michigan Journal of International Law 495, 511–512 (2000); Simon SC Tay, *Southeast Asian Fires: The Challenge for International Environmental Law and Sustainable Development*, 11 Georgetown International Environmental Law Review 241, 266–267 (1999).

the preamble to the United Nations Framework Convention on Climate Change (UNFCCC).[6]

Critics challenging the doctrinal significance of the *Trail Smelter* arbitration rightly warn against making too much of such manifestations of the *Trail Smelter* principles. Indeed, *Trail* probably had very little influence on the substantive norms forming the global climate change regime. But, as John Wirth noted in *Smelter Smoke in North America*, the *Trail Smelter* arbitration is about much more than the heavily debated principles of limited territorial sovereignty and polluter liability. "The main issues of the case," Wirth complained, "have been otherwise ignored or obscured by the conventional wisdom about *Trail*'s [doctrinal] significance."[7] Several facets of the *Trail Smelter* arbitration, beyond the eponymous principles, suggest surprising and important links between *Trail*, the global climate change regime, and international environmental law in general. Two of these neglected facets strike me as particularly significant: the extensive role played by nonstate actors in the case, and the dynamic regime created by the Tribunal, which permitted a resolution of the *Trail Smelter* dispute in the face of scientific uncertainty.

NONSTATE ACTORS

International environmental law has come of age at the very time that global civil society and transnational enterprises have taken international law by storm. The rise of these nonstate actors suggests a new world order in which the nation state's Westphalian prerogative is increasingly suspect. The literature is right to remind us that nonstate actor involvement in international affairs is nothing new.[8] But the nature and degree of the contemporary involvement of nonstate actors is a genuine phenomenon.[9] International environmental lawyers have found it necessary to consult and coordinate with, not to mention confront, NGOs and the representatives of global industrial interests. The global climate change regime is certainly no exception to this rule. Remarkably, it also was true of the *Trail Smelter* arbitration, which played out long before the terms "civil society" and "globalization" were coined, bearing all their counter-Westphalian potential.

6 "States have, . . . the sovereign right to exploit their own resources pursuant to their own environmental and developmental policies; and the responsibility to ensure that activities within their jurisdiction or control do not cause damage to the environment of other States . . ." Framework Convention on Climate Change, preamble, *reprinted in* 31 International Legal Materials 849 (1992).

7 John D. Wirth, Smelter Smoke in North America 80 (2000).

8 *See, e.g.*, Jordan J. Paust, *The Reality of Private Rights, Duties, and Participation in the International Legal Process*, 25 Michigan Journal of International Law 1229 (2004).

9 *See, e.g.*, Harold Hongju Koh, *Bringing International Law Home*, 35 Houston Law Review 623 (1998).

Arising at a time when NGOs did not enjoy their now ubiquitous status in international law and international affairs,[10] the *Trail Smelter* arbitration owes its existence to the persistence of the Citizens' Protective Association (CPA). This committee of "smoke farmers," Washington residents who began complaining of smoke damage from the Canadian smelter in the early 1920s, was nothing less than an early, single-issue, environmental NGO. To be sure, there is little evidence that the CPA raised its challenge to the Canadian smelter's pollution out of a concern for the environment for its own sake. Rather, the farmers were chiefly interested in the negative effects the Canadian pollution was having on their own industrial activities, namely farming and logging.[11] But the CPA was advancing an international environmental agenda: the Canadian smelter was emitting harmful transboundary pollutants and the CPA wanted it to stop. Underscoring the CPA's grass-roots character, and thus, further aligning the CPA with today's civil society movements that play such an integral role in international environmental law, Wirth reported that CPA Chairman John Leaden was a veteran local activist who viewed the CPA and its challenge to the smelter as "an effective organization of the people."[12]

To an astonishing degree, the plucky CPA came to dictate and control the American side of this international dispute. The CPA's bitter rejection of the $350,000 judgment of the International Joint Commission stirred political pressure that forced the State Department to reopen the case in 1933,[13] leading to the now-famous arbitration. The State Department lawyer handling the matter felt obliged to consult the CPA in the run-up to the formation of the arbitration agreement between the U.S. and Canada.[14] Wirth suggested that loyalty to the CPA on the part of the American diplomatic and scientific agents involved in the case ran so high as to endow the farmers with a *de facto* veto over the government's handling of the matter.[15] Ultimately, the local CPA attorney assumed control of the presentation of the American case, including the hearings before the Tribunal, after the surprising, untimely death of the State Department lawyer who had been handling the case.[16] Throughout the evolution of the case, the local CPA attorney had been "skilled at focusing and holding Washington, D.C.'s attention on the farmers' (and other landholders') plight in the intensely local worlds [in which the case played out], as well as on the larger implications of the Trail case."[17]

[10] WIRTH, *supra* note 7, at 7 ("By today's standards, the absence of nongovernmental organizations . . . that have operated in Canada and the United States since the 1970s is striking.").

[11] *See* Allum in this volume.

[12] WIRTH, *supra* note 7, at 18.

[13] *Id.* at 37. *See Trail Smelter* (1939), *supra* note 1, at 192 ("[The IJC] recommendation, apparently, did not commend itself to the interested parties.").

[14] WIRTH, *supra* note 7, at 38. *See* Convention on Damages Resulting from Operation of Smelter at Trail, British Columbia, June 5, 1935, U.S.-Canada, 49 Stat. 3245. *See* Annex to this volume.

[15] WIRTH, *supra* note 7, at 38.

[16] *Id.* at 39–40.

[17] *Id.* at 20.

The CPA's "extraordinary and unusual influence"[18] in the *Trail Smelter* arbitration was matched on the Canadian side of the dispute by the powerful, international industrial interests represented in the case by the Consolidated Mining and Smelter Company. After all, "Trail was the largest smelting complex of its kind in the British Empire, and the visibility and power of Canadian Pacific Railroad, the parent corporation, assured that the [smelter's] interests would be considered national interests in Ottawa."[19] Certainly, a bald assertion of the power of industry in domestic and international politics was nothing new. What gives Consolidated's commanding role in the *Trail Smelter* arbitration special resonance for contemporary international environmental law, and the global climate change regime in particular, was the globalized nature of the industrial interests that coalesced around the smelter's defense. Consolidated did not stand alone as a Canadian corporation, hoping to capitalize on its political clout to ensure Canada's vigorous defense of its interests in the international arena. Instead, the company served as the proxy for the defense of the broader mining and smelting industry's interests, regardless of where the industry's distinct actors made their home.[20] In this sense, Canada served as a front for the advancement of the global mining and smelting industry's agenda every bit as much as the United States served as the vehicle for advancing the interests of the CPA.

The globalized nature of the industrial interests at stake in the *Trail Smelter* case exposed the diminishing relevance of national identity to the industry's managing elite in a way that must have seemed revolutionary at the time. "To be in mining and smelting," Wirth explained, "was almost by definition to have an international career. These executives did not cease to be citizens [of their respective countries], but the nature of their business led them to derive perspectives that were regional and continental."[21] Under these conditions it can hardly be surprising that U.S. smelting interests sought leave (unsuccessfully) to join the matter on the Canadian side, and that these mining and smelting interests pursued every other available avenue for asserting their influence on behalf of Consolidated.[22] Recognizing the uniquely confounding influence this transnational industrial bloc was having on the case, the State Department lawyer managing the U.S. case went so far as to threaten that "immunity from the Logan Act [an obscure antisedition statute from

18 *Id.* at 42.

19 *Id.* at 2.

20 *Id.* at 40, 90. *See* Allum in this volume.

21 *Id.* at 90–91 ("...[T]he smelters acted continentally, across national borders. Only recently have we come to realize the significance of the fact that certain businesses were thinking and acting along continental lines, and such thinking and acting occurred well before the passage of NAFTA. Students of North American business should pay more attention to these earlier networks of executives in the transportation and natural resource sectors, men whose careers in railroads, mining, smelting, electric power, oil, and gas developed a transnational point of view with respect to sharing technology and expertise.").

22 Sulfur producers from Texas and Louisiana were able to secure the support of Texas senators who strenuously advocated against a vigorous prosecution of the case in Washington, D.C. *Id.*, at 94–95. They also sought to have the State Department lawyer handling the case fired. *Id.*

the 1790s] could not be assured if the [American] smelters persisted in supporting [the Canadian case on behalf of Consolidated], a foreign corporation."[23]

Thus, the *Trail Smelter* arbitration transcended its formal framework, cast as a dispute between the United States and Canada, to encompass the clash of interests between nonstate actors: an environmental NGO facing off against global industrial interests on the international plane. This clash of nonstate actors is paradigmatic of contemporary international environmental law.[24]

Nonstate actors play a no-less-prominent role in the global climate change dispute and the legal responses thereto. Jessica Mathews, in her influential *Foreign Affairs* article *Power Shift*, described the post–Cold War "redistribution of power among states, markets, and civil society."[25] As a paramount example of the increasing importance of NGOs in this shift, Mathews cited the comprehensive involvement of NGOs in the negotiation of the global climate change treaty:

> NGOs set the original goal of negotiating an agreement to control greenhouse gases long before governments were ready to do so, proposed most of its structure and content, and lobbied and mobilized public pressure to force through a pact that virtually no one else thought possible when the talks began.[26]

In capacities reminiscent of the role played by the Citizens' Protective Association in the *Trail Smelter* arbitration, NGOs advised and even participated as members of government delegations to the negotiations.[27] Much as the United States eventually handed the *Trail Smelter* case over to the local CPA attorney, Mathews noted that "the tiny nation of Vanuatu turned its delegation over to an NGO with expertise in international law . . . , thereby making itself and the other sea-level island states major players in the fight to control global warming."[28]

NGOs have continued to play a major role in the global climate change regime, surpassing the sweeping involvement of the CPA in the *Trail Smelter* arbitration. First, the sheer number of civil society participants in the Kyoto Protocol negotiations is suggestive of the broad responsibility NGOs have come to assume in international environmental law since the *Trail Smelter* case. "By the time COP 6 to the UNFCCC convened in The Hague, Netherlands, in November 2000, representatives of NGOs outnumbered representatives of states."[29] Dating

[23] *Id.* at 93.

[24] Edith Brown Weiss, *The Changing Structure of International Law*, GEORGETOWN LAW – *RES IPSA LOQUITUR* 52 (Spring 1997), *reprinted in* EDITH BROWN WEISS ET AL., INTERNATIONAL ENVIRONMENTAL LAW AND POLICY 50 (1998) ("Particularly in international environmental law, nonstate actors are now prominent participants, either formally or informally.").

[25] Jessica T. Mathews, *Power Shift*, 76 FOREIGN AFFAIRS 50, 55 (Jan./Feb. 1997).

[26] *Id.* at 55. *See* Daniel Bodansky, *The United Nations Framework Convention on Climate Change: A Commentary*, 18 YALE JOURNAL OF INTERNATIONAL LAW 451, 482 and 486 (1993).

[27] Mathews, *supra* note 25, at 55.

[28] *Id.*

[29] Michele Bestill, *Environmental NGOs Meet the Sovereign State: The Kyoto Protocol Negotiations on Global Climate Change*, 13 COLORADO JOURNAL OF INTERNATIONAL ENVIRONMENTAL LAW & POLICY 49 (2002).

back to the negotiations for the UNFCCC, the 280-plus NGO members of the Climate Action Network (CAN) continued to coordinate civil society's global involvement with the issue of climate change during the Kyoto negotiations.[30] Second, NGO participation in the global climate change regime has been formalized. Mathews noted that NGOs were long relegated to seeking to influence international law and international affairs from the hallway. This is exactly the posture from which the CPA exerted its influence in the *Trail Smelter* arbitration. "Even when [NGOs like the CPA] were able to shape government's agendas, . . . their influence was largely determined by how receptive their own government's delegation happened to be."[31] Thus, Wirth noted that the U.S. engagement with the CPA and the rigor with which the United States prosecuted the *Trail Smelter* case was contingent on the sympathy and motivation of the State Department attorney who was originally assigned to the matter.[32] The UNFCCC and the Kyoto Protocol, by contrast, invite NGOs to join the Parties to these instruments "around the table."[33] Both the UNFCCC and the Kyoto Protocol explicitly provide for the participation of civil society as observers at sessions of the Conference of the Parties (COP).[34] "Such observers may, upon invitation of the President [of the COP], participate without the right to vote in the proceeding of any session in matters of direct concern to the body or agency they represent. . . ."[35] At least one commentator has underscored the valuable contribution NGOs can make to the enforcement of the Kyoto Protocol's binding emission's targets.[36]

In *Red Sky at Morning*, James Gustave Speth lists the economic implications of legal and policy responses to global environmental threats among his *Ten*

[30] Bestill reports that CAN "has more than 280 members. CAN is a loose organization, divided into eight regions. Each region has its own coordinator. . . . CAN was the voice of the environmental community during the Kyoto Protocol negotiations." *Id.*, at 53.

[31] Mathews, *supra* note 25, at 55.

[32] Wirth described the successor as a "lawyer doing a job he neither relished nor particularly believed in." Wirth, *supra* note 7, at 41 and 81. He bungled the presentation of the American case before the Tribunal, inflicting damage that couldn't be repaired after civilian attorney John Raftis was appointed U.S. agent on the matter. *Id.* at 42. "This," Wirth explained, "marked the end of the CPA's extraordinary and unusual influence in the Trail smelter case." *Id.*

[33] Mathews, *supra* note 25, at 55.

[34] "Any body or agency, whether national or international, governmental or non-governmental, which is qualified in matters covered by the Convention, and which has informed the secretariat of its wish to be represented at a session of the Conference of the Parties as an observer, may be so admitted . . ." Article 7, Paragraph 6, UNFCCC; Article 13, Paragraph 8, Kyoto Protocol.

[35] Rule 7.2, *Draft Rules of Procedure of the Conference of the Parties and its Subsidiary Bodies*, UNFCCC, 2nd Sess., Agenda Item 4(b), at 4, FCCC/CP/1996/2 (May 22, 1996), *available at* http://unfccc.int/cop4/02-5.pdf.

[36] "There will be potentially significant opportunities for public participation in compliance proceedings. Intergovernmental and nongovernmental organizations will be entitled to submit technical and factual information to the committee's relevant branch, that is, either the facilitative or enforcement branch." Glenn Wiser, *Kyoto Protocol Packs Powerful Compliance Punch*, 25 International Environmental Report – Current Reporter No. 2, p. 86 (2002), *available at* http://www.ciel.org/Publications/INER_Compliance.pdf.

Drivers of Environmental Deterioration. "Here as elsewhere," he explained, "economic interests are typically pitted against environmental ones."[37] Echoing the role played by international industrial interests in the *Trail Smelter* arbitration, Speth observes that "[t]here is often a seamless link between economic interests and the positions governments take in all negotiations." Indeed, many scholars have observed that "government negotiating positions in Europe and the United States have tended to track the stances of major industries active on key issues, such that achievement of global environmental accords is impossible if important economic sectors are unified in opposition."[38]

It has become impossible to engage international environmental issues without, at the same time, engaging international industrial interests. After all, the active cooperation of large multinational corporations is needed to implement the international environmental regimes agreed to by Westphalian states. "[T]hese large companies are the 'street level bureaucrats' on whom policy makers rely, like it or not, for successful implementation."[39] This short comment is the wrong vehicle for a broad survey of the interrelationship between globalization and the environment, which has acquired an independent doctrinal significance in international environmental law in the form of the principle of sustainable development. It is adequate for my thesis that this interrelationship, graphically described by Speth and undeniably central to the *Trail Smelter* case, is also distinctly present in the context of the dispute over global climate change.

As it turns out, the participation of these corporate actors, with the aim of empowering sympathetic governments and persuading world public opinion, is a facet of the global climate change dispute that is difficult to understate. As Perry Wallace observed: "In the current debate over 'global climate change,' . . . corporate actors are both prominent and controversial."[40] Not unlike the efforts of international mining and smelting corporations in the *Trail Smelter* arbitration, international industrial interests in the global climate change dispute have largely focused their energies on challenging the scientific basis for the harm their economic activities is alleged to have caused. To stave off undesired regulation, these corporate actors continually reiterate that answering several key questions should be a precondition for implementation of Kyoto, including: "the extent of global warming, the proportionate contribution (and hence the significance) of anthropogenic GHG sources to the warming process, and, in any event, whether the consequences of present global warming patterns are actually so great a threat to the world."[41]

[37] James Gustave Speth, Red Sky at Morning 107 (2004).

[38] *Id.* (quoting David Lecy and Peter Newell, *Oceans Apart? Business Responses to Global Environmental Issues in Europe and the United States*, 42 Environment No. 9, p. 9 (2000)).

[39] *Id.*

[40] Perry E. Wallace, *Global Climate Change and the Challenge to Modern American Corporate Governance*, 55 Southern Methodist Law Review 493, 494 (2002) (citations omitted).

[41] *Id.* at 510–511. *See* Bratspies in this volume.

To this end, the international fossil fuel industry, in conjunction with other industrial interests, commenced an aggressive campaign of financial support for "a small but vocal group of 'green-house skeptics'" following the release in 1995 of a report from the Intergovernmental Panel on Climate Change's (IPCC), in which the IPCC noted "a consensus among scientists that 'the balance of evidence suggests that there is a discernable human influence on global climate.'"[42] This industry campaign, which greatly enhanced "the core of shared [American] governmental and corporate attitudes about climate change" was managed through the Global Climate Coalition and the Global Climate Information Project,[43] two industry-funded NGOs. Industry representatives countered the presence of environmental NGOs at the negotiations for the Kyoto Protocol by attending, consulting and lobbying in significant force.[44]

What was true of the *Trail Smelter* case is true of the dispute over global climate change. Nonstate actors have played a prevalent role in both: NGOs and globalized industrial interests battling, largely but not exclusively *via* their state sponsors, in the international arena. The international realm was once thought the preserve of state interests. The paradigmatic role of nonstate actors in international environmental law and policy, as exemplified by the global climate change dispute, seems to represent another, long-overlooked *Trail Smelter* "principle."

DYNAMIC REGIMES

At the core of most international environmental issues lies a set of evolving and often strenuously contested scientific claims. Edith Brown Weiss explained:

> Scientific uncertainty is inherent in all international environmental law. We do not have a full understanding of the natural system or of our interactions with it. Our scientific understanding is always changing, as is our technological knowledge and know-how.[45]

The *Trail Smelter* Tribunal's approach to the scientific uncertainty clouding the question of how best to respond to the harm caused in the United States by the smelter's smoke, is remarkably similar to international law's most prevalent contemporary approach to the problem, namely evolving or "dynamic legal regimes" that are "sufficiently flexible to enable the parties to respond to changes in scientific knowledge."[46]

In fact, the whole of the *Trail Smelter* Tribunal's work was infused with scientific inquiry. The first question posed to the Tribunal by the 1935 Convention

42 David Hunter et al., International Environmental Law and Policy 626 (2d ed. 2002).

43 Wallace, *supra* note 40, at 495.

44 Hunter et al, *supra* note 42, at 629.

45 Edith Brown Weiss, *International Environmental Law: Contemporary Issues and the Emergence of a New World Order*, 81 Georgetown Law Journal 675, 688 (1993).

46 *See* Thomas Gehring, *International Environmental Regimes: Dynamic Sectoral Legal Systems*, International Yearbook of International Environmental Law 35 (1990).

was almost exclusively a scientific matter: "Whether damage caused by the Trail Smelter in the State of Washington has occurred since the first day of January, 1932 . . . ?"[47] To address this scientific issue, and building on the extensive but contentious research instigated by the earlier IJC inquiry into the dispute, the two sides built their cases around complex and contradictory scientific theories. Wirth catalogued the distinct scientific claims advanced by the Americans and the Canadians in the arbitration, and the integrity of those claims in light of subsequent research.[48] Whatever scientific posterity might have to say about the matter, the Tribunal sided with the Canadian government, accepting the Canadian refutation of the American "invisible injury" thesis, pursuant to which the Americans sought to establish that "long-term, cumulative (or 'invisible') injury"[49] had occurred, requiring broader indemnification and more strict abatement.

The *Trail Smelter* Tribunal expressed some frustration over the conflicting and incomplete scientific evidence presented. With respect to crops and timber, however, the Tribunal found the evidence sufficient to permit it to conclude, in Canada's favor, that the smelter had only the slightest causal link to damage in Washington State,[50] and, further, that the degree and range of any damage caused by the smelter to crops and timber were extremely limited. On this distinct claim, the Tribunal's response constituted a clear rejection of the American "invisible injury" theory. The Tribunal concluded: "The contention made by the United States that fumigation prevents germination of seed is, in the opinion of the Tribunal, not sustained by the evidence. . . . The Tribunal is of the opinion that it is not proved that fumigation prevents trees from producing sufficient seeds, . . ."[51] For claimed damages to crops and timber in Washington State between 1932 and 1937, the Tribunal ordered a minimal indemnity of $78,000.[52] As regards other damages claimed by the United States, the Tribunal resorted to a variation on the conclusion that "the United States has failed to prove that the presence of fumes from the Trail Smelter has" caused the injury, and no further indemnification was ordered.[53]

Determining the cause and extent of damage between 1932 and 1937 was not, however, the end of the *Trail Smelter* Tribunal's scientific inquiry. The Tribunal was charged by the third question in the 1935 Convention with deciding on a permanent regime that would reduce, to an acceptable level, the damage in Washington State resulting from the smelter's fumigations. Confounded as it was by the shortcomings of the science presented to it, the Tribunal refrained from issuing a final, discrete and strictly formalistic judgment. Instead, the Tribunal

47 1935 Convention, *supra* note 14, at III, para. 1.

48 See the second chapter of Wirth's book: "'*This Most Illusive of Problems*': *The Scientific Dispute*." Wirth, *supra* note 7, at 45–79.

49 *Id.* at 39.

50 *Trail Smelter* (1939), *supra* note 1, at 198.

51 *Id.* at 204.

52 *Id.* at 206.

53 *Id.* at 206.

opted for a radically different exercise of its legal authority, ordering a series of procedures under a temporary regime which aimed primarily at pursuing a "more adequate and intensive study . . . of meteorological conditions in the valley."[54] The Tribunal's reply to the Convention's third question was a dynamic and evolving application of the law by which it recognized that its judgment regarding the operating regime for the smelter would be most effectively shaped if it were responsive to the dynamic and evolving science of the case. Thus, the Tribunal ordered a minutely detailed series of scientific investigations, the results of which it would finally review at a second round of proceedings it ordered for October 1940.[55] The Tribunal appointed "Technical Consultants" who were to head this effort. It also ordered the commissioning of a "consulting meteorologist" and outlined the specific studies (including the dates and methods) this expert was to undertake. The Tribunal ordered the smelter to install the relevant equipment and to grant the Technical Consultants freedom to pursue their investigation. Finally, the Tribunal scheduled the interim and final proceedings at which the Technical Consultants were expected to report on their findings.[56]

The resulting reports were authoritatively referenced by the Tribunal in its second opinion, first in summarily dismissing American claims that acute damage had been suffered in Washington State between 1938 and 1940.[57] More significantly, the Tribunal's resolution of the Convention's third question, regarding the imposition of a permanent regime, begins with an exhaustive survey of the information developed by the Technical Consultants.[58] By the end of the Tribunal's scrupulous accounting of the scientific evidence, there can be little doubt about the determinative weight the Tribunal had given to the new science developed under its temporary regime. And, after imposing the detailed "permanent" regime, which should have been the end of the matter, the Tribunal declared:

> While the Tribunal refrains from making the following suggestion a part of the regime prescribed, it is strongly of the opinion that it would be to the clear advantage of the Dominion of Canada, if during the interval between the date of filing of this Final Report and December 31, 1942, the Dominion of Canada would continue, at its own expense, the maintenance of the experimental and observational work by two scientists similar to that which was established by the Tribunal under its previous decision, and has been in operation during the trial period since 1938. It seems probable that a continuance of investigations until at least December 31, 1942, would provide additional valuable data both for the purpose of testing the effective operation of the regime now prescribed and for the purpose of obtaining information as to the possibility of the necessity of improvements in it.[59]

It is my thesis that the process-oriented approach to the law adopted by the *Trail Smelter* Tribunal in response to the Convention's third question parallels the

[54] *Id.* at 209.
[55] *Id.* at 184.
[56] *Id.* at 210–211.
[57] *Trail Smelter* (1941), *supra* note 1, at 707–709.
[58] *Id.* at 719–726.
[59] *Id.* at 731.

dynamic regimes that now predominate as the legal response to transboundary and international environmental issues, dramatically exemplified by the UNFCCC. It is a pragmatic response to the scientific complexity raised by these issues.[60] More important, however, it is representative of the realist jurisprudence that has prevailed in international law since the period during which the *Trail Smelter* case was decided. In fact, pragmatism is welcomed as a legitimate basis for the exercise of legal authority in realist jurisprudence in ways that it cannot be embraced by formalist and naturalist jurisprudence. The dynamism of the *Trail Smelter* Tribunal's temporary regime was a product of an emerging realist jurisprudence in international law that, when it later took hold, would facilitate the current widespread use of framework agreements in international environmental law.

Early international environmental law reflected the prevailing formalist jurisprudence and took the form of discrete, supposedly impartial rules set down in treaties.[61] It "had no special processes for adjusting to changes in scientific understanding of the problem."[62] However, precisely at the time the *Trail Smelter* case arose, this formalism was coming under siege from legal realists such as Frank and Llewellyn in the American, domestic jurisprudential discourse.[63] The realist debate also was beginning to trickle into international law. Martti Koskenniemi explained that:

> American international lawyers were faced with three responses to the decline of inter-war formalism. First, legal realism was overwhelming domestic academic law and cultivated an image of the lawyer as a policy-maker.... Second, domestic social science was being instrumentalized into "policy science."...Third was the activity of the German émigré internationalists working on international relations as well as international law.[64]

60 *See* Brown Weiss, *supra* note 45. It is also a concession to a gradualist approach necessitated by the previously noted economic interests implicated by these issues.

61 Admittedly, the proposed parallel suffers from the fact that the *Trail Smelter* arbitration is an arbitral decision, whereas the UNFCCC and most modern international environmental law is treaty-based. It would be possible to argue that this distinction is so comprehensive as to be determinative of the result in *Trail Smelter*, thus severely undermining the proposed parallel. Although I cannot dedicate the space necessary to exploring this potential criticism in the present project, I would suggest at least two responses. First, as a treaty institution, it is possible to characterize the work of the Tribunal as a kind of vicarious treaty undertaking between the United States and Canada. Second, in light of the general lack of either form of international environmental norm elucidation (whether treaty or decisional) at the time, it is fair to consider the *Trail Smelter* arbitration alongside treaty-based law, if only as a matter of ancestry.

62 Brown Weiss, *supra* note 45, at 689.

63 Pound and Llewellyn engaged in their seminal debate over legal realism in the 1931 volume of the Harvard Law Review. *See* Roscoe Pound, *The Call for a Realist Jurisprudence*, 44 HARVARD LAW REVEIW 697 (1931); Karl N. Llewellyn, *Some Realism About Realism – Responding to Dean Pound*, 44 HARVARD LAW REVIEW 1222 (1931). *See* JEROME FRANK, LAW AND THE MODERN MIND (1930).

64 MARTTI KOSKENNIEMI, THE GENTLE CIVILIZER OF NATIONS 475 (2001).

Koskenniemi went on to describe the effect of realism on international law jurisprudence as a "convergence of realism" on international law "from three sides."[65] The Yale School (led by Myres McDougal and Harold Laswell) and the Columbia School (led by Wolfgang Friedmann and Louis Henkin) made their distinctive realist impression on American international law thinking, emphasizing rule skepticism and interdisciplinarity.[66] In opting for process over a formalistic resolution of the case, and by explicitly making its eventual resolution of the case dependent on the incorporation of science, the *Trail Smelter* Tribunal took up both of these facets of international legal realism. As Llewellyn noted, some of the "common points of departure" for the legal realists were the "conception of law in flux" and the "conception of law as a means to a social ends . . . so that any part needs constantly to be examined for its purpose, and for its effect, . . ."[67] It is a rejection of the formalism of "disembodied rules,"[68] an embrace of a living, sociological jurisprudence that welcomes "the management of government by experts."[69] Thus, the *Trail Smelter* Tribunal was assisted in its dynamic approach to resolving the case by appointed Technical Consultants.[70] Koskenniemi closed the circle on my thesis when he concluded his survey of the international legal realism movement by explaining that:

> A relevant law . . . would mean a shift of emphasis from formal obligations to informal understandings and "*regimes*," with the acknowledgment that violations could be of different degrees. Such policy *pragmatism* [preferred] an "essentially modest, low-level, empirically-based, step-by-step approach."
>
> * * *
>
> Realist insights have been used to project an interdependent world of co-operation beyond the nation-States. As a consequence, an intellectual alliance has been proposed between international lawyers and international relations scholars advocating *regime theory* – that is, a theory about the effects of informal norms in constructing collaborative "*regimes*."[71]

And so it is that, long ago, the *Trail Smelter* arbitration pursued a realist response to scientific uncertainty, in much the same manner that the UNFCCC grapples with the scientific controversy swirling around the issue of global climate change. Of course, scientific uncertainty and contentious debate have featured more prominently in the efforts to address the global climate change issue than

65 *Id.* at 475.

66 *Id.* at 475–476.

67 Llewellyn, *supra* note 63.

68 Stephen B. Presser and Jamil S. Zainaldin, Law and Jurisprudence in American History 806 (4th ed. 2000).

69 G. Edward White, *From Sociological Jurisprudence to Realism: Jurisprudence and Social Change in Early Twentieth-Century America*, 58 Virginia Law Review 999, 1003–1004 (1972).

70 *Trail Smelter* (1939), *supra* note 1, at 210.

71 Koskenniemi, *supra* note 64, at 478–479 (emphasis added) (citations omitted).

in many other international environmental issues.[72] Consensus was only possible under a regime that is frequently cited as a leading example of the framework agreements in international environmental law "to which protocols and annexes may subsequently be negotiated as scientific understanding advances or political consensus is reached."[73] Patricia Birnie and Alan Boyle explained that "the most notable feature of environmental treaties over [the period since the Stockholm Conference] has been their increasing sophistication, characterized by the greater attention now paid to . . . the problems of amendment and flexibility." They concluded that the "1992 Convention on Climate Change, [represents one] of the most developed examples of this sort of international regulatory regime."[74] Brown Weiss identified the specific features of the UNFCCC that serve to promote flexibility. In particular:

> [The UNFCCC] provides for a standing body to provide scientific and technological advice on a timely basis. This body will provide scientific assessments of climate change and its effects, and the impact of implementing measures under the Convention. It will also identify relevant new technologies, assist in building local capacity for scientific research and assessment, and respond to the scientific inquiries of the parties. In sum, this body establishes a process for integrating scientific and technological advances into the operation of the Climate Framework Convention.[75]

For an example of the widely held view that international law should be understood as a process "rather than an impartial set of rules," W. Bradnee Chambers also cited the UNFCCC.[76] The evolutionary capacity of modern treaties is facilitated by "mechanisms such as framework and protocol approaches, learning systems such as education clauses, science and technology mechanisms that review

[72] It is frequently noted that the Second Assessment (1995) of the Intergovernmental Panel on Climate Change (IPCC) went a great distance to resolving the scientific uncertainty surrounding the anthropocentric causes of Climate change. It concluded that observable warming trends were "unlikely to be entirely natural in origin" and that there was a "discernible human influence" on the climate. IPCC, Working Group I, The Science of climate Change, 3–5 (Second Assessment Report, 1995). *See* HUNTER ET AL., *supra* note 45, at 593; EDITH BROWN WEISS ET AL., INTERNATIONAL ENVIRONMENTAL LAW AND POLICY 681–684 (1998). But the IPCC's report follows the scientific tradition of avoiding absolutism. The resulting conditional language, and its own admissions of empirical shortcomings in the field of climate change, have fueled vigorous questioning of these conclusions. "Disbelief is partly due to the historical fluctuations in temperature." BROWN WEISS ET AL., *id.*, at 684. Brown Weiss and her coauthors provide a useful survey of the persistent claims of scientific uncertainty in the field. *See id.*, at 684–695.

[73] EDITH BROWN WEISS ET AL., *supra* note 72, at 615 (1998).

[74] PATRICIA BIRNIE AND ALAN BOYLE, INTERNATIONAL LAW & THE ENVIRONMENT 152 (2nd ed. 2002).

[75] Brown Weiss, *supra* note 45, at 689.

[76] W. Brandee Chambers, *Towards an Improved Understanding of Legal Effectiveness of International Environmental Treaties*, 16 GEORGETOWN INTERNATIONAL ENVIRONMENTAL LAW REVIEW 501, 525–526 (2004).

progress in knowledge and advancement on the issue area."[77] Chambers, like others, remarked the UNFCCC's "review system and scientific mechanism that ensured [that] new knowledge made its way into the membership of the parties." The Kyoto Protocol, Chambers concluded, was a consequence of these mechanisms and processes, and, true to the realist mandate, the Kyoto Protocol "itself also has built-in systems that allow it to adapt as norms may change again in the future."[78]

CONCLUSION

There can be no doubt that, "because the *Trail Smelter* arbitration is a rare example of international environmental adjudication [from an] early period, it has acquired an unusually important place in the jurisprudence of international environmental law."[79] I think the case has much to tell us about modern international environmental law, but perhaps for reasons that have long escaped international environmental commentators, who have focused on the doctrine the case may or may not have generated. The parallels between *Trail Smelter* and the global climate change regime suggest that the *Trail Smelter* arbitration was important, indeed radically innovative, for surprising reasons. It is possible that *Trail Smelter's* real significance is that, all those years ago, it presaged the contextual and jurisprudential template for international environmental law's approach to a wide range of distinct issues, including global climate change.

77 *Id.* at 526.
78 *Id.* at 526.
79 Brown Weiss, *supra* note 45, at 677.

15 Sovereignty's Continuing Importance: Traces of *Trail Smelter* in the International Law Governing Hazardous Waste Transport

Austen L. Parrish

INTRODUCTION

The *Trail Smelter* arbitration[1] occupies an ambiguous place in history. On the one hand, scholars revere it for being the first of only a few landmark cases that address transboundary pollution.[2] On the other hand, scholars are prone to dismiss the case as a relic of a bygone era.[3] The arbitration resolved itself thirty years before international environmental law was even in its nascent stages, and the case arose under unique circumstances when science had yet to reveal the threat human activities pose to the environment.[4] Although praised for acknowledging the "no harm" principle,[5] environmentalists lament the decision's implicit recognition of a right to pollute, "so long as the polluting activities do not cause 'damage' in the sense of direct injury . . . to a second state."[6]

Those scholars whose assessment of the *Trail Smelter* arbitration's place in history is less than generous, explain that times have changed. We are told that

[1] *Trail Smelter Arbitral Decision*, 33 AMERICAN JOURNAL OF INTERNATIONAL LAW 182 (1939) [hereinafter "*Trail Smelter* (1939)"]; *Trail Smelter Arbitral Decision*, 35 AMERICAN JOURNAL OF INTERNATIONAL LAW 684 (1941) [hereinafter "*Trail Smelter* (1941)"]. *See* Annex to this volume.

[2] Alfred P. Rubin, *Pollution by Analogy: The Trail Smelter Arbitration*, 50 OREGON LAW REVIEW 259, 259 (1971) (explaining that "[e]very discussion of general international law relating to pollution starts, and must end, with a mention of the *Trail Smelter* arbitration"). Republished in this volume. *See* JAN SCHNEIDER, WORLD PUBLIC ORDER OF THE ENVIRONMENT: TOWARDS AN INTERNATIONAL ECOLOGICAL LAW AND ORGANIZATION 50 (1979) (describing the *Trail Smelter* arbitration as a "milestone" in the development of international environmental law).

[3] *See* Knox in this volume.

[4] Karin Mickelson, *Rereading* Trail Smelter, 31 CANADIAN YEARBOOK OF INTERNATIONAL LAW 219, 222–23, n.12 (1993) (discussing the limited "precedential value" of the *Trail Smelter* arbitration "because of the unique circumstances that surround both the decision to submit the dispute to arbitration and the decision of the Tribunal"). Republished in this volume.

[5] *See generally* EDITH BROWN WEISS ET AL., INTERNATIONAL LAW AND POLICY 257 (1998) (listing writers who have treated the *Trail Smelter* as an expression of the "no harm" principle of customary international law).

[6] Rubin, *supra* note 2, at 272; *see also* Mickelson, *supra* note 4, at 219–24 (explaining the precedential limitations of the *Trail Smelter* case).

the *Trail Smelter* arbitration has limited modern applicability because many of the world's environmental problems no longer fit into *Trail Smelter's* traditional conception of transboundary pollution.[7] States and other actors have turned away from resolving environmental disputes through adjudication, and instead have created elaborate regulatory regimes designed to address environmental concerns. The troubling aspects of the *Trail Smelter* arbitration, which appear to endorse a right to pollute, have been forgotten. For many commentators then, the basic teachings of the *Trail Smelter* arbitration are of marginal relevance to modern international environmental law.

But these commentators underestimate the *Trail Smelter* arbitration's continuing importance, at least insofar as the case reflects how solutions to international environmental problems are tied to preserving state sovereignty. Notwithstanding its development and growing importance, international environmental law has, in many ways, departed little from the basic legal principles that were articulated in the *Trail Smelter* arbitration. One poignant example, on which this chapter focuses, is the international law governing the transboundary movement and disposal of hazardous waste. Although that law is certainly different, both substantively and procedurally, from the rules applied in the *Trail Smelter* arbitration, the principles framing the international law regulating the transportation of hazardous waste reflect *Trail Smelter's* legacy and show how its notion of state sovereignty, for good or bad, continues to be important. Although the *Trail Smelter* arbitration and international hazardous waste laws both demonstrate a recognition of the continuing importance of territorial integrity, they also reveal the willingness of states to transfer some control to the international system to, paradoxically, protect sovereignty.

This chapter does three things. First, it describes how the *Trail Smelter* arbitration embraced the concept of permanent sovereignty over natural resources: the idea that nations have a right to exploit their resources so long as that exploitation does not cause serious transboundary harm.[8] The chapter draws from scholarship showing that the *Trail Smelter* arbitration rejected the notion of absolute sovereign integrity, yet also was unconcerned with environmental damage. Second, the chapter describes the problems surrounding the transboundary movement of hazardous waste and how the world community has chosen to address this

7 Rubin, *supra* note 2, at 272; *see, e.g.*, Michael Bothe, *The Responsibility of Exporting States, in* TRANSFERRING HAZARDOUS TECHNOLOGIES AND SUBSTANCES: THE INTERNATIONAL LEGAL CHALLENGE 160 (Günther Handl & Robert E. Lutz eds. 1989) (explaining the "fundamental difference" between transfrontier pollution and hazardous waste transportation); KATHARINA KUMMER, INTERNATIONAL MANAGEMENT OF HAZARDOUS WASTES: THE BASEL CONVENTION AND RELATED LEGAL RULES 19 (1995) (noting that traditional concepts of transboundary pollution do not apply in situations where the source of the pollution is transferred to an area beyond the jurisdiction of the state of origin). *See also* Knox in this volume.

8 FRANZ XAVER PERREZ, COOPERATIVE SOVEREIGNTY: FROM INDEPENDENCE TO INTERDEPENDENCE IN THE STRUCTURE OF INTERNATIONAL ENVIRONMENTAL LAW 96–108 (2000).

particular environmental challenge. Specifically, it discusses the Basel Convention and the world's unwillingness either to enact a total ban on hazardous waste transport or to set specific standards for environmentally safe hazardous waste generation and disposal. Finally, it addresses the linkages between the international law governing hazardous waste transport, and the sovereignty principles on which the *Trail Smelter* arbitration decision is grounded. The chapter then concludes by suggesting that the international community's motivation to address the problem of hazardous waste was not a concern over environmental harm, but a desire to reclaim sovereign integrity.

BACKGROUND: THE *TRAIL SMELTER* ARBITRATION AND PERMANENT SOVEREIGNTY

In the *Trail Smelter* arbitration, a specially appointed arbitral tribunal held Canada liable for property damage in the United States caused by the Trail smelter's release of sulfur dioxide from its tall smoke-stacks.[9] In its now famous proclamation of the "no-harm principle," the Tribunal explained that:

> no State has the right to use or permit the use of its territory in such a manner as to cause injury by fumes in or to the territory of another or the properties or persons therein, when the case is of serious consequence and the injury is established by clear and convincing evidence.[10]

The *Trail Smelter* arbitration occupies a special place in the history of international environmental law because it remains the "only decision of an international court or tribunal that deals specifically, and on the merits, with transfrontier pollution."[11] Scholars routinely cite and discuss the case and its no-harm principle.[12]

Mentioned less, however, is how the *Trail Smelter* arbitration was first and foremost a case about the limits of sovereignty. The dispute arose from the exercise of sovereign rights: Canada's right to carry out lawful activities in its own territory (to smelt ore), and the U.S. right to determine what acts may take place within its territory (to harvest apples without interference from Canadian smelter smoke).[13] In fact, the arbitral decision especially focused on sovereignty. The no-harm principle the Tribunal articulated reflected an "obligation of all states to protect within their territory the rights of other states, especially the rights

[9] *Trail Smelter* (1941), *supra* note 1, at 687–88, 733–34.

[10] *Id.* at 716.

[11] Weiss et al., *supra* note 5, at 257.

[12] Tuomas Kuokkanen, International Law and the Environment: Variations on a Theme 89 (2002) (stating that "[t]he Trail Smelter case is one of the landmarks of the traditional period to which scholars constantly refer"); Mickelson, *supra* note 4, at 219–20 (observing that the Trail Smelter arbitration, although often "more an object of reverence than a subject of analysis" is "one of the best known and most frequently cited international decisions").

[13] For a discussion of territorial sovereignty, see Günther Handl, *Territorial Sovereignty and the Problem of Transnational Pollution*, 69 American Journal of International Law 50, 51, 60–61 (1975).

to national integrity and inviolability during peace and war."[14] The Tribunal rejected other principles inconsistent with this notion of permanent sovereignty. It neither accepted the Harmon doctrine and its reliance on unfettered, absolute territorial sovereignty,[15] nor prohibited pollution as a constraint on sovereignty.[16] The Tribunal instead held that nations may use their natural resources as they see fit, so long as serious transboundary harm does not occur.[17] Both countries favored that approach, as did the "smelter industries on both sides of the border, [who] shared a common interest in adjudicating the case in a manner that would allow them to continue polluting the atmosphere with minimal regulatory restrictions."[18]

Notably absent from the *Trail Smelter* decisions was any notion that individuals have a right to be free from environmental harm. Equally missing was any suggestion that environmental preservation is an end unto itself. *Trail Smelter* focused on state rights – the rights and responsibilities of Canada and the United States – and not individual, natural or environmental rights.[19] The arbitration protected pollution: "despoiling the environment of one's own territory so as to endanger the ecology of the planet" did not violate international legal obligations.[20] Because of sovereignty, "international environmental law provides no basis for external intervention when [environmental] harm is purely domestic," no matter how severe or irreparable that harm may be.[21] Thus, unlike international human rights law, which "challenges traditional notions of sovereignty by viewing a state's treatment of its citizens as of international rather than merely domestic concern,"[22] *Trail Smelter* kept state sovereignty and equality as its cornerstones.

[14] Franz X. Perrez, *The Relationship between 'Permanent Sovereignty' and the Obligation not to Cause Transboundary Environmental Damage*, 26 Environmental Law 1187, 1198 (1996).

[15] *Id.*; *see also* Weiss et al., *supra* note 5, at 867–69 (discussing the Harmon Doctrine); Stephen McCaffrey, *The Harmon Doctrine One Hundred Years Later: Buried, Not Praised*, 36 Natural Resources Journal 549 (1996) (same).

[16] Rubin, *supra* note 2, at 275.

[17] Perrez, *supra* note 14, at 1199

[18] John D. Wirth, Smelter Smoke in North America xiv, 89–90 (2000); *see also* Weiss et al., *supra* note 5, at 259 (noting that the "United States in effect gave the tribunal the authority to legalize future damage if that was necessary to the establishment of an effective regime of control"); John E. Read, *The Trail Smelter Dispute*, 1963 Canadian Yearbook of International Law 213, 224–25 (1963) (explaining how neither party wanted to stop the pollution entirely, as "absolute cessation of damage might have shut down the Trail Smelter; but it would also have brought Detroit, Buffalo and Niagara Falls to an untimely end"). Republished in this volume.

[19] For a good discussion of environmental rights, see Dinah Shelton, *Human Rights, Environmental Rights, and the Right to Environment*, 28 Stanford Journal of International law 103, 117 (1991).

[20] Rubin, *supra* note 2, at 275.

[21] Hari M. Osofsky, *Learning from Environmental Justice: A New Model for International Environmental Rights*, 24 Stanford Environmental Law Journal 71, 80 (2005).

[22] *Id.* at 83; *see generally* Louis Henkin, *Keynote Address, Sibley Lecture (March 1994): Human Rights and State Sovereignty*, 25 Georgia Journal of International & Comparative Law 31 (1995).

Not an environmentally focused decision, the *Trail Smelter* arbitration is best understood as extending the principle of good-neighborliness.[23] Yet the good-neighborliness principle is derived directly from the permanent sovereignty concept because it requires that a state consider the interests, and respect the rights, of other states.[24] Even early cases, like the *Corfu Channel* case and the *Lac Lanoux* arbitration, recognized that cooperation is sovereignty's necessary corollary.[25] Permanent sovereignty over natural resources "generates a corresponding duty to recognize and respect other states' sovereignty over their natural resources and it thus includes a duty to avoid actions which are contrary to the similar rights of other states."[26] Or as Günther Handl has explained:

> [T]he notion of "neighborliness" simply implies . . . the exercise of sovereign territorial rights. . . . The concept of "neighborliness" is thus but the factual background against which the exercise of territorial rights must be seen. It does not constitute an independently existing body of specific legal rules imposing restraints on the exercise of territorial rights but merely represents an expression of the principle of abuse of rights.[27]

The sovereign right to territorial integrity – tempered by the no-harm principle – was thus the starting and ending point to the *Trail Smelter* arbitration.

The *Trail Smelter* arbitration eventually developed into a paradigm for resolving environmental challenges. A state would agree to impose restraints on unilateral sovereign action only because by so agreeing other states would do the same, thus better preserving sovereign interests in the long run. The primary benefit of entering into environmental agreements (or, in the case of *Trail Smelter*, agreeing to international arbitration) would be to preserve permanent sovereignty; the environmental benefits being collateral. Indeed, the concept of permanent sovereignty over a nation's own natural resources – as conceived in the *Trail Smelter* arbitration – is now a firmly entrenched principle of international law.

[23] *See* Ellis in this volume.

[24] *See generally* Island of Palmas Case (U.S. v. Neth.), 2 R.I.A.A. 829 (Perm. Ct. Arb. 1928); *see also* Handl, *supra* note 13, at 55–57 ("The emerging principle of *sic utere tuo ut alienum non laedas* constituted recognition of the fact that territorial sovereign rights in general were correlative and interdependent and were consequently subject to reciprocally operating limitations.").

[25] Corfu Channel Case (United Kingdom v. Albania), 1949 I.C.J. Rep. 4 (1949) (holding that it is "every state's obligation not to allow knowingly its territory to be used for acts contrary to the rights of other states"); Lake Lanoux Arbitration (Fr. v. Spain), 12 R.I.A.A. 281 (1957), *reprinted in* 53 American Journal of International Law 156 (1959) (holding that a state can lawfully utilize the waters of an international river in its territory so long as it takes into account the interests of coriparian states).

[26] Perrez, *supra* note 14, at 1209; *see also* Franz Xaver Perrez, *The Efficiency of Cooperation: A Functional Analysis of Sovereignty*, 15 Arizona Journal of International and Comparative Law 515, 515–16 (1998) (arguing that "according to a modern understanding of international law, permanent sovereignty indeed includes a responsibility to cooperate and to prevent environmental damage with negative external effects").

[27] Handl, *supra* note 13, at 56.

"More than eighty resolutions have covered the right to permanent sovereignty."[28] It is believed to be essential to the development of underdeveloped countries,[29] reflects the right of self-determination,[30] and is considered by some international tribunals to constitute customary international law.[31]

INTERNATIONAL ENVIRONMENTAL LAW ON TRANSBOUNDARY HAZARDOUS WASTE TRANSPORT

That the *Trail Smelter* arbitration was unconcerned with environmental harm *per se* is hardly surprising. At the time, nations were only beginning to understand the necessity of limiting the exploitation of natural resources.[32] The arbitration occurred thirty years before the 1972 Stockholm Conference, which is widely regarded as the birthplace of modern international environmental law.[33] At the time, few international organizations existed, let alone organizations with competence in environmental matters.[34] What is surprising is that seventy years later, and after appreciable growth in the understanding of the dangers facing the international environment,[35] permanent sovereignty continues to play a more important role in solving environmental challenges than a global concensus or a perceived moral obligation to protect and preserve the environment. A good example of how permanent sovereignty, as opposed to environmental concerns, drives international environmental law is found in the international laws governing transboundary hazardous waste transport.

The Problem: The Hazardous Waste Challenge

Transfrontier movements of hazardous waste present a significant environmental challenge.[36] Hazardous waste that crosses national borders include "a range

[28] Melissa A. Jaminson, *Rural Electric Cooperative: A Model for Indigenous Peoples' Permanent Sovereignty Over Their Natural Resources*, 12 TULSA JOURNAL OF COMPARATIVE & INTERNATIONAL LAW 401, 428–29 (2005); *see also* PHILIPPE SANDS, PRINCIPLES OF INTERNATIONAL ENVIRONMENTAL LAW 237 (2d ed. 2004) (listing conventions recognizing a nation's right to permanent sovereignty over its natural resources).

[29] PERREZ, *supra* note 8, at 72.

[30] *Id. See also* Nico J. Schrijver, *Permanent Sovereignty Over Natural Resources Versus the Common Heritage of Mankind: Complementary or Contradictory Principles of International Economic Law? in* INTERNATIONAL LAW AND DEVELOPMENT 87, (de Waart et al. eds. 1988).

[31] SANDS, *supra* note 28, at 237.

[32] *Id.* at 25

[33] Report of the Stockholm Conference on the Human Environment, U. N. Doc. A/CONF. 48/14, *reprinted in*, 11 INTERNATIONAL LEGAL MATERIALS 1416 (1972); *see* Edith Brown Weiss, *International Environmental Law: Contemporary Issues and the Emergence of a New World Order*, 81 GEORGETOWN LAW JOURNAL 675, 678–79 (1993); *see also* Marc Pallemaerts, *International Environmental Law from Stockholm to Rio: Back to the Future?, in* GREENING INTERNATIONAL LAW 2 (Philippe Sands ed., 1994) ("[The Stockholm Declaration] is generally regarded as the foundation of modern international environmental law.").

[34] SANDS, *supra* note 28, at 25.

[35] *Id.*

[36] For a comprehensive and systematic study of the global hazardous waste challenge, *see* GÜNTHER HANDL & ROBERT E. LUTZ, TRANSFERRING HAZARDOUS TECHNOLOGIES AND SUBSTANCES: THE

of materials, from chemical and radioactive wastes to municipal solid waste, asbestos, incinerator ash and old tires."[37] The global generation of these kinds of waste has increased dramatically over the last three decades. Depending on how hazardous waste is defined, between three hundred and five hundred metric tons are generated each year[38] and the amount continues to grow.[39] Concomitant with the increased generation of hazardous waste has been the growing cost of safe disposal. Industrialized nations generate ninety percent of all hazardous waste.[40] Over the last twenty-five years, these industrialized nations have faced exponentially increased waste disposal costs given strict domestic environmental legislation.[41] As a result, industrialized nations shipped staggering amounts of hazardous waste to developing nations in the 1980s and early 1990s.[42] Cost savings to industrial nations are significant: the disposal of hazardous waste can cost $2,000 per ton or more in a developed country, compared to as little as $40 per ton in Africa.[43] Developing nations are attractive places for the disposal of

International Legal Challenge (1989); *see also* Robert E. Lutz, *The Export of Danger: A View From The Developed World*, 20 New York University Journal of International Law & Policy 629 (1988).

37 David Hunter et al., International Environmental Law & Policy 832 (2d ed. 2002).

38 Karen Dawson, *Wag the Dog: Towards a Harmonization of the International Hazardous Waste Transfer Regime*, 19 Canadian Journal of Law & Society 1, 3 (2004) (citing 1999 statistics); *see also* Alexandre Kiss & Dinah Shelton, International Environmental Law 601 (3d ed. 2004) (noting that "UNEP estimates that more than 400 million tons of hazardous waste is generated annually throughout the world . . . "); Jonathan Krueger, International Trade and the Basel Convention 14–22 (1999) (listing statistics); Kummer, *supra* note 7, at 5 ("Estimates of the global volume of hazardous waste generated each year range from 300 to 500 million tons."). The exact amount of hazardous waste generated is unknown because of disagreement as to what constitutes hazardous waste and illegal trafficking. *Id.* at 4–5.

39 Hunter, *supra* note 37, at 832 (explaining that the "[w]orldwide generation of hazardous waste is estimated to have increased from approximately 5 million metric tons in 1945 to 300 million in 1988 – an increase of 60-fold since the end of World War II").

40 Dawson, *supra* note 38, at 753.

41 Sean D. Murphy, *The Prospective Liability for the Transboundary Movement of Hazardous Waste*, 88 American University Journal of International Law & Policy 24, 30–31 (1994); *see also* Donna Valin, *Comment, The Basel Convention on the Control of Transboundary Movements of Hazardous Waste and Their Disposal: Should the United States Ratify the Accord?*, 6 Indiana International & Comparative Law Review 267, 269–70 (1995) (discussing the "soaring costs of hazardous waste disposal").

42 Cyril Uchenna Gwam, *Travaux Preparatoires of the Basel Convention on the Control of Transboundary Movements of Hazardous Wastes and their Disposal*, 18 Natural Resources & Environmental Law Journal 1, 7–8 (2004); *see also* Kiss & Shelton, *supra* note 38, at 607–08 (3d ed. 2004); Vep P. Nanda and Bruce C. Bailey, *Export of Hazardous Waste and Hazardous Technology: Challenge for International Environmental Law*, 17 Denver Journal of International Law & Policy 155, 155–56, 161–79 (1988). For a general discussion of the amount of hazardous waste crossing national borders, *see* J.W. MacNeill, *Policy Issues Concerning Transfrontier Movements of Hazardous Waste*, *in* Transfrontier Movements of Hazardous Wastes 7–8 (OECD 1985).

43 Jennifer Kitt, Note, *Waste Exports to the Developing World: A Global Response*, 7 Georgetown International Environmental law Review 485, 488 (1995); Hunter, *supra* note 37, at 833.

hazardous waste because, even if those countries have environmental laws, they often do not have the administrative infrastructure necessary to enforce those laws.[44] The incentives flow both ways. Despite potential hazards, developing nations have strong economic incentives to accept hazardous waste from other countries.[45]

The potential detrimental effects of hazardous waste transport and disposal raise three concerns. First, significant, irreparable short-term and long-term environmental damage may result from the transport, storage, disposal, or treatment of hazardous waste.[46] Second, hazardous waste disposal has had a disparate impact on developing nations. Because developing nations often lack the technology and infrastructure to properly manage hazardous wastes,[47] "the recipient country and bordering nations experience soil contamination, ground water pollution, air pollution, and threats to natural resources and biodiversity."[48] Third, illegal dumping and trafficking of hazardous waste, sometimes characterized as "toxic terrorism" or "garbage imperialism," has become prevalent.[49] This problem received significant worldwide media attention with a number of high profile cases.[50]

44 Günther Handl and Robert E. Lutz, *An International Policy Perspective on the Trade of Hazardous Materials and Technologies*, 30 HARVARD INTERNATIONAL LAW JOURNAL 351, 362 (1989); David P. Fidler, *Challenges to Humanity's Health: The Contributions of International Environmental Law to National and Global Public Health*, 31 ENVIRONMENTAL LAW REPORTER 10048, 10066 (2001).

45 HUNTER, *supra* note 37, at 834, 846; Handl and Lutz, *supra* note 44, at 355–57.

46 Dawson, *supra* note 38, at 3; *see also* KRUEGER, *supra* note 38, at 9 (discussing the long-term and short-term environmental impacts from improper hazardous waste disposal); Handl and Lutz, *supra* note 44, at 354 (explaining why waste trading and the hazards it poses is a matter of international concern).

47 Sejal Choksi, *Comment, The Basel Convention on the Control of Transboundary Movements of Hazardous Wastes and Their Disposal: 1999 Protocol on Liability and Compensation*, 28 ECOLOGY LAW QUARTERLY 509, 514 (2001).

48 Valin, *supra* note 41, at 270.

49 Choksi, *supra* note 47, at 515 (citing William N. Doyle, *United States Implementation of the Basel Convention: Time Keeps Ticking, Ticking Away*, 9 TEMPLE INTERNATIONAL AND COMPARATIVE LAW JOURNAL 141, 142 n.13 (1995)); KISS & SHELTON, *supra* note 38, at 607 (describing "garbage imperialism"); *see also* Kimberly K. Gregory, Note, *The Basel Convention and the International Trade of Hazardous Waste: The Road to Destruction of Public Health and the Environment Is Paved with Good Intentions*, 10 CURRENTS: INTERNATIONAL TRADE LAW JOURNAL 80, 81 (2001) (discussing "toxic terrorism" and "garbage imperialism").

50 HUNTER, *supra* note 37, at 834 (discussing how "in return for paying $100 monthly rent to a Nigerian national for use of his farmland, five ships transported 8,000 barrels of Italian hazardous waste to the small river town of Koko, Nigeria."); KUMMER, *supra* note 7, at 8 (describing well-published cases such as the "Philadelphia fly ash case"; the Koko, Nigeria case; and the *Khian Sea* incident); Nanda & Bailey, *supra* note 42, at 155 (explaining how the Seveso accident, the 1984 Bhopal, India disaster, the Chernobyl catastrophe, and the 1986 Basel fire "pushed to the forefront of public attention" the problems with hazardous technology transfers); *see generally* WEISS, *supra* note 5, at 779–81; *cf.* Gregory, *supra* note 49, at 80 (describing problems illicit toxic dumping caused in Haiti, Paraguay, and Costa Rica).

The International Law: From Cairo to Basel

The growing recognition in the early 1980s that the transport and disposal of hazardous waste had become an international problem spurred action. In 1982, the United Nations Environment Programme (UNEP) established a working group to develop guidelines governing the hazardous waste trade.[51] In 1987, this group developed the Cairo Guidelines. The Cairo Guidelines established a working group with a mission to create a global regime addressing "the threat posed to the environment of ill-equipped developing countries by the illegal import of hazardous wastes from industrialized nations."[52] The global regime came to fruition in the Basel Convention that 116 nations adopted on March 22, 1989,[53] and which entered into force on May 5, 1992.[54] Currently, 165 states have ratified the Convention, with the United States notably absent from the list.[55]

The Basel Convention established rules designed to regulate hazardous waste trade rather than prohibit it.[56] Broadly, the Basel Convention did three things. First, it recognized that states have the sovereign right to ban hazardous waste imports.[57] Second, it attempted to minimize the generation of hazardous waste and promote disposal at its source (the proximity principle) by permitting transboundary movement only "in circumstances where the state of export does not have the capacity or facilities to dispose of the wastes in an environmentally sound manner."[58] Third, for trade that does occur, the Convention required the prior informed consent of both the transit and importing states.[59] The prior-informed-consent concept is the heart of the Basel Convention.[60] A signatory state cannot send hazardous waste to: (1) a nonparty state; (2) a signatory state that bans the import of that waste; or (3) a signatory that does not have adequate facilities to

[51] Choksi, *supra* note 47, at 516; HUNTER, *supra* note 37, at 835–36. For a good discussion of the basic international legal framework existing before the Basel Convention, see Lutz, *supra* note 36, at 659–69.

[52] KUMMER, *supra* note 7, at 43.

[53] The Basel Convention on the Control of Transboundary Movements of Hazardous Wastes and Their Disposal, adopted Mar. 22, 1989, *reprinted in* 28 INTERNATIONAL LEGAL MATERIALS 649 (1989) [hereinafter "Basel Convention"]; *see generally* KISS & SHELTON, *supra* note 38, at 608.

[54] *Id.*

[55] Statistics available at http://www.basel.int/ratif/frsetmain.php (for current list of ratifications); *see also* Jutta Brunnée, *The United States and International Environmental Law: Living With An Elephant*, 15 EUROPEAN JOURNAL OF INTERNATIONAL LAW 617, 620–29 (2004) (discussing the United States' unwillingness to engage in international environmental law, including its failure to ratify the Basel Convention).

[56] SANDS, *supra* note 28, at 692.

[57] PATRICIA BIRNIE & ALAN BOYLE, INTERNATIONAL LAW & THE ENVIRONMENT 430 (2d ed. 2002).

[58] *Id. See also* SANDS, *supra* note 28, at 692.

[59] BIRNIE & BOYLE, *supra* note 57, at 430; SANDS, *supra* note 28, at 692.

[60] David A. Wirth, *Trade Implications of the Basel Convention Amendment Banning North-South Trade in Hazardous Wastes*, 7 REVIEW OF EUROPEAN COMMUNITY & INTERNATIONAL ENVIRONMENTAL LAW 237, 237–38 (1998).

dispose of the waste in an environmentally sound manner.[61] Hazardous waste traffic which contravenes the notification and consent requirements of the Basel Convention is illegal and considered criminal.[62]

Beyond Basel: The Basel Ban and Liability Protocol

Since its enactment, several amendments have attempted to "give teeth" to the Basel Convention and ensure compliance; but none has been successful. The first, known as the Basel Ban, attempted to prohibit hazardous waste exports from industrialized nations to the developing world.[63] The ban is essentially a "North/South ban" because it "does not affect the movements between developing countries themselves."[64] The ban was formally proposed for incorporation into the Basel Convention in September 1995.[65] The Basel Ban has lacked broad-based support, and has been criticized as an "exercise in futility."[66] A decade later, the Basel Ban has not yet entered into force. The Secretariat has received only fifty-six of the required sixty-two ratifications.[67]

The second amendment attempted to create a liability regime. In December 1999, the parties adopted a Protocol on Liability and Compensation.[68] The Protocol intended to "provide a comprehensive regime for liability and for adequate and prompt compensation for damage, defined to include damage to persons and property and loss of income."[69] The Protocol applies to damage caused during transboundary movement of hazardous wastes,[70] and imposes a strict liability regime.[71] Under this regime, the notifying state is generally liable for damage until the disposer takes possession of the waste, at which point liability shifts to the disposer.[72] Like the Basel Ban, the Protocol also lacked broad-based support. To date, only thirteen parties have signed the Protocol and only five of the required twenty have ratified it.[73]

[61] Basel Convention, Art. 4(2)(b), (e) and (g).

[62] Basel Convention, Arts. 4(3) & 9; *see also* SANDS, *supra* note 28, at 692.

[63] Basel Convention, Art. 4A and Annex VII; SANDS, *supra* note 28, at 694–95.

[64] Muthu S. Sundram, *Basel Convention on Transboundary Movement of Hazardous Wastes: Total Ban Amendment*, 9 PACE INTERNATIONAL LAW REVIEW 1, 3–4 (1997).

[65] *Id.*

[66] Wirth, *supra* note 60, at 245; *see also* William Schneider, Note, *The Basel Convention Ban on Hazardous Waste Exports: Paradigm of Efficacy or Exercise in Futility*, 20 SUFFOLK TRANSNATIONAL LAW REVIEW 247, 247–48, 278–86 (1996) (arguing that the Basel Ban is unlikely to be successful).

[67] For a current list of ratifications, see http://www.basel.int/ratif/frsetmain.php (status of ratifications as of April 8, 2005).

[68] Choksi, *supra* note 47, at 522–24; *see generally* HUNTER, *supra* note 37, at 854–55.

[69] SANDS, *supra* note 28, at 924.

[70] *Id.* at 925.

[71] *Id.*

[72] Basel Convention, Art 4(1); SANDS, *supra* note 28, at 924.

[73] For a current list of ratifications, see http://www.basel.int/ratif/frsetmain.php#protocol (status of ratifications as of April 8, 2005); *see also* Choksi, *supra* note 47, at 524 (noting that "environmentalists have branded the Protocol as text with as many holes and exclusions as Swiss cheese.").

TRACES OF THE *TRAIL SMELTER* ARBITRATION IN THE BASEL CONVENTION

Just as the *Trail Smelter* arbitration can be viewed as a case primarily concerned with protecting sovereignty and only secondarily concerned with protecting the environment, the Basel Convention is also best seen as a reflection of permanent sovereignty's enduring importance. States agreed to the Basel Convention not for its environmental protections, but for its preservation of sovereignty in the face of a burgeoning waste trade that threatened state independence and self-determination.

The Basel Convention does not impose conditions that would be expected if the Convention was principally concerned with preventing environmental harm. The Basel Convention does not ban hazardous waste trade and "a policy ending all trade in hazardous wastes has [not] prevailed at a global level."[74] Except for requiring that importing states dispose of waste in an "environmentally sound" manner – a vague and easily circumvented standard[75] – the Basel Convention does not set specific standards, establish targets, or create quantitative restrictions on the generation and disposal of hazardous wastes.[76] The treaty does not address waste management within states, nor does it create any fund for cleanup of hazardous waste accidents.[77] All of this is disconcerting from an environmental perspective as near unanimity exists in the assessment that "limiting the avalanche of waste which is now threatening to engulf industrialized countries . . . requires the development of strategies and legal rules which address the waste problem at its source by preventing its generation."[78]

Instead, the Basel Convention is aimed at preserving the right to self-determination and ensuring that nations can use their resources as they see fit, so long as other nations' interests are not infringed. This is evident in the Convention's provisions themselves, including:

- The definition of the term "transboundary" requiring the involvement of at least two states, which effectively excludes the protection of the global commons from hazardous waste contamination.[79]

[74] Birnie & Boyle, *supra* note 57, at 430; *see also* Sundram, *supra* note 64, at 3 (noting that "[t]he Basel Convention is as much of a Convention affecting the international trade, as it is to protect the global environment.").

[75] Choksi, *supra* note 47, at 519.

[76] Basel Convention, Art 4(2)(g); Elli Louka, Overcoming National Barriers to International Waste Trade: A New Perspective on the Transnational Movements of Hazardous and Radioactive Waste xiii–xiv, 49 (1994) (noting "lack of definition of sound waste management" in Basel Convention).

[77] Choksi, *supra* note 47, at 518–19.

[78] Birnie & Boyle, *supra* note 57, at 432.

[79] Kummer, *supra* note 7, at 78.

- The express recognition of a state's sovereign right to ban hazardous waste imports.[80]
- The right of states to enter into separate agreements with nonparties and thus exempt themselves from the Convention's provisions.[81]
- The requirement of prior informed consent and notification, which itself "is simply an expression of the sovereignty of a state over the use of its territory and resources."[82]

The importance of sovereignty is also evident in the Convention's failure to define hazardous wastes, leaving the definition of "wastes" to national governments.[83]

That the Basel Convention is a law embracing the notion of permanent sovereignty can also be seen in the contrary positions countries took during the treaty negotiations. The first position was taken by developing nations.[84] Fueled by recent high-profile, public incidents involving waste trafficking, those countries saw the Basel Convention as an opportunity to stop rich industrialized nations from using their territories as toxic dumps.[85] They sought a complete worldwide ban of all transboundary movements of hazardous wastes.[86] But this position again was taken not to stop environmental harm, but by the desire to control their own nation's destiny and prevent so-called garbage imperialism. The opposing position, championed by developed countries, was not entirely environmentally rooted either. Those nations urged the world to minimally control and regulate waste traffic, and opposed "any proposed measures that would put too many restrictions on the trade in wastes – especially recyclable materials with an economic value – amongst industrialized nations."[87] Some hazardous wastes were valuable tradable commodities.[88] Although the "vast majority of international waste transport takes place between industrialized nations," this problem was widely ignored."[89]

80 Basel Convention, Preamble & Art. 4(1), (6) & (13). For a discussion of the impact international trade rules have on the right to curtail hazardous waste transfers in their country, see Harold H. Koh, *The Responsibility of the Importer State*, *in* TRANSFERRING HAZARDOUS TECHNOLOGIES AND SUBSTANCES: THE INTERNATIONAL LEGAL CHALLENGE 170–96 (Günther Handl & Robert E. Lutz eds. 1989).

81 Basel Convention, Art. 11; *see also* Choksi, *supra* note 47, at 519.

82 BIRNIE & BOYLE, *supra* note 57, at 432; Cyrus Mehri, Note, *Prior Informed Consent: An Emerging Compromise for Hazardous Exports*, 21 CORNELL INTERNATIONAL LAW JOURNAL 365, 387 (1988) (explaining that prior informed consent "preserves the sovereignty and self-determination of importing states . . . and avoids infringing state sovereignty as do bans.").

83 LOUKA, *supra* note 76, at 49; Paul E. Hagen, *Update on the Basel Convention and Other Agreements Governing the Transboundary Movement of Hazardous Wastes*, SG056 ALI-ABA 331, 336 (2002).

84 KUMMER, *supra* note 7, at 43–45.

85 *Id.* at 43.

86 *Id.*

87 *Id.*; *see also* Gwam, *supra* note 42, at 33–34, 45 (explaining that "industrialized countries sought only minimal regulation.").

88 Gwam, *supra* note 42, at 9.

This is not to condemn the Basel Convention. The Basel Convention, as Katharina Kummer has carefully detailed, "represented the maximum degree of consensus that was politically possible at that time."[90] But the Convention certainly betrays that the global community is still – as it was as the time of the *Trail Smelter* – willing to solve environmental challenges only so long as the law acts as a means to reclaim sovereignty rather than erode it. Perhaps predictably then, the Basel Convention has been only partially successful in stopping environmental degradation as a result of hazardous waste transport and disposal. Many believe the international laws have facilitated waste transfers rather than waste minimization,[91] "spurred illegal waste transfers,"[92] and "failed to adequately address some of the most critical hazardous waste problems underdeveloped nations now confront."[93] This is not surprising. If the world wishes to fully solve the environmental challenges that hazardous waste poses, it must be willing to enact a treaty that addresses the environmental problem head on, rather than enacting one that pays only lip-service to the problem, while primarily ensuring that there is no encroachment on territorial sovereignty.

CONCLUSION

It would be a mistake to suggest that the *Trail Smelter* arbitration was essential in the creation of international law governing hazardous waste transport and disposal. And yet it would be equally naïve to ignore the case, for the permanent sovereignty principles (constrained by the no-harm principle) on which that case relied, remain pertinent. As in the *Trail Smelter* arbitration, states are willing to surrender sovereignty in an effort to solve environmental challenges only when by doing so they restore some control lost to external forces. This is certainly true with the Basel Convention. The Basel Convention's focus on prior informed consent and self-determination, and the world's unwillingness to ban or set specific environmentally sound disposal standards, has meant the hazardous waste generation and disposal problem remains largely unsolved. Traces of the *Trail Smelter*

[89] KUMMER, *supra* note 7, at 43.

[90] *Id.* at 79.

[91] Choksi, *supra* note 47, at 520 (citing assessment of Greenpeace, the Basel Action Network, and other monitoring organizations).

[92] LOUKA, *supra* note 76, at xiv; *see also* KUMMER, *supra* note 7, at 81–82 (noting that "illegal schemes for the transfer of hazardous wastes have tended to become more ingenious and elaborate since the inception of the Basel Convention.").

[93] B. John Ovink, *Transboundary Shipments of Toxic Waste: The Basel and Bamako Conventions – Do Third World Countries Have A Choice?*, 13 DICKINSON JOURNAL OF INTERNATIONAL LAW 281, 294 (1995); *see also* Alexandre Kiss, *The International Control of Transboundary Movement of Hazardous Waste*, 26 TEXAS INTERNATIONAL LAW JOURNAL 521, 539 (1991) ("Some states have criticized the Convention. African states in particular have found it too lenient. Other states have characterized the stringent requirements and procedures as inefficient."); *see* Dawson, *supra* note 38, at 20–21 (discussing the limitations and failings of the Basel Convention from a Critical Race Theory perspective).

arbitration can be seen in many modern international laws like the Basel Convention and, accordingly, scholars are premature in their attempt to resign the case to international environmental law history. The *Trail Smelter* arbitration reflects a continuing paradigm used to solve environmental challenges. Environmentalists and ecology-minded persons can only hope that the next generation of international environmental laws will move beyond sovereignty and meet environmental challenges on their own terms. But that day has not yet come.

16 The Legacy of *Trail Smelter* in the Field of Transboundary Air Pollution

Phoebe Okowa

INTRODUCTION

When I began my research in this field a few years ago I assumed, like many writing about international environmental law, that the *Trail Smelter* dispute would provide a useful starting point from which to assess the framework of accountability that existed in international law for transboundary air pollution damage. The Tribunal's famous pronouncement, "that under the principles of international law, as well as the law of the United States, no state has the right to use or permit the use of its territory in such a manner as to cause injury by fumes in or to the territory of another or the properties of persons therein, when the case is of serious consequence and the injury is established by clear and convincing evidence,"[1] had in many ways irrevocably confirmed that transboundary environmental damage, and specifically transboundary air pollution, entailed state responsibility.

The precedential value of this passage had been acknowledged in diverse contexts such as water disputes,[2] pollution of the seas, and in various neighbourhood disputes between European States,[3] as well as in the extensive literature[4] that proliferated in the period after the Stockholm Conference on the Human Environment.[5] It substantially informed the work of the International Law Commission

[1] *Trail Smelter Arbitral Decision*, 35 AMERICAN JOURNAL OF INTERNATIONAL LAW 684, 716 (1941) [hereinafter "*Trail Smelter* (1941)"]. *See* Annex to this volume.

[2] For a helpful discussion, *see* STEPHEN MCCAFFREY, THE LAW OF INTERNATIONAL WATERCOURSES: NON-NAVIGATIONAL USES 354 (2001).

[3] For an extensive survey of European practice reflecting the *Trail Smelter* principle, *see* JOHAN G. LAMMERS, POLLUTION OF INTERNATIONAL WATERCOURSES (1984).

[4] Karin Mickelson, *Rereading Trail Smelter*, 31 CANADIAN YEARBOOK OF INTERNATIONAL LAW 219 (1993); Arthur P. Rubin, *Pollution by Analogy: The Trail Smelter Arbitration*, 50 OREGON LAW REVIEW 259 (1971). Both republished in this volume. *See also* JOHN D. WIRTH, SMELTER SMOKE IN NORTH AMERICA (2000).

[5] Report of the Stockholm Conference on the Human Environment, U.N. Doc. A/CONF.48/14, *reprinted in* 11 INTERNATIONAL LEGAL MATERIALS 1416, at 1420 (1972).

on the Non-Navigational Uses of International Watercourses,[6] and on Liability for Injurious Consequences of Acts not Prohibited by International Law.[7] It also was unequivocally endorsed, either directly or in form, in the many environmental treaties that were enacted in the wake of the Stockholm Conference. The preamble to the air pollution treaties equally affirmed the normative quality of the *Trail Smelter* principle.[8]

But why had the *Trail Smelter* arbitral model and its jurisprudence not been used to resolve the new wave of far more serious air pollution problems in Europe and North America?[9] Why, given *Trail's* seemingly clear articulation of rules regulating air pollution, had states failed to rely on the regime of state responsibility for their enforcement? And why had arguments for the imposition of responsibility been largely absent in the conventional regimes on air pollution as well as the diplomatic exchanges between states affected by acid rain? Why, in the face of the Chernobyl nuclear disaster (an often forgotten case of aerial pollution), had states failed to rely on the *Trail Smelter* precedent in calling for the responsibility of the Soviet Union?

It has been repeatedly pointed out in the literature that the dispute remains one of the few environmental controversies to have been resolved directly through the medium of litigation.[10] Nonetheless, far from endorsing the *Trail Smelter* model, the emerging air pollution treaties eschewed any firm commitment to the liability model as a method of resolving these disputes. Most notable in this respect was the footnote to Article 8 of the 1979 Convention on Transboundary Air Pollution, which in a controversial provision purported to exclude any form of state

6 *See* McCaffrey in this volume.

7 On the relevance of *Trail Smelter* to the International Law Commission's work on liability for injurious consequences of acts not prohibited by international law, *see* Drumbl in this volume.

8 ECE Convention on Long-Range Transboundary Air Pollution,1302 U.N.T.S. 217, *reprinted in* 18 INTERNATIONAL LEGAL MATERIALS 1442 (1979). Among its related protocols are: 1985 Helsinki Protocol on the Reduction of Sulfur Emissions or their Transboundary Fluxes by at least 30 Percent, July 18, 1985, *reprinted in* 27 INTERNATIONAL LEGAL MATERIALS 698 (1988); Protocol to the 1979 Convention on Long-Range Transboundary Air Pollution on Further Reduction of Sulfur Emissions, June 14, 1994, *reprinted in* 33 INTERNATIONAL LEGAL MATERIALS 1540 (1994) (not yet entered into force); Protocol to the Convention on Long- Range Transboundary Air Pollution on Persistent Organic Pollutants, March 31, 1998, *reprinted in* 37 INTERNATIONAL LEGAL MATERIALS 505 (1998); Protocol Concerning the Control of Emissions of Nitrogen Oxides, October 31, 1988, *reprinted in* 28 INTERNATIONAL LEGAL MATERIALS 212 (1989); 1991 Protocol on the Reduction of Volatile Organic Compounds, *reprinted in* 31 INTERNATIONAL LEGAL MATERIALS 568 (1992). *See also* Canada – United States Agreement on Air Quality, March 13, 1991, *reprinted in* 30 INTERNATIONAL LEGAL MATERIALS 676 (1991); Vienna Convention for the Protection of the Ozone Layer, March 22, 1985, 1513 U.N.T.S. 293, *reprinted in* 26 INTERNATIONAL LEGAL MATERIALS 1529 (1987); Montreal Protocol on Substances that Deplete the Ozone Layer, September 16, 1987, 1522 U.N.T.S. 3, *reprinted in* 26 INTERNATIONAL LEGAL MATERIALS 1541 (1987).

9 On the outcome of the research, *see* PHOEBE OKOWA, STATE RESPONSIBILITY FOR TRANSBOUNDARY AIR POLLUTION IN INTERNATIONAL LAW (2000)

10 For a history of adjudication of environmental disputes, *see* Ellis and Knox in this volume.

liability.[11] *Trail Smelter's* lasting legacy was surely bound to be marginal if it could not be relied on to provide a dependable set of precepts for resolving the seemingly insurmountable difficulties presented by acid rain.

The first part of this chapter will explore a number of the salient features of the *Trail Smelter* dispute, which, arguably, limited its potential utility in the resolution of the new wave of air pollution problems. The second part will emphasise those aspects of the award that have proved of lasting value.

BILATERAL CHARACTER OF THE DISPUTE

On close analysis a number of unique features of the dispute emerge that were clearly bound to restrict *Trail Smelter's* wider application beyond the confines of that specific dispute.

First, the dispute was only marginally international; the pollutants in question crossed the U.S.-Canadian border by only a few miles. It is probably best characterised as a localized case of industrial nuisance,[12] which in other circumstances would hardly engage the legal departments of most states.[13] In the *Trail Smelter* dispute the source of harm and its recipients were clearly identifiable. It was incontrovertible that the source of harm was the Canadian company and, although the extent and scale of injury was controversial, there was no doubt at all that a causal link existed between the harm and the damage suffered by U.S. farmers.

Three questions were submitted to the tribunal, the first of which required it to determine: "whether damage caused by the Trail Smelter in the State of Washington has occurred since the first day of January 1932, and if so what indemnity should be paid therefore?"[14] Answering this question was arguably quite straightforward. The pollution was notorious, and in a previous claim referred to the International Joint Commission (IJC), the IJC reached the conclusion that the pollution from the smelter was responsible for damage in Washington state.[15] There also was evidence before the Tribunal that the damage caused by the emissions had been the subject matter of previous litigation in the local county courts, forcing the smelter owners to purchase smoke easements from local farmers. Moreover, as

[11] The footnote was inserted at the insistence of the United Kingdom which did not want duties pertaining to the exchange of information and consultation to be expressly linked to the idea of responsibility; UN Doc. ECE/HLM.1/2/Add 1, Annex IV (1979) at 104. I have argued elsewhere that a better interpretation of this provision is to see it as leaving questions of liability to be resolved by reference to the applicable standards of customary law rather than excluding responsibility altogether. The *Trail Smelter* experience in this context would remain of continuing relevance in the determination of state responsibility issues under the Convention. *See* OKOWA, *supra* note 9.

[12] *See* Allum in this volume.

[13] This is not the place to discuss in any detail why a problem of local nuisance effectively became an international dispute. For helpful accounts of this aspect of the dispute, *see* WIRTH, *supra* note 4. *See also* Knox in this volume.

[14] *Trail Smelter* (1941), *supra* note 1, at 686.

[15] *Id.* at 693–694; WIRTH, *supra* note 4.

John Wirth has eloquently demonstrated in a masterly survey of the history of the dispute, Trail was not the only smelter on the border between the two states; the United States was deeply aware that an unfavourable regulatory or compensatory regime could rebound and lead to similar disputes against U.S.–based industry.[16] There was concern in particular that residents of Ontario, Canada, could putatively challenge American industries in Detroit on the same terms. Thus, from the outset both governments were committed to a "negotiated settlement" that would cause minimal interference to industry.[17]

In the new problems of long-range transboundary pollution that emerged in the decades since the *Trail Smelter* dispute, a direct causal nexus of the kind present in *Trail Smelter* have been impossible to establish. The complex web of relationships, in which states affected were both sources and recipients of pollution, made it practically impossible to create a workable liability framework. Air pollution monitoring programs in Europe have long pointed out that the problem of transboundary air pollution was a regional one in which all European states were both exporters and importers of pollutants.[18] This was further compounded by evidence that the science was in any event far more complex than had been the case at the time of the *Trail Smelter* dispute.[19] A distinctive feature of long-range transboundary air pollution is the inability to trace the precise sources of pollutants that eventually cause damage. These pollutants undergo a number of complex atmospheric processes making it practically impossible, at least for the purpose of imposing liability, to identify the responsible state.[20] Furthermore, in a departure from the cooperative climate that prevailed in the *Trail Smelter* arbitration, it appears that the states affected by long-range transboundary air pollution do not want a litigated settlement creating normative precedents, which could form the basis of any future claims against them.

The general consensus thus appears to have moved away from state responsibility towards a regulatory model in which all affected states have undertaken a reduction in the amount of pollutants.

Second, the damage from the new wave of transboundary air pollution is attributable, in part, to pollution that is delayed, sometimes taking several decades before its effects are manifested. This is another feature of long–range transboundary air pollution that has made its resolution through the medium of

16 WIRTH, *supra* note 4, at xv, 1, 22, 89–90 and 92. 17 WIRTH, *supra* note 4.

18 OECD, The OECD Programme on Long Range Transport of Air Pollutants—Measurements and Findings, 3 (2nd ed. 1979); Sophia Mylona, Detection of Sulphur Emissions Reductions in Europe During the Period 1979–1986, EMEP/MSC-W Report 1/89 (1989); UN ECE, Effects and Control of Long-Range Transboundary Air Pollution, No. 10 (1994).

19 *See* Miller in this volume.

20 Royal Swedish Ministry of Agriculture, Acidification Today and Tomorrow, (B. Aniasson, ed.1982); *see also* Bert Bolin et al., Air Pollution Across Boundaries: The Impact on the Environment of Sulphur in Air and Precipitation, Sweden's Case Study for the United Nations Conference on the Human Environment (1971); IRENE H. VAN LIEV, ACID RAIN AND INTERNATIONAL LAW 1–10 (1981).

litigation particularly difficult. There is incontrovertible evidence that acidification of lakes and forests, although in part attributable to transboundary air pollution, also was a result of natural causes. A number of studies have in fact assigned a secondary role to acidic inputs in assessing the acidification of lakes.[21] Direct accountability in the face of these uncertainties is, to say the least, problematic.

RADIOACTIVE CONTAMINATION

In the *Trail Smelter* arbitration, the Tribunal was effectively called upon to strike a balance between the interests of what were arguably two equally viable economic activities. The balance between the interests of the Canadian smelter industry and the American farming community called for mutual toleration of a certain amount of inconvenience.[22] The risks of damage through pollution caused by the continuing operation of the smelter, although not insignificant, were generally perceived as benign, which could be easily accommodated in a satisfactory compensation regime. The Tribunal saw its task as one of arriving at a just solution;

> which would allow the continuance of the operation of the *Trail Smelter* but under such restrictions and limitations as would, as far as foreseeable, prevent damage in the United States, and as would enable indemnity to be obtained, if in spite of such restrictions and limitations, damage should occur in the future in the United States.[23]

Some of the new dangers attributed to aerial pollution after 1945 were, however, not so benign, and did not involve a balancing of equally important interests.

From the 1950s onward, right up to the 1980s, a significant problem of radioactive contamination caused by atmospheric nuclear tests became a major concern. These atmospheric nuclear tests were initially carried out by the United Kingdom and the United States, mostly in the South Pacific, without protest and with the cooperation of some states.[24] However, by the late 1950s, the dangers inherent in the conduct of such tests from the perspective of nuclear proliferation as well as environmental contamination had become apparent. France initially carried out tests in the Algerian Sahara and, from 1966 onwards, in the two uninhabitable islands of Mururoa Atoll and Fangataufa, both forming part of French Polynesia in the South Pacific. These tests were conducted in the face of protracted protests

[21] Frank A. Record et al., Acid Rain Information Book 16, 142 (1982).

[22] *See* Jacobson in this volume.

[23] *Trail Smelter* (1941), *supra* note 1, at 685.

[24] The United Kingdom had conducted such tests on the Australian territory with the cooperation of the Australian Government. *See* Keesings, Vol, XXXII 3471(1986).

by the affected states,[25] culminating in the cases brought by Australia and New Zealand in the 1973 *Nuclear Tests Cases*.[26]

Although such tests were conducted underground as well as in the atmosphere, the most significant amounts of radioactive contamination were attributed to atmospheric nuclear tests. A limited amount of local dispersion results from nuclear explosions, but the bulk of radioactive properties is injected into the upper troposphere and stratosphere, allowing their dispersion and subsequent deposition on to the earth's surface on a global scale.[27] Once in the stratosphere, the injected radioactive materials may constitute a reservoir from where they may contaminate the earth's atmosphere through fallout for many years.[28] Not only was there a marked sensitivity to this form of contamination, but it was also increasingly argued that that the resulting contamination was global and that all states had a legitimate interest in being protected from it. Arguably if the interests affected transcended narrow national concerns, it also followed that their resolution could not be achieved in a bilateral dispute settlement framework. Moreover, unlike the case of industrial pollution, many states took the view that there were no thresholds below which exposure to radiation could be deemed acceptable.[29] The tests in the view of these States were taken in pursuit of military and political aspirations of the participating states, with no compensating benefits for those states not involved in the activity.

[25] The dangers to humanity as a result of radioactive contamination caused by nuclear tests was noted in the following General Assembly Resolutions, all of which are available from http://www.un.org/documents/resga.htm: G.A. Res. 1379 (XIV), U.N. GAOR, 14th Sess., Supp. No. 65, at 6, U.N. Doc. 4280 (1959); G.A. Res.1629, U.N. GAOR, 16th Sess., Supp. No. 16A, at 241–42, U.N. Doc. A/4684/Add.1 (1961); G.A. Res. 2828, U.N. GAOR, 26th Sess., Supp. No. 29, at 33, U.N. Doc. A/8429 (1971); G.A. Res. 2934, U.N. GAOR, 27th Sess., (1972); G.A. Res. 3078, U.N. GAOR, 28th Sess., Supp. No. 30, at 17, U.N. Doc. A/9030 (1973); G.A. Res. 3257, U.N. GAOR, 29th Sess., at 20 (1974) *available at* http://daccessdds.un.org/doc/Resolution/ G.A. Res. 3460, 30th Sess., at 30 (1975) *available at* http://daccessdds.un.org/doc/Resolution/ On November 24, 1961, the General Assembly passed Resolution 1652 recommending that Africa should be a denuclearized zone. The resolution was adopted by 50 votes to none with 44 abstentions.

[26] Nuclear Tests (Aust. v. Fr.) 1974 I.C.J. 253 (Dec. 20); Nuclear Tests (N.Z. v. Fr.) 1974 I.C.J. 457 (Dec. 20); For a full record of New Zealand's opposition to French nuclear tests, *see* New Zealand Ministry of Foreign Affairs and Trade, New Zealand at the International Court of Justice: French Nuclear Testing in the Pacific (1996).

[27] Alexander Leaf, *Intermediate and Long-term Health Effects*, *in* WHO Report: Effects of Nuclear War on Health and Health Services, Annex 7 (2nd ed. 1987).

[28] For a summary of the effects of these tests on the global environment, *see* Report of the UNSCEAR, U.N. GAOR, 37th Session, Supp. No 45, at 6, U.N. Doc. A/37/45 (1982).

[29] Nuclear Tests, *supra* note 26, paras. 2–20 and para 49, and at 479–90 (argument of Mr. Byers). Like other substantive aspects of the application, the issue was never addressed by the Court on the merits. Instead, the Court found that the French unilateral undertaking to cease nuclear testing had rendered the claim obsolete. *Id.* In his dissenting opinion Judge Barwick was of the view that such tests were "intrinsically harmful and that their harmful effect is neither capable of being prevented nor indeed capable of being ascertained with any degree with any degree of certainty." *Id.* at 433.

Another new form of aerial contamination presented itself in the form of dangers arising out of nuclear reactor accidents. The benefits associated with nuclear energy resulted in the proliferation of nuclear reactors with very little specific attention being paid to the likely environmental consequences in the event of an accident.[30] The normal operation of nuclear reactors involves a number of steps from mining, preparation of fissionable material, and production of power to the disposal of radioactive wastes. In theory, there is a risk of radioactive contaminants escaping into the atmosphere at each of these stages. In practice, the normal operation of nuclear reactors raises very little concern of environmental contamination except for persons living within the immediate vicinity of nuclear power plants.[31] Consequently, as far as transboundary air pollution is concerned the main potential danger lies in the risk of accidental radioactive contamination.[32]

The Three-Mile Island accident in the United States and, more fundamentally, the accident at the Chernobyl nuclear plant in the Soviet Union, have demonstrated the very serious nature of the radioactive contamination that may occur in the event of a nuclear accident. The accident at Chernobyl caused a prolonged release of large quantities of radioactive material in the form of gases, halogens and particulate matter throughout the northern hemisphere but mainly across Europe. Radioactive contaminants were detected in Japan and countries as distant as the United States.[33]

Given the scale of affected states and the wide range of economic activities that were disrupted by the radioactive contamination, it is immediately apparent that a *Trail Smelter*-like adjudicatory framework was unlikely to be of much use in the resolution of any resulting disputes. Adjudication would have remained problematic even if claims were advanced not at the intergovernmental level (by way of diplomatic protection), but in form of private law actions by the affected farmers. As almost all European states and their farming sectors were extensively affected, discrete bilateral settlements would be not only cumbersome but also expensive and impractical. As all the states concerned were in any case unlikely to concede to arbitration before a common forum, it is clear that there was not much prospect for *ex post facto* accountability. Even assuming the existence of a forum, a further difficulty arises in the application of the relevant legal standards and principles. It would have been extremely difficult to establish a causal relationship between

30 Alan Boyle, *Nuclear Energy and International Law: An Environmental Perspective*, 60 British Yearbook of International Law 257 (1989).

31 C. Nick Hewitt, *Radioactivity in the Environment*, *in* Pollution, Causes, Effects and Control 343, 357–8 (Roy M. Harrison ed., 1988).

32 Ionizing Radiation: Sources and Biological Effects, Report of the UNSCEAR to the General Assembly, U.N. GAOR, 37th Session, Supp. No 45, U.N. Doc. A/37/45 (1982).

33 John H. Gittus et al., *The Chernobyl Accident and its Consequences*, UKAEA Report No. NOR 4200 (1987); OECD, The Radiological Impact of the Chernobyl Accident in OECD Countries, (1987). *See also* Report of the President's Commission on the Accident at Three Mile Island – The Need For Change: the Legacy of TMI (Oct. 1979).

exposure to radiation and any particular injury suffered by these states. This is due to the long interval of time that may elapse between irradiation and clinical manifestations, making it difficult to establish the connection between the effect and its cause. In relation to damage to human health, the United Nations Scientific Committee on the Effects of Atomic Radiation (UNSCEAR) has observed that the late effects are in fact usually indistinguishable from diseases induced by other causes and radiation only increases their incidence in the population.[34] Moreover, genetic damage may not be manifest for many generations. In any event, most of the immediate losses suffered by the farming industry as a result of the Chernobyl accident were a result of precautionary measures adopted by governments, the scientific basis of which was at best controversial.[35] Again, recovery for this type of harm in a liability framework would have been problematic.

In conclusion, it is apparent that for these new radiation-related dangers the *Trail Smelter* precedent, with its narrow definition of protected interests and heavy evidentiary burden expected of the plaintiff, was unlikely to be of much practical use in the resolution of the resulting conflicts.

PROCEDURAL DIFFICULTIES IN ADJUDICATING A MULTIPARTY PROBLEM

There are other features of the *Trail Smelter* arbitration, which made it of limited relevance to problems of transboundary air pollution that emerged in the years after the arbitral award. The dispute was between two states with a long history of cooperation in transboundary matters. In fact, as is apparent from the diplomatic history of the dispute, the parties had made good use of the dispute settlement provisions provided for in the IJC.[36] The new problems of transboundary air pollution do not, however, arise between similarly like-minded states with a history of bilateral cooperation in dispute settlement matters. Moreover, as is well known, dispute settlement in the international system remains distinctly consensual, and the treaties on long-range transboundary air pollution have stoically refrained from imposing any compulsory dispute settlement mechanisms.[37]

An additional, practical complication involved the issue of how to adjudicate and apportion responsibility between states in the position of multiple tortfeasors, in the absence of developed procedures for dealing with multiple party claimants. Issues of responsibility in international jurisprudence have largely been concerned with bilateral disputes between no more than two parties. This is apparent, for instance, from the wording of the Statute of the International Court of Justice

34 1982 UNSCEAR Report, *supra* note 32, at 11–15.

35 Philippe J. Sands, Chernobyl: Law and Communication 17 (1988).

36 Treaty Relating to the Boundary Waters and Questions Arising Along the Boundary Between the United States and Canada, Jan. 11, 1909, U.S.-Gr. Brit., 36 Stat. 2448.

37 *See, e.g.*, the 1979 ECE Convention and its 8 Protocols, *supra* note 8.

(ICJ), which only envisages disputes between two states.[38] Moreover, as the ICJ has pointed out in its recent jurisprudence, it lacks a procedure known to municipal systems whereby parties may be compulsorily joined to a dispute if they are implicated in the conduct in dispute.[39] Added to this is an almost complete absence of rules for the apportionment of responsibility or for compelling contribution should a multiplicity of states be found to have acted in concert. Furthermore, in the literature, there is hardly any discussion of the principles to be applied in those cases where impermissible harm is the result of delictual conduct on the part of a number of states.[40]

The problem is compounded by the fact that, in many cases, the victim of pollution is also invariably a contributor to the harm complained of. A further complication is the fact that, although the states whose delictual conduct is in question may have acted independently, the harm suffered is unitary. A state victim of pollution has the unenviable task of finding a jurisdictional basis for bringing a claim against each one of the parties. Failing that, it may only bring a claim against those states which have accepted the jurisdiction of the arbitral forum[41] Serial litigation in this form is not only cumbersome for the parties, it also unnecessarily burdens the adjudicating body, which, in turn, may lead to conflicting judgments on the same facts. The adjudicator may, in any case, decline jurisdiction under the indispensable parties rule, if it concludes that the dispute relates to the interests of a number of parties not before it and cannot therefore be determined without adjudicating on the rights and obligations of third states.[42]

What emerges from this brief account is that long-range transboundary air pollution disputes raise large questions of public order, which cannot be satisfactorily resolved through adjudication, at least in the international legal order as it presently stands. The *Trail Smelter* precedent will remain

[38] Statute of the International Court of Justice, June 25, 1945, arts. 43, 44, 59 Stat. 1055, 33 U.N.T.S. 993.

[39] *See* Concerning Continental Shelf (Libyan Arab Jamahiriya/Malta),Italy's Application for Permission to Intervene, Judgment, 1984 I.C.J. 25, para. 40 (March 28); see also Concerning Military and Paramilitary Activities in and Against Nicarauga (Nicar. v. U.S.), Jurisdiction and Admissibility of Judgment, 1984 I.C.J. 431, para 88 (Nov. 26); Land, Island and Maritime Frontier Dispute (El Salvador/Honduras), Petition of Nicaragua to Intervene, Judgment, 1990 I.C.J. 135 (September 13); Case Concerning Certain Phospate Lands in Nauru (Nauru v. Aus.) (Preliminary Objections), 1992 I.C.J. 260 (June 26).

[40] Of the general texts only the following advert to the problem under discussion, *see* CHRISTINE CHINKIN, THIRD PARTIES IN INTERNATIONAL LAW 15 (1993); IAN BROWNLIE, STATE RESPONSIBILITY – PART I 198 (1983); *See also* LAMMERS, *supra* note 3, at 603–61.

[41] The problems raised by multiple party cases in the specific context of air pollution are discussed by the INTERNATIONAL LAW ASSOCIATION, THIRD REPORT ON LEGAL ASPECTS OF LONG-DISTANCE AIR POLLUTION – 61ST CONFERENCE 294, 297 (1990).

[42] Monetary Gold Removed from Rome in 1943 (Ital. v. Fr.), 1954 I.C.J. 19, 32 (June 15); Chinkin, *supra* note 40, at 199.

relevant in small transfrontier claims where private law actions of a civil nature would, in any case, be more suitable than diplomatic protection at the interstate level.

RECOVERABLE CATEGORIES OF LOSS

The approach of the *Trail Smelter* Tribunal as to what items of loss merited indemnification has been roundly criticised as too narrow even for the time.[43] The Tribunal emphasized that the injury had to be "serious" and be established by "clear and convincing evidence."[44] The catastrophic but subtle effects on the environment not computable in property terms received scant attention in the Tribunal's final judgment. In the years that followed, there has been a broad consensus on the need for caution if irreversible environmental damage is to be prevented and this may require action in the face of scientific uncertainty.[45] However, the problem of quantifying what amounts to detrimental alteration of environmental quality and the mechanisms for its evaluation has remained problematic.[46] It is nevertheless unlikely that a tribunal faced with the same set of facts would impose such a heavy evidentiary burden on a litigant, or be so dismissive of invisible environmental damage. In the *Gabčikovo-Nagymaros* case, Hungary argued that the application of the principle of precaution required states to take measures to prevent pollution even if the deleterious effects of such pollution remain unproven, and that the burden is on the state indulging in the polluting activity to prove that its actions are in fact harmless.[47]

Although the *Trail Smelter* Tribunal's outlook on compensable items of loss was conservative, one of the more lasting contributions of the *Trail Smelter* litigation is the effort undertaken by the Tribunal, in spite of intractable difficulty, to determine what kinds of property-related harm should trigger state responsibility as well as the methods for assessing such harm. The Tribunal's approach in defining acceptable categories of loss has generally been accepted as a reliable guide as to what categories of property-related loss are compensable in a liability framework.[48] The categories of harm claimed before the Tribunal included

43 *See generally* WIRTH, *supra* note 4, at 1–2; McCaffrey in this volume.

44 *Trail Smelter* (1941), *supra* note1, at 716.

45 *See* Bratspies and McCaffrey in this volume.

46 *See, e.g.*, Francisco Orrego Vicuña, *State Responsibility, Liability and Remedial Measures under International Law: New Criteria for Environmental Protection*, *in* ENVIRONMENTAL CHANGE AND INTERNATIONAL LAW 124–28 (Edith Brown Weiss ed., 1992); Gabčikovo-Nagymaros (Hungary v. Slovakia), 1997 I.C.J. 7 (Sept. 25).

47 Gabčikovo-Nagymaros, *supra* note 46.

48 *See, e.g.*, IAEA Protocol to Amend the 1963 Vienna Convention on Civil Liability for Nuclear Damage, September 27, 1997, IAEA Doc. INFCIRC/566, *reprinted in* 36 INTERNATIONAL LEGAL MATERIALS 1454 (1997); IAEA Convention on Supplementary Compensation for Nuclear Damage, September 27, 1997, IAEA Doc. INFCIRC/567 *reprinted in* 36 INTERNATIONAL LEGAL MATERIALS 1473 (1997).

damage to persons and property. The Tribunal accepted as proved the existence of a causal relationship between the sulfur emissions and the reduction in crop yield, as well as damage to timber stands. The claims of the United States that the sulfur emissions had resulted in the impairment of the soil contents through increased acidity, reduction of livestock, milk and wool productivity were rejected as having an insufficient causal connection with the sulfur fumigations. The Tribunal also rejected claims by the United States, which, although causally linked to the sulfur fumigations, were nevertheless too remote a consequence. The United States had claimed damages for "loss of business and impairment of the value of good will because of the reduced economic status of the residents of the damaged area."[49]

It is also important to note that the Tribunal did not insist on absolute certainty of proof, but was instead prepared to draw liberal inferences from the available facts.[50] It quoted with approval the following passage from the decision of the U.S. Supreme Court in *Story Parchment Company v. Paterson Parchment Paper Company*, where the Court had observed that: "Where the tort itself is of such a nature as to preclude the ascertainment of the amount of damages with certainty. . . . it will be enough if the evidence shows the extent of the damages as a matter of just and reasonable inference, although the result be only proximate."[51]

The Tribunal accepted that, in principle, precautionary measures and cleanup costs also were recoverable items of loss. It observed that expenses incurred in mending the consequences of an injury were recoverable as an appropriate item of damage.[52] It nevertheless rejected the U.S. claim for recovery of monies spent on investigation, preparation, and proof of its case. The Tribunal observed that these expenses were incidental to the normal conduct of litigation and could not therefore be treated as separate items of loss.[53]

Although the treaties on transboundary air pollution have eschewed any direct consideration of recoverable items of loss, the *Trail Smelter* approach is bound to be invaluable in any bilateral resolution of a transfrontier air pollution problem. The Nuclear Civil Liability Conventions have broadly endorsed a similar approach and the Tribunal's jurisprudence is likely to prove invaluable in that context as well.[54]

49 *See Trail Smelter Arbitral Decision*, 33 AMERICAN JOURNAL OF INTERNATIONAL LAW 182 (1939) [HEREINAFTER "*Trail Smelter* (1939)"], at 206.

50 This approach of the Tribunal to evidentiary matters was endorsed without qualification by the INTERNATIONAL LAW ASSOCIATION, *supra* note 41.

51 *Trail Smelter* (1939), *supra* note 49, at 193.

52 *See* Craik in this volume.

53 *Trail Smelter* (1941), *supra* note 1, at 709–712.

54 *See, e.g.*, Vienna Convention on Civil Liability for Nuclear Damage, May 21, 1963, 1063 U.N.T.S. 265, 2 I.L.M. 727; Paris Convention on Third Party Liability in the Field of Nuclear Energy, July 29, 1960, 956 U.N.T.S. 251; Alan E. Boyle, *Nuclear Energy and International Law: An Environmental Perspective*, 1989 BRITISH YEARBOOK OF INTERNATIONAL LAW 257, 276–77.

A REGULATORY REGIME?

In much of the literature, the *Trail Smelter* decision has been derided as providing for no more than a regime of rudimentary *ex post facto* accountability.[55] For this reason, it is widely assumed that the decision provides an inadequate framework for protecting the environment from incidents that ought not to happen in the first place. The fact that the Tribunal also established a regulatory regime of prevention, which went hand in hand with the imposition of compensation, is frequently lost in the many commentaries on the case. In addition to asking the Tribunal to establish whether damage had been caused in the state of Washington and to what extent, the two subsequent questions in the *Compromis* were:

> 2. In the event of the answer to the first part of the preceding Question being in the affirmative, whether the Trail Smelter should be required to refrain from causing damage in the state of Washington in the future and if so to what extent?
> 3. In the light of the answer to the preceding Question, what measures or regime, if any should be adopted or maintained by the Trail Smelter?[56]

In answering Questions 2 and 3 in its final judgment the Tribunal declared that:

> (2) So long as the present conditions in the Columbia River Valley prevail, the Trail Smelter shall be required to refrain from causing any damage through fumes in the State of Washington.[57]

In answering the third question, the Tribunal imposed a technical regime designed to control the emission of sulfur dioxide fumes. The Tribunal anticipated the possibility that it might prove inadequate, and as a supplementary requirement prescribed a regime of amendment to be supervised by a commission of scientists. The regulatory regime, which included fume abatement at the stack and acid recovery measures, substantially reduced the cross-border pollution. In light of the now prevailing scientific evidence regarding dispersal of long-range air pollutants, the regulatory regime prescribed by the Tribunal in 1938 and 1941 would, however, be regarded as environmentally unsound today. The point of principle is nevertheless commendable: industry was expected to adopt economically feasible measures, which, in light of the scientific evidence before the Tribunal, would substantially reduce the emissions.[58]

55 Peter H. Sand, Transnational Environmental Law: Lessons in Global Change 97–98 (1999); McCaffrey in this volume. For a more general criticism of the *ex post facto* character of the liability approach, *see* Andrew Hurrell and Benedict Kingsbury, International Politics of the Environment 26 (1992); Patricia Birnie and Alan E. Boyle, International Law and the Environment 111 (2002). For a more balanced account of the award, *see* Cesare P.R. Romano, The Peaceful Settlement of International Environmental Disputes 261–278 (2000).

56 *Trail Smelter* (1941), *supra* note 49, at 183.

57 *Trail Smelter* (1941), *supra* note 1, at 717.

58 *See* Bratspies in this volume.

The regulatory model has proved particularly favorable in the network of framework conventions regulating transboundary air pollution.[59] In each case, the parties are required to set in place a regulatory structure designed to reduce the amount of pollutants. The logic is apparent; it is far more sensible to prevent the harm from occurring than to offer monetary compensation afterward. The structures in place in these treaties are, however, substantially weaker than those put in place by the *Trail Smelter* Tribunal. There are no provisions for mandatory external review of the regulatory schemes by teams of technical experts should they prove inadequate.

For instance, in the 1979 ECE Convention for Long-range Transboundary Air Pollution,[60] the parties are required to do no more than endeavor to limit and to gradually reduce the amount of air pollution emanating from their territories. The regulatory framework of the Convention is to be found in three key provisions. Article 2 provides that:

> The contracting parties, taking due account of the facts and the problems involved are determined to protect man and his environment against air pollution and shall endeavour to limit, and, as far as possible, gradually reduce and prevent air pollution including long range transboundary air pollution.[61]

This provision is supplemented by Article 6, which requires the parties to adopt emission control policies and strategies to the extent that they are economically feasible having regard to the best available technology. It provides that:

> Taking into account Articles 2 to 5, the on-going research, exchange of information and monitoring and the results thereof, the cost and effectiveness of local and other remedies and, in order to combat air pollution, in particular that originating from new or rebuilt installations, each contracting party undertakes to develop the best policies and strategies including air quality management systems and, as part of them, control measures compatible with balanced development, in particular by using the best available technology which is economically feasible and low- and non-waste technology.[62]

In North America, the 1991 Air Quality Agreement between Canada and the United States laid down specific targets and schedules to be observed by the two parties as part of the framework for regulating transboundary air pollution. The key provision in Article II (a) requires the parties "to . . . establish specific objectives for emission limitations or reductions of air pollutants and adopt the necessary programs and other measures to implement such specific objectives."[63]

59 *See* Miller in this volume.

60 ECE Convention on Long-Range Transboundary Air Pollution, *supra* note 8.

61 *Id.* at art. 2.

62 *Id.* at art. 6.

63 United States-Canada Air Quality Agreement, *supra* note 8. For a very useful commentary on the treaty, *see* Jeffrey L. Roelofs, "*United States-Canada Air Quality Agreement: A Framework for Addressing Transboundary Air Pollution Problems*, 26 CORNELL INTERNATIONAL LAW JOURNAL 421 (1993).

Article IV reiterates this commitment, and provides that the specific objectives to be achieved by the parties are to be set in an Annex to the Convention. The detailed goals to be realized by the parties are also set in Annexes to the Convention and relate principally to the reduction of sulphur and nitrogen oxides.

The subsequent Protocols on Sulfur and Nitrogen emissions imposed more concrete obligations on the parties. The Protocols were an attempt to lay down concrete emission targets based on sound scientific principles. However, like in the main Framework Convention, the parties are left a wide margin of appreciation in deciding on the exact methods to be employed in complying with the obligation. The parties, like in the Framework Convention itself, eschewed any commitment to a mandatory review process. Again, analogies with the review process prescribed by the Tribunal in the *Trail Smelter* arbitration may be somewhat misplaced. To require a technical committee to review compliance with regulatory regimes, which relate to the conduct of a wide range of industrial sectors in a multiplicity of states, is financially and logistically unrealistic.

CONCLUSION

On close analysis it appears that much of the negative commentary on the *Trail Smelter* arbitration is misplaced. Viewed in its context, in light of both the science of the day and the declared wishes of the states involved, the decisions were both innovative and forward-looking. The Tribunal did not, and could not realistically be expected to, anticipate the radical shift in the science of air pollution and the dramatic sensitivity to the importance of environmental values that crystallized in the period after the 1972 Stockholm Conference. That the Tribunal paid scant attention to invisible environmental damage seems glaringly myopic today, but even the most forward-looking scientists would not have been unduly disturbed by this decision at the time. The problems that emerged in the field of aerial contamination from 1945 onward were so radically different that it is unrealistic to expect the modest jurisprudence of a localized air pollution problem to provide a solution. Confined to the narrow range of problems it was designed to address, the decision remains relevant. With modification, it might even be applied to solve similar local problems of industrial pollution today.

17 The Impact of the *Trail Smelter* Arbitration on the Law of the Sea

Stuart B. Kaye

INTRODUCTION

The 1941 *Trail Smelter* arbitral decision[1] is often described as a landmark case in the development of international environmental law.[2] Although not a case dealing with the marine environment, the same claim has been advanced in the context of the law of the sea, with the case exerting a significant impact on aspects of the development of the Law of the Sea Convention,[3] albeit indirectly. *Trail Smelter* certainly informed the deliberations of delegates to the Third United Nations Conference on the Law of the Sea in the 1970s and 1980s, as well as a number of other diplomatic endeavors aimed at providing greater protection to the marine environment. Some of the principles underlying the provisions in the Convention, particularly with respect to marine pollution, clearly have their antecedents in the 1941 *Trail Smelter* decision. By contrast, there is a dearth of references to *Trail Smelter* in international cases on the law of the sea, and the development of marine environmental law seems to have occurred without direct reliance on the decision. Indeed, an argument could be made that shipping disasters such as the *Torrey Canyon*,[4] *Amoco Cadiz*,[5]

[1] *Trail Smelter Arbitral Decision*, 35 American Journal of International Law 684 (1941) [hereinafter "*Trail Smelter* (1941)"]. *See* Annex to this volume.

[2] Handl described the *Trail Smelter* arbitration as the "*locus classicus* of international legal principles on transnational pollution." Günther Handl, *Territorial Sovereignty and the Problem of Transnational Pollution*, 69 American Journal of International Law 50, 60 (1975).

[3] United Nations Convention on the Law of the Sea, Dec. 10, 1982, 1833 U.N.T.S. 397 [hereinafter LOSC].

[4] The *Torrey Canyon* foundered on Seven Stones Reef near the Scilly Isles in March 1967, losing 860,000 barrels of crude oil into the sea. It is regarded by some as the first large oil tanker environmental disaster. *See* NOAA, *Oil Spill Case Histories 1967–1991: Summaries of Significant US and International Spills*, HMRAD Report 92–11 (1992), *available at* http://response.restoration.noaa.gov/oilaids/spilldb.pdf [hereinafter "NOAA Report"].

[5] The *Amoco Cadiz* ran aground off the coast of Brittany in March 1978 spilling 1.6 million barrels of oil. *See* NOAA Report, *supra* note 4.

and *Exxon Valdez*,[6] as well as the emergence of the precautionary principle,[7] have had a greater impact on the development of international marine environmental law than the 1941 *Trail Smelter* decision. This chapter will explore the relationship between the 1941 *Trail Smelter* decision and the Law of the Sea Convention and evaluate the influence it has exerted.

DEVELOPMENT OF INTERNATIONAL MARINE ENVIRONMENTAL LAW

After tentative steps before World War II, the first substantial attempt by the international community to address marine pollution was OILPOL 54,[8] which was a limited attempt at preventing discharges of oil from ships in certain circumstances, and other measures to limit oil discharges. Although groundbreaking, OILPOL 54 was criticized as too limited in its effect and practically unenforceable.[9] Further measures followed in the 1960s,[10] culminating in MARPOL 73/78,[11] which established a wide-ranging regime for dealing with pollution of the marine environment from shipping.[12] MARPOL's structure reflected the incremental nature of its approach to marine environmental protection, with many of its substantive provisions contained in a series of annexes. Each annex covered a distinct area of marine pollution from ships, and could be implemented by State parties independently of the other

6 The *Exxon Valdez* ran aground on Bligh Reef in King William Sound, Alaska, in March 1989, spilling approximately 240,000 barrels of oil. *See* NOAA Report, *supra* note 4.

7 For a discussion of the relationship between *Trail Smelter* and the precautionary principle, *see* Bratspies in this volume.

8 International Convention for the Prevention of the Pollution of the Sea by Oil, May 12, 1954, 327 U.N.T.S. 3 [hereinafter "OILPOL 54"].

9 *See* Edgar Gold, *Pollution of the Sea in International Law: A Canadian Perspective*, 3 JOURNAL OF MARITIME LAW & COMMERCE 13 (1972); Norman A. Wulf, *Contiguous Zones for Pollution Control*, 3 JOURNAL OF MARITIME LAW & COMMERCE 537, 541 (1972); Emeka Duruigbo, *Reforming the International Law and Policy on Marine Pollution*, 31 JOURNAL OF MARITIME LAW & COMMERCE 65, 69 (2000).

10 In addition to amendments to OILPOL 54, two other conventions were concluded in the 1960s: International Convention Relating to Intervention on the High Seas in Cases of Oil Pollution Casualties, Nov. 29, 1969, 26 U.S.T. 765, 970 U.N.T.S. 211; International Convention on Civil Liability for Oil Pollution Damage, Nov. 29, 1969, 973 U.N.T.S. 3 [hereinafter "Civil Liability Convention"].

11 Protocol of 1978 Relating to the International Convention for the Prevention of Pollution from Ships, 1973, Feb. 17, 1978, 1340 U.N.T.S. 61 [hereinafter "MARPOL"].

12 The *Torrey Canyon* disaster was a significant impetus for the reform of existing international law with respect to pollution from ships at sea. *See* Alan E. Boyle, *Marine Pollution under the Law of the Sea Convention* 79 AMERICAN JOURNAL OF INTERNATIONAL LAW 347, 349 (1985).

annexes.[13] In addition, the London Convention[14] was negotiated as an attempt to regulate the dumping of wastes and pollutants into the world's oceans.[15]

In recent times, more progress has been made in marine environmental law at a regional level than a global level A number of regional marine pollution agreements have been concluded, most notably in European waters although by no means restricted to that part of the world.[16] Some additional wide-ranging

[13] The MARPOL Annexes are:

Annex I: Prevention of pollution by oil

Annex II: Control of pollution by noxious liquid substances

Annex III: Prevention of pollution by harmful substances in packaged form

Annex IV: Prevention of pollution by sewage from ships

Annex V: Prevention of pollution from garbage from ships

Annex VI: Prevention of air pollution from ships

[14] Convention on the Prevention of Marine Pollution by Dumping of Wastes and other Matter, Dec. 29, 1972, 26 U.S.T. 2403, 1046 U.N.T.S. 137 [hereinafter "London Convention"].

[15] The London Convention was substantially overhauled in 1996. 1996 Protocol to the Convention on the Prevention of Marine Pollution by Dumping of Wastes and other Matter, Nov. 7, 1996, *reprinted in* 36 INTERNATIONAL LEGAL MATERIALS 1 (1997). One major change was the shift from a black list of materials the dumping of which was prohibited, to a white list of substances approved for ocean dumping. This shift reflects a substantial change in approach, bringing the London Convention more in accord with developments at the United Nations Conference on the Environment and Development in Rio in 1992, and with a precautionary approach, as indicated at Article 3(1). For a discussion of *Trail Smelter* and the precautionary principle, *see* Bratspies in this volume.

[16] *See, e.g.*, Convention on the Protection of the Marine Environment of the North-East Atlantic, Sep. 22, 1992, *reprinted in* 32 INTERNATIONAL LEGAL MATERIALS 1069 (1993) (combining the 1974 Paris Commission on Land-Based Sources of Marine Pollution and the 1972 Oslo Commission on Dumping Wastes at Sea); Convention on the Protection of the Marine Environment of the Baltic Sea Area, Mar. 22, 1974, 1507 U.N.T.S. 167; Convention for the Protection of the Mediterranean Sea against Pollution, Feb. 16, 1976, ECE Decision 77/585 (OJ L 240, 1) *reprinted in* 15 INTERNATIONAL LEGAL MATERIALS 290 (1976); Convention for the Protection and Development of the Marine Environment of the Wider Caribbean Region, Mar. 24, 1983, T.I.A.S. No. 110,85, 1506 U.N.T.S. 157; Convention for the Protection of the Marine Environment and the Coastal Area of the South-East Pacific, Nov. 12, 1981, U.N. Doc. UNEP-CPPS/I.G. 32/4 (1981); Convention for the Conservation of the Red Sea and Gulf of Aden Environment, Feb. 14, 1982, *reprinted in* 9 ENVIRONMENTAL POLICY & LAW 56 (1982); Convention for the Co-operation in the Protection and Development of the Marine and Coastal Environment of the West and Central African Region, Mar. 23, 1981, *reprinted in* 20 INTERNATIONAL LEGAL MATERIALS 746 (1981); Convention on the Protection of the Black Sea against Pollution, Apr. 21, 1992, *reprinted in* 32 INTERNATIONAL LEGAL MATERIALS 110 (1993); Convention for the Protection, Management and Development of the Marine and Coastal Environment of the Eastern African Region, Jun. 21 1985, *available at* http://www.unep.ch/seas/main/eaf/eafconv.html; Convention for the Protection of the Natural Resources and the Environment of the South Pacific Region, Nov. 24, 1986, 1990 Austl.T.S. No. 31, *reprinted in* 26 INTERNATIONAL LEGAL MATERIALS 38 (1987); Agreement Establishing the South Pacific Regional Environment Programme, Jun. 16, 1993, 1995 Austl. T.S. No. 24; Kuwait Regional Convention for Co-operation on the Protection of the Marine Environment from Pollution, Jun. 24, 1978, 1140 U.N.T.S.

international pollution control measures have been negotiated, including agreements dealing with the transboundary movement of hazardous waste,[17] the use of antifouling paints,[18] oil spill preparedness,[19] and most recently ballast water management.[20]

All these measures exist within the framework of the Law of the Sea Convention. Some, such as MARPOL, existed before the Law of the Sea Convention was concluded. The Convention recognised that substantial progress had already been made in some areas, and it was not the intention of delegates to reprise or replace those provisions. Rather, the Law of the Sea Convention was to act as a unifying framework, with a series of more specific agreements dealing with discreet areas.

The Law of the Sea Convention deals with environmental matters principally in Part XII. It imposes a general duty on states to protect and preserve the marine environment,[21] and then expands on this by indicating some additional measures that are within the competency of a state. These include: the instituting of measures within their capabilities to prevent, reduce, and control marine pollution; a duty not to transfer or transform marine pollution from one area to another: and a duty to cooperate with other states to deal with marine pollution.[22] Those duties relevant to the 1941 *Trail Smelter* decision will be considered further later.

The remainder of Part XII links the Law of the Sea Convention to more specialised international regulatory instruments in order to cover all types of pollution that can affect the sea including: land-based sources of marine pollution;[23] pollution from the seabed subject to national jurisdiction;[24] pollution from activities in the Area;[25] pollution by dumping;[26] pollution from vessels;[27] and pollution through the atmosphere.[28] Although pollution from ships and from ocean dumping is the subject of wide-ranging protection regimes under MARPOL and under the London Convention, the other areas lack analogous overarching international

133; Convention for the Cooperation in the Protection and Sustainable Development of the Marine and Coastal Environment of the Northeast Pacific, Feb. 18, 2002, *available at* http://www.cep.unep.org/services/nepregseas/Convention_English_NEP.doc.

17 Basel Convention on the Control of the Transboundary Movements of Hazardous Wastes and their Disposal, Mar. 22, 1989, 1673 U.N.T.S. 125. *See* Parish in this volume.

18 IMO International Convention on the Control of Harmful Anti-fouling Systems on Ships, Oct. 5, 2001, *available at* http://www.austlii.edu.au/au/other/dfat/treaties/notinforce/2002/18.html.

19 IMO International Convention on Oil Pollution Preparedness, Response and Cooperation, Nov. 30, 1990, *reprinted in* 30 INTERNATIONAL LEGAL MATERIALS 773 (1991).

20 International Convention for the Control and Management of Ships' Ballast Waters and Sediments, Feb. 13, 2004, *available at* http://globallast.imo.org/mepc.htm.

21 LOSC, *supra* note 3, art. 192.

22 *Id.* sec. 1 and 2, Part XII.

23 *Id.* art. 207.

24 *Id.* art. 208.

25 *Id.* art. 209.

26 *Id.* art. 210.

27 *Id.* art. 211.

28 *Id.* art. 212.

agreements. The Law of the Sea measures themselves provide little detail beyond a requirement that states legislate to implement protection of the marine environment, and cooperate with an appropriate international organization[29] or diplomatic effort.

The Law of the Sea Convention also deals with liability under Part XII. Article 235 provides that States are responsible for fulfilling their obligations to preserve and protect the marine environment, and can be liable in accordance with international law.[30] Article 235 recognized that further development of this area of state liability was ongoing in 1982,[31] and the same observation would be as accurate almost twenty-five years later. The fact that after decades of negotiation, the Articles on Responsibility of States for Internationally Wrongful Acts prepared by the International Law Commission are still some way from being incorporated into a binding international instrument indicates that issues surrounding state liability are still contentious.[32]

In a development parallel to the ongoing discussions about state liability, a number of specific liability conventions for certain types of pollution have been negotiated. These conventions are limited to pollution from ships, particularly, although not exclusively, in the context of oil pollution.[33] These instruments have dealt with mechanisms to provide for shipowner liability, and the adequacy of insurance coverage to deal with marine pollution from ships.[34] These mechanisms have been favoured over regimes to determine state liability, and are designed to provide some insurance coverage to deal with the consequences of damage from marine pollution.

[29] For example, in the context of pollution from ships, the International Maritime Organization would be the appropriate international body, as it administers MARPOL.

[30] LOSC, *supra* note 3, art. 235(1).

[31] *Id.* art. 235(3).

[32] *See* Draft Articles on Responsibility of States for Internationally Wrongful Acts, in Report of the International Law Commission on the Work of Its Fifty-third Session, UN GAOR, 56th Sess., Supp. No. 10, UN Doc. A/56/10 (2001). *See* Drumbl in this volume.

[33] *See* Civil Liability Convention, *supra* note 10; Protocol to Amend the International Convention on Civil Liability for Oil Pollution Damage, May 30, 1992, IMO Doc. LEG/CONF. 9/15, Dec. 2, 1992, *available at* http://www.jus.uio.no/lm/imo.civil.liability.oil.pollution.damage.protocol.1992/certificate; International Convention on Civil Liability for Bunker Oil Pollution Damage, Mar. 23, 2001, *available at* http://www.imo.org/Conventions/mainframe.asp?topic_id=256&doc_id=666 [hereinafter "Bunkers Convention"]; International Convention on Liability and Compensation for Damage in Connection with the Carriage of Hazardous and Noxious Substances by Sea, May 3, 1996, *available at* http://www.jus.uio.no/lm/imo.carriage.by.sea.liability.compensation.damage.connected.to.hazardous.and.noxious.substances.convention.1996/index [hereinafter "HNS Convention"].

[34] For example, *see* Bunkers Convention, *supra* note 33, art.3, and HNS Convention, *supra* note 33, chap. II and Civil Liability Convention, *supra* note 10, art.3. *See generally*, Edgar Gold, *Liability and Compensation for Ship-Source Marine Pollution: The International System*, 1999 YEARBOOK OF INTERNATIONAL COOPERATION ON ENVIRONMENT AND DEVELOPMENT 31, 31–37. *See also* Drumbl in this volume.

MARINE ENVIRONMENTAL PROTECTION AND THE *TRAIL SMELTER* ARBITRATION

The first necessary task in exploring the links between the *Trail Smelter* arbitration and the Law of the Sea Convention is to identify the key principles arising out of the arbitration. While this has been explored in other chapters in this volume, for the purposes of this analysis, the core principle deriving from the case is neatly summarized in the oft-cited passage from the 1941 judgment:

> . . . under principles of international law, as well as the law of the United States, no State has the right to use or permit the use of its territory in such a manner as to cause injury by fumes in or to the territory of another or the properties or persons therein, when the case is of serious consequence and the injury is established by clear and convincing evidence.[35]

Essentially, liability flows where one state permits an activity that causes harm that extends beyond its territory into the territory of another state. The principle has been described as lying at the heart of international environmental law,[36] creating what amounts to an international law equivalent of the Latin maxim *sic utere tuo* or not allowing activities on your property to harm another's property.[37]

The first observation that can be made on the *Trail Smelter* arbitration is that the case had very little to do with marine pollution. Although the smelter itself was located in the valley of the Columbia River, the focus of the case was the atmospheric pollution that the smelter caused, not on degradation of the river itself.[38] The physical location of the smelter was well inland, and marine areas were unaffected. Even if contamination of the river had been a significant concern, the Law of the Sea Convention clearly distinguishes rivers from ocean areas, and does not deal with the former.[39]

As much was acknowledged by the Tribunal in the 1941 *Trail Smelter* decision itself, which noted:

> No case of air pollution dealt with by an international tribunal has been brought to the attention of the Tribunal nor does the Tribunal know of any such case. The nearest analogy is that of water pollution. But here also, no decision of an international tribunal has been cited or has been found.[40]

35 *Trail Smelter* (1941), *supra* note 1, at 716.

36 *See* Alfred P. Rubin, *Pollution by Analogy: The Trail Smelter Arbitration*, 50 Oregon Law Review 259 (1971). Republished in this volume.

37 Karin Mickelson, *Rereading Trail Smelter*, 31 Canadian Yearbook of International Law 219, 220 (1993). Republished in this volume.

38 For a discussion of the current water pollution dispute surrounding the Trail smelter, *see* Craik in this volume.

39 The Law of the Sea Convention only deals with rivers in the context of territorial sea baselines, confirming that rivers are entirely under the sovereignty of the coastal State in the absence of any other arrangement.

40 *Trail Smelter* (1941), *supra* note 1, at 714.

The principle that was ultimately used was not, therefore, drawn from a marine context, although that omission should not, of itself, prevent the application of the principle more widely. Nevertheless, it is necessary to consider what issues might arise in removing the arbitration from its wholly terrestrial context.

Because the 1941 *Trail Smelter* decision was not concerned with the marine environment, the Tribunal did not have to deal with the same range of jurisdictional issues that a case concerning ocean areas would inevitably have raised. The pollution produced at the Trail smelter clearly emanated from a factory located in an area under Canadian sovereignty. Similarly, the pollution that was produced crossed the 49th parallel and adversely affected lands under the sovereignty of the United States.

An equivalent case dealing with damage to ocean areas would have to contend with a lesser degree of clarity on the issue of sovereign responsibility. Only the territorial sea, internal waters, and, if applicable, archipelagic waters, are part of the sovereignty of a coastal state. A coastal state has sovereign rights within its Exclusive Economic Zone (EEZ) only over resources within the EEZ, not full sovereignty over the EEZ itself.[41] Whether or not this point of difference is significant enough to justify distinguishing the principles derived from the 1941 *Trail Smelter* decision alone, it nevertheless remains a point of difference.

More significantly, the jurisdiction possessed by a coastal state over its adjacent waters is different from the control it exercises over land. States must permit the vessels of other states to pass through their territorial sea, archipelagic waters, and EEZ, and cannot generally exclude rights of passage from these areas.[42] Canada was responsible for the Trail smelter's pollution because the smelter was contained in its territory, as domestic law could have been applied to restrict the operation of the smelter. Such an option would not necessarily be available with respect to harm emanating from ocean areas. Although the territorial sea is within the sovereignty of a coastal state, it is not always able to exert direct control over the activities taking place there. For example, a ship polluting the area around it with toxic emissions might cause substantial harm to areas in its vicinity. The ship might be located within the territorial sea of one state and the pollution from it might drift over the territorial sea and land territory of another state. Applying *Trail Smelter*, there would be liability attaching to the state in whose territorial sea the offending ship was located. Such a result would not be consistent with contemporary international law, where liability for the pollution would attach to the ship, rather than the coastal state.[43] The coastal state would

[41] LOSC, *supra* note 3, art. 54.

[42] Only a temporary suspension of the right of innocent passage in the territorial sea or archipelagic waters from essential defense purposes is permissible. Such a closure must be only temporary and applied in a nondiscriminatory fashion. *See* LOSC, *supra* note 3, art. 25(3).

[43] There is a positive obligation on the flag state to legislate with respect to marine pollution from their ships: LOSC, *supra* note 3, art. 211(2).

have jurisdiction to deal with the pollution, but is not under an obligation to do so.[44]

As can be seen, there is no positive obligation on a coastal state to exercise its environmental jurisdiction in its adjacent waters, although it possesses the ability to do so. If it chooses to exercise such jurisdiction, it is obliged not to hamper freedom of navigation. It is therefore very clear that the coastal state does not have the same degree of control over its adjacent waters as its land territory, where it clearly could exert unfettered control over activities taking place.

TRAIL SMELTER WITHIN THE LAW OF THE SEA

One key impact on the Law of the Sea Convention possibly attributable to the *Trail Smelter* arbitration is in the context of the environmental jurisdiction of the coastal state in the exclusive economic zone (EEZ). This is found in Article 56 of the Convention, which deals in part with the marine environment. Article 56(1)(b)(iii) gives the littoral state jurisdiction over marine environmental protection to a distance of at least two hundred nautical miles from the coast, as well as over living and non-living resources and marine scientific research.[45]

The motivation for this provision has more to do with its other elements than with environmental protection. The utility and value of sovereign rights over marine living resources would be substantially diminished if pollution of the EEZ by other states could take place without restriction. States dumping pollution could damage or destroy fish stocks in parts of the EEZ, rendering jurisdiction over such stocks worthless in the absence of a jurisdiction to prevent such activities.[46]

The theoretical underpinning of this concept owes something to the *Trail Smelter* arbitration. There, Canada was liable because it allowed environmental harm to damage property and interests across the border in the United States. States, although not possessing more than sovereign rights in the EEZ, still possess jurisdictional control over the resources of the EEZ. Having an international law right of action for harm caused by another state to such resources is clearly analogous to the circumstances of the *Trail Smelter* arbitration.

44 LOSC, *supra* note 3, art. 211(4).

45 For a broad discussion of the nature of the EEZ in the context of the environment *see* FRANCISCO ORREGO VICUNA, THE EXCLUSIVE ECONOMIC ZONE: REGIME AND NATURE UNDER INTERNATIONAL LAW 83–89 (1989).

46 As much is evident from the development of Article 56, which in the earlier drafts stressed the importance of State jurisdiction over marine pollution. *See* 2 UNITED NATIONS CONVENTION ON THE LAW OF THE SEA 1982: A COMMENTARY 521–544 (Myron H. Nordquist ed., 1993). In addition, the IMO examined the question of the applicability of State jurisdiction over marine dumping under the then-London Dumping Convention. The United Nations Secretary-General, noting the IMO's work, reported that, for the purposes of dumping. State jurisdiction should be considered equivalent to waters under territorial jurisdiction. *See* Law of the Sea: Report of the Secretary-General, UN GAOR, 44th Sess. Agenda Item 30, para. 45, U.N. Doc. No. A/44/461 (1989).

In addition to the jurisdictional linkage that might derive through Article 56, the environmental provisions in Part XII of the Law of the Sea Convention also may have links to the principle derived from *Trail Smelter*. The provision with the most substantial apparent linkage is Article 194(2) of the Convention. It provides:

> States shall take all measures necessary to ensure that activities under their jurisdiction or control are so conducted as not to cause damage by pollution to other States and their environment, and that pollution arising from incidents or activities under their jurisdiction or control does not spread beyond the areas where they exercise sovereign rights in accordance with this Convention.

This Article is most closely related to the *Trail Smelter* arbitration, as it deals explicitly with a situation where the activities of a coastal state damage the marine environment in such as way as to allow that damage to escape beyond the EEZ.[47] Although distinct from *Trail Smelter* in that the areas in question are not necessarily part of the sovereignty of a coastal state or its neighbours, the thread linking this to the *Trail Smelter* arbitration is clear. Allowing harm to spread beyond national jurisdiction is not permissible, with the implicit corollary that a failure to discharge the obligation will give rise to liability. Interestingly, under Part XV of the Law of the Sea Convention, an offending state would, in the absence of reaching agreement with its neighbor, be obliged to accept the compulsory jurisdiction of a judicial body to determine the dispute.[48]

Other provisions within the Law of the Sea Convention also contain echoes of the *Trail Smelter* arbitration. Article 195 of the Convention provides that states have a duty not to transfer damage or harm caused by pollution from one area to another. Similarly Article 196 provides for measures to prevent, reduce or control pollution of the marine environment resulting from the use of technologies or of alien species.

An excellent example of the use of *Trail Smelter* can be seen in Article 195 of the Law of the Sea Convention. As considered above, Article 195 draws out a principle for the prohibition of the movement of pollution similar to that expounded in the *Trail Smelter* arbitration. However, it is apparent from the text of the article that it draws heavily from Principle 13 of the Principles of Assessment and Control of Marine Pollution endorsed at the United Nations Conference on the Human Environment, which uses very similar language.[49] The relationship to the *Trail Smelter* arbitration can be seen in the Report of the International Law

47 It is clear that the State's obligation extends to pollution from land-based sources, ships, and platforms at sea: LOSC, *supra* note 3, art. 194(3).

48 Part XV of the Law of the Sea Convention provides for dispute resolution by the International Court of Justice, an Arbitral Tribunal, Special Arbitral Tribunal or the International Tribunal on the Law of the Sea. *See generally* NATALIE KLEIN, DISPUTE SETTLEMENT AND THE UNITED NATIONS CONVENTION ON THE LAW OF THE SEA (2004).

49 *See* Report of the United Nations Conference on the Human Environment, at Annex III UN Doc. A/conf.48/14/rev.1 (1972), *reprinted in* 11 INTERNATIONAL LEGAL MATERIALS 1416 (1972).

Commission's Working Group on International Liability for Injurious Consequences arising out of Acts not Prohibited by International Law.[50] That report draws heavily on the *Trail Smelter* arbitration in the context of general principles for liability arising out of transboundary harm,[51] but prefers more recent materials in the context of its discussion of marine pollution.

However, in Part XII of the Law of the Sea Convention, many of the obligations concerning protection of the marine environment would not appear to have their antecedents in the *Trail Smelter* arbitration alone. The general duty in Article 192 frames a duty to preserve and protect the marine environment in much wider terms than the notion of the escape of harm.[52] *Trail Smelter* allows the conclusion that a state may do as it pleases with its own environment, so long as the harm it does will not extend its consequences to another state.[53] Article 192 places a stricter obligation on state parties, as their duty is to protect the marine environment, not merely the marine environment beyond the coastal state's own jurisdiction.

The tenor of other duties in Part XII also go beyond that of the *Trail Smelter* arbitration. Several articles require that cooperative arrangements be negotiated between states to deal with marine pollution.[54] Exchange of scientific data,[55] the development of contingency plans,[56] and the notification of actual or imminent damage to another's environment[57] are all required. *Trail Smelter* provides guidance for liability arising out of an incident, providing a response as to an indication of liability in an adversarial process. The scope of duties under Part XII takes this notion much further, providing for measures to mitigate harm before damage can occur. Although *Trail Smelter* has provided some foundation to these provisions, subsequent cases and other developments in international environmental law have probably taken a greater role in fleshing out the present provisions.[58] Certainly these provisions are far more sophisticated in their approach to dealing with marine pollution than the application of liability under the *Trail Smelter* arbitration.

A major point of difference between the *Trail Smelter* arbitration and contemporary marine environmental protection is the basic approach to liability.

[50] *See* International Law Commission, Working Group on International Liability for Injurious Consequences Arising out of Acts not Prohibited by International Law, 48th Sess., U.N. Doc. A/CN.4/L.533 and Add.1 (1996)

[51] *Id.* at para. (2) of the General Commentary, and para. (10) of the Commentary to Article 1. *See* Drumbl in this volume.

[52] LOSC, *supra* note 3, art. 192 ("States have the obligation to protect and preserve the marine environment.").

[53] *See* LOSC, *supra* note 3, art. 193.

[54] *Id.* art. 197.

[55] *Id.* art. 200.

[56] *Id.* art. 199.

[57] *Id.* art. 198.

[58] For example, the requirement to notify of actual or imminent damage probably owes more to the duty to notify imposed by the International Court of Justice on Albania in the *Corfu Channel Case* than it does to the *Trail Smelter* arbitration. *See* Corfu Channel (UK v. Albania), 1949 I.C.J. 3 (April 9).

International environment marine environmental law has been focused on pollution deriving from ships rather than on pollution originating on land. The emergence of MARPOL and a number of other regimes in the late 1960s and early 1970s were more likely directly a result of the international reaction to the *Torrey Canyon* disaster in 1967, than the continued development of *Trail Smelter*. In the years since, the development of MARPOL and other regimes since 1973 has been, in part, hastened on its way as a result of other international shipping disasters.

Furthermore, these developments can be distinguished from the *Trail Smelter* arbitration in that the focus has been on dealing with pollution from a ship, which generally attracts the liability of the shipowner and the ship's insurers rather than liability at a state level. State responsibility attaching to a flag state has been seen as undesirable, possibly because so-called flag-of-convenience states would not be in a position to guarantee that payment for damages could be made. For example, it would be financially impossible for Liberia, Vanuatu, or Tuvalu, all flag-of-convenience states, to cover the cost of environmental damage on the scale of the damage possible from a major oil spill, such as the harm to King William Sound in the case of the *Exxon Valdez*.[59]

The *Trail Smelter* arbitration was concerned with the liability of Canada for the harm it permitted to escape from its territory. International marine pollution law is directed at prevention of harm, and the creation of insurance and liability schemes that seek to ensure, that in the event of harm, shipowners will be liable. There appears to be little impetus to pursue flag states, and even less to pursue of coastal states who have an environmental incident in their EEZ, which is where *Trail Smelter*'s reasoning might take matters. The connection between the flag and coastal states to the offending ship may be too tenuous to bring about an effective solution to the generated environmental harm.

A more elemental difference also exists between contemporary law and the approach in *Trail Smelter*. As already noted, much of the development of international law looking at the protection of the marine environment has been directed towards pollution from ships. MARPOL is entirely concerned with pollution from ships, either directly in terms of engine emissions, garbage, sewage, and the like, or pollution from ships carrying cargoes that escape because of misadventure, such as oil spills. Similarly, instruments like the Basel Convention on the Transboundary Movement of Hazardous Waste or the London Convention, address pollution from cargoes or the disposal of pollutants at sea from ships. As a result of shipping accidents like the *Torrey Canyon*, the international community has

59 Exxon was ultimately fined US$25 million, paid US$100 million in criminal restitution to the Federal and Alaskan governments, and US$900 million in payments as a civil settlement to governments. Exxon has indicated it spent US$2.1 billion on the cleanup. *See* Exxon Valdez Oil Spill Trustee Council, Oil Spill Facts: Settlement, *available at* http://www.evostc.state.ak.us/facts/settlement.html.

moved beyond looking to simply deal with liability for harm. Most measures are directly designed to minimize the likelihood of harm occurring in the first place. An example of this can be seen in MARPOL measures to phase out single-hulled oil tankers, particularly those that are over fifteen years old, as these vessels pose the greatest risk of substantial harm in the event of a marine accident.[60] Similarly, Article 221 of the Law of the Sea Convention legitimizes action to minimize harm,[61] provided there is imminent danger of damage, and the measures are in response to the magnitude of the threat posed.[62]

This is also borne out in the powers available to port states to limit access to their port facilities. Port states clearly have the power under Article 219 of the Law of the Sea Convention, to impose requirements that visiting vessels meet minimum standards of seaworthiness and safety. These extend not merely to the denial of access, but also to the holding of noncompliant vessels in port until remedial action is taken.[63] Again, these are measures that go far beyond anything envisaged in the *Trail Smelter* arbitration.

Although the *Trail Smelter* arbitration would seem to have been surpassed in some areas, it may be of substantial utility in those areas of marine environmental law where less headway toward regime-building has been made. In one such area, the issue of land-based sources of marine pollution, very little progress has been made, in spite of decades of diplomatic effort.[64] This is one area in which the *Trail Smelter* arbitration may still exert some influence in the protection of the marine environment. Where such pollution extends into the EEZ or territorial sea of another state and causes harm, the *Trail Smelter* Arbitration would provide clear authority to the effect that the polluting state was liable for the damage caused.

If land-based sources of marine pollution are not substantially regulated, aside from the general duties in the Law of the Sea Convention, already largely considered, then there is a substantial lacuna in marine environmental protection. Although *Trail Smelter* is not a solution to this lack of progress in creating a binding instrument, it presents a limited way of dealing with substantial

60 MARPOL is implementing a phase-out of single-hulled oil tankers. *See* MEPC Resolution 111(50), *available at* http://www.imo.org/Environment/mainframe.asp?topic_id=1046.

61 *See* LOSC, *supra* note 3, art. 211(1).

62 This would conceivably cover a situation such as the *Torrey Canyon* disaster, where the Royal Navy bombed the ship in order to set fire to its oil and minimize the harm it would cause the Cornish Coast: *See* Nordquist (ed.), *supra* note 46, Vol. 4, 305.

63 LOSC, *supra* note 3, art. 219.

64 Although the framework of the Law of the Sea Convention could accommodate such an instrument, the creation of a global regime for dealing with land-based sources of marine pollution has yet to occur. Diplomatic effort spanning a number of decades has made little progress. *See* Daud Hassan, *International Conventions Relating to Land-Based Sources of Marine Pollution Control: Applications and Shortcomings*, 16 GEORGETOWN INTERNATIONAL ENVIRONMENTAL LAW REVIEW 657 (2004). The weakness in this area can be seen in the Law of the Sea Convention itself, where states are obliged to "take into account" international developments rather than meet international standards. *See* Boyle, *supra* note 12, 354.

land-based sources of marine pollution. Admittedly, the liability principle from *Trail Smelter* would only come into play where land-based pollution from a coastal state escaped beyond its national jurisdiction, potentially a substantial distance, but the application of the principle would seem clear.

CONCLUSION

The legacy of the *Trail Smelter* arbitration in marine environmental law would appear to be muted. The principle that a state is under a duty not to allow harm to escape from its jurisdiction owes much to *Trail Smelter*, and, as has been indicated, this principle resonates within parts of the Law of the Sea Convention. However, the issues surrounding protection of the marine environment are more complex than the factual situation faced in the *Trail Smelter* arbitration, and the means available to tackle the problems must equally be more sophisticated.

It is significant that in the one excursion to date into a marine pollution dispute before the International Tribunal on the Law of the Sea, in the *MOX Plant Case*, none of the judges in the case referred to *Trail Smelter* in their judgments.[65] Whereas the case was dealt with at a preliminary measures stage, and seems unlikely to be heard to its conclusion by the Tribunal, it would seem strange that if *Trail Smelter* occupies an august place in the lexicon of marine environmental law, none of the Tribunal's members felt obliged to refer to it.

Ultimately, the *Trail Smelter* arbitration can be seen as a starting point in the development of marine environmental protection, but by no means the end of the story. Much of this area of international law is about prevention of harm before it ever occurs, rather than attributing liability. Furthermore, the jurisdictional differences between the marine and terrestrial environments would seem to indicate a mismatch in the direct transposition of the principles out of the Arbitration.

[65] The MOX Plant Case (Ir. v. U.K.), ITLOS Permanent Court of Arbitration, *reprinted in* 42 International Legal Materials 1187 (2003).

PART THREE

Trail Smelter and Contemporary Transboundary Harm – Beyond the Environment

18 *Trail Smelter* and Terrorism: International Mechanisms to Combat Transboundary Harm

Pierre-Marie Dupuy and Cristina Hoss

INTRODUCTION

This chapter compares and analyzes the international legal issues arising from transboundary environmental pollution and international terrorism. At first glance, it might appear to be somewhat peculiar to consider such disparate topics alongside one another. However, the potential cross-border effects of terrorism and pollution, as well as the fact that, in the vast majority of cases, the resulting harm is the result of nonstate actors, makes such an examination worthwhile. Moreover, both of these transboundary phenomenons share one vitally important characteristic: they cannot be left to the respective domestic legal orders alone, but need common action in order to raise an effective response.

This examination is facilitated by considering the features of international environmental law arising out of the historic *Trail Smelter* arbitration and the emerging international law on transboundary terrorism, which allows for a consideration of international law's general mechanisms for combating all manners of transboundary harm. Such a consideration gives rise to the interesting question of whether these legal mechanisms constitute special international law, or whether they are, in fact, merely applications of the general law of state responsibility.

The legal mechanisms developed by international law to combat transboundary pollution and international terrorism share two important features. First, international law has placed the prevention of transboundary harm at the very center of both, whereas the responsibility of states, the "curative" side of international law, does not seem to be the preferred strategy to combat transboundary harm.[1] Nevertheless, the law of state responsibility remains applicable in the event of a violation of the obligations to prevent transboundary harm.

[1] Pierre-Marie Dupuy, A *propos des mésaventures de la responsabilité internationale des Etats dans ses rapports avec la protection internationale de l'environnement, in* Les Hommes et L'environnement, En Hommage a Alexandre Kiss 269 (M. Prieur ed., 1998).

The emergence and strengthening of the obligation of due diligence is at the center of the development of international law on transboundary harm. The content of the general obligation of due diligence can differ according to the issue under consideration; indeed, the very nature of the general due diligence obligation is to offer the necessary flexibility to respond to most of the potential conflicts between states.

The second notable commonality between the law governing transboundary pollution and international terrorism is the focus on nonstate actors.[2] International treaties as well as the United Nations Security Council have recognized that nonstate actors can be "partial" subjects of the international legal order in the context of environmental law, as well as in the context of terrorist activities. The expansion of the range of subjects of international law is an important and effective means for combating transboundary harm arising out of private activities,[3] as it enables the international community to impose legal obligations directly on nonstate actors, which would not be possible under the classic, Westphalian view of international law. This second commonality did not, however, predominate in the *Trail Smelter* arbitration, where strictly construed state liability controlled. Given the topic of the present chapter, the move to a focus on private actors will therefore not be envisaged here.

The *Trail Smelter* arbitration offers an appropriate starting point for the consideration of the general international law on transboundary harm, as it was from this arbitration that a central principle of international law was introduced. "Invented," might be a better description. The *Trail Smelter* Tribunal did not provide a detailed justification or legal reasoning in support of its now-famous "crystallization" of the primary obligation, which held that states were not allowed "to use or permit the use of [their] territory in such a manner as to cause injury by fumes in or to the territory of another or the properties of persons therein, when the case is of serious consequence and the injury is established by clear and convincing evidence."[4] The International Court of Justice similarly recognized, in its judgment on the *Corfu Channel* case, that every state has an obligation, "not to allow knowingly its territory to be used for acts contrary to the rights of other States."[5]

These two cases are the starting point for a considerable strengthening of the international obligation of due diligence. However, as will be seen in the next section, this development has brought with it a certain degree of confusion.

2 *See* Zumbansen and Miller in this volume.

3 Pierre-Marie Dupuy, *State Sponsors of Terrorism: Issues of International Responsibility*, *in* ENFORCING INTERNATIONAL LAW NORMS AGAINST TERRORISM 3–11 (Andrea Bianchi ed., 2004).

4 *See Trail Smelter Arbitral Decision*, 33 AMERICAN JOURNAL OF INTERNATIONAL LAW 182 (1939) [hereinafter "*Trail Smelter* (1939)"]; *Trail Smelter Arbitral Decision*, 35 AMERICAN JOURNAL OF INTERNATIONAL LAW 684 (1941) [hereinafter "*Trail Smelter* (1941)"]. *See* Annex to this volume.

5 Corfu Channel (UK v. Albania) (Merits), 1949 I.C.J. 4, 21 (Apr. 9).

INTERNATIONAL LAW AND TRANSBOUNDARY ENVIRONMENTAL HARM

Responsibility Versus Liability

An overly broad interpretation of the *Trail Smelter* arbitration[6] lies at the center of several attempts to establish a new type of state responsibility for transboundary environmental damage called "responsabilité objective," "responsabilité pour risque," or "liability for injurious consequences arising out of acts not prohibited by international law."[7] However, far from being a new manifestation of state responsibility, this type of liability is better described as the responsibility of states for the violation of due diligence obligations. It therefore still fits within existing concepts of responsibility for wrongful acts, and attempts to view it as a new kind of liability are misplaced and, in our view, doomed to fail.

The failure of these novel attempts is particularly evident in the codification work of the International Law Commission (ILC), which, despite more than twenty-five years of effort, has yet to reach a final conclusion.[8] In 1997, the ILC established a working group that determined that "prevention" and "international liability" are distinct, yet related, topics. Thus, although they were related, the ILC determined that the topics should be considered separately.[9] This ILC strategy is supported by the widely-held view that it is much more effective to protect the environment through a proactive or preventive approach, rather than a reactive approach.[10] Despite there having been many recent instances of transboundary pollution, there have been only a small number of international decisions which deal with reparations for transboundary environmental harm.[11]

After considering the first report of the newly appointed Special Rapporteur Pemmaraju Sreenivasa Rao, and after the adoption of 17 articles upon a first reading, the ILC decided in 2001 to recommend that the United Nations General

6 For a critique, *see* Pierre-Marie Dupuy, La Responsabilité Internationale Des Etats Pour Les Dommages D'Origine Technologique et Industrielle 32 (1976).

7 C. Wilfred Jenks, *Liability for Ultra-Hazardous Activities in International Law*, 117 Recueil des Cours 99, 102 (1966); L. Frederick E.. Goldie, *Liability for Damage and the Progressive Development of International Law*, 14 International and Comparative Law Quarterly 1189 (1965).

8 Draft Articles on Responsibility of States for Internationally Wrongful Acts, *in* Report of the International Law Commission on the Work of Its Fifty-third Session, UN GAOR, 56th Sess., Supp. No. 10, UN Doc. A/56/10 (2001) at 366–436 [hereinafter "2001 ILC Report"]. *See* the Reports of Special Rapporteurs Quentin Quentin-Baxter and Julio Barboza, *in* 2 Yearbook of the International Law Commission, part I (1980 to 1996); *see also* Pierre-Marie Dupuy, *Quarante ans de codification du droit de la responsabilité internationale. Un bilan*, 107 Revue Générale de Droit International Public 305 (2003); *See also* Drumbl in this volume.

9 Report of the International Law Commission, U.N. GAOR 51st Sess., Supp. No. 10, UN Doc. A/52/10 (1997) para. 168.

10 Dupuy, *supra* note 1, at 270, *See also* Bratspies in this chapter.

11 In particular, the Chernobyl incident or the pollution of the river Rhine by Sandoz did not lead to any jurisdictional dispute resolution.

Assembly (GA) negotiate a convention on the basis of the Draft Articles on Prevention of Transboundary Harm from Hazardous Activities (Draft Articles).[12] It remains unclear if the resulting Draft Articles will ever become legally binding on states.

The GA recommended using the Draft Articles as a basis for negotiating a multilateral convention.[13] So far, no such regime exists although some of the principles of the liability regime under consideration by the ILC are already enshrined in specific international agreements. Therefore the future convention may be formulated by extracting general principles from existing agreements in order to provide for their general applicability.

The concept of liability in international law is premised on the assertion that a state implicitly endorses any "hazardous" activities taking place on its territory and accepts responsibility for any resulting harm.[14] A state's responsibility is engaged, irrespective of whether it breached an international obligation.[15] Furthermore, the state would only be liable if "transboundary harm" exists.[16] The latter requirement is always fulfilled in the case where the activity giving rise to the damage is performed within the territory of one state and the damage is caused within the territory of another state.

However, the Draft Articles only cover activities that involve the risk of causing "significant transboundary harm,"[17] the meaning of which is limited in the Draft Articles to the "physical consequences" of activities involving risks.[18] The risk of causing harm is given a flexible definition: the greater the potential harm, the lower its probability need be.

The Content of Due Diligence Obligations in Environmental Law

According to the Draft Articles, states have a duty to prevent or minimize risks of harm. The Draft Articles contain detailed procedural rules on the requirement of authorization for activities involving risks, the need for environmental impact assessments, notification of the project to the states potentially affected, exchange of information, and the duty to consult.[19]

The general duty to prevent and to minimize the risk of transboundary harm[20] is derived from the fundamental principle *sic utere tuo ut alienum non laedas*,

[12] 2001 ILC Report, *supra* note 8, at 370. [13] *Id.*

[14] *See* Anderson in this volume.

[15] 2001 ILC Report, *supra* note 8, at 386. Article 1 (Scope) of the Draft Articles reads: "The present articles apply to activities not prohibited by international law which involve a risk of causing significant transboundary harm through their physical consequences."

[16] *Id.* [17] *Id.*

[18] *Id.*

[19] *Id.* at 372–72 (arts. 3–9). *See* Handl in this volume.

[20] *Id.* at 372. Article 3 states: "The State of origin shall take all appropriate measures to prevent significant transboundary harm or at any event to minimize the risk thereof."

which was enshrined in the *Trail Smelter* award and was, more recently, reflected in Principle 21 of the Stockholm Declaration[21] and endorsed by the Rio Declaration.[22] In its advisory opinion on *The Legality of Use by a State of Nuclear Weapons in Armed Conflicts*,[23] the ICJ emphasized that the obligation of prevention formed part of the corpus of international principles. The ILC considers the prevention obligation to be a primary one[24] and further imposes on states the obligation to minimize the risk of harm in cases where complete prevention is "not fully possible." Article 3 of the Draft Articles also imposes a due diligence requirement,[25] and gives the states discretion to adopt administrative, legislative, or executive measures in respect to hazardous activities.

The standards and parameters of "due diligence," however, are not uniform and depend on the circumstances of each case. Whether a measure was appropriate may depend on several factors such as the economic resources of the respective state,[26] or the technological level of its industries. The ILC has explained that the due diligence requirement should be measured by a standard of care that can be expected of a "good government," noting that, "It should possess a legal system and sufficient resources to maintain an adequate administrative apparatus to control and monitor the activities."[27]

States also have an obligation to cooperate in order to prevent significant transboundary harm or, as the case may be, to minimize the risk thereof.[28] Perhaps the requirement to cooperate could have been integrated into the general obligation of "due diligence," but as cooperation is a central element of prevention it is contained in a separate article of the Draft Articles. The obligation of cooperation is to be fulfilled by "all states concerned," which encompasses both the state of origin, as well as any state likely to be affected by the significant transboundary harm.[29]

[21] Report of the Stockholm Conference, U.N. Doc. A/CONF.48/14, at 7, *reprinted in* 11 International Legal Materials 1416, at 1420 (1972). Principle 21 of the Stockholm Declaration reads:

> States have, in accordance with the Charter of the United Nations and the principles of international law, the sovereign right to exploit their own natural resources pursuant to their own environmental policies, and the responsibility to ensure that activities within their jurisdiction or control do not cause damage to the environment of other States or of areas beyond the limits of national jurisdiction.

[22] Report of the United Nations Conference on Environment and Development, Annex I, U.N. Doc. A/CONF.151/26 (Vol. I), *reprinted in* 31 International Legal Materials 874 (1992) [hereinafter "Rio Declaration"].

[23] Legality of the Threat or Use of Nuclear Weapons, Advisory Opinion, 1996 I.C.J. 226, 242 (July 8), *reprinted in* 35 International Legal Materials 1343 (1996).

[24] 2001 ILC Report, *supra* note 8, at 391.

[25] *Id.* at 392 (citing the dispute between Germany and Switzerland relating to the Sandoz accident, where Switzerland recognized that it had not acted with "due diligence.")

[26] *Id.* at 394.

[27] *Id.* at 395.

[28] *Id.* at 396. Article 4 (Co-operation) provides: "States concerned shall cooperate in good faith and, as necessary, seek assistance of one or more competent international organizations in preventing significant transboundary harm or at any event minimizing the risk thereof."

[29] *Id.* at 397.

States also must implement legislative, administrative, or other instruments, in order to comply with their duties.[30]

Going far beyond the reactionary approach to due diligence mapped out by the *Trail Smelter* arbitration, the ILC has introduced a proactive state obligation of prior authorization for certain activities carried out within a state's territory.[31] The governmental authorities, in order to fulfill the state's obligation "not to allow knowingly its territory to be used for acts contrary to the rights of other States,"[32] would require private operators to request prior authorization for "hazardous activities" covered by the scope of the Draft Articles. This obligation requires that the state take special legislative and administrative measures, and further requires that the responsible government organ conduct a risk assessment before the grant of authorization.[33] The latter requirement derives from the general obligation of cooperation. States likely to be affected must be notified of the result of the risk-assessment.[34]

In order to ensure the aim of an "equitable balance of interests,"[35] the Draft Articles would require states to enter into good faith consultations[36] on preventive measures. Based on the jurisprudence of the International Court of Justice, the ILC has established a list of factors that, *inter alia*, should be considered in order to reach the desired "equitable balance of interest."[37]

The state of origin also incurs an obligation to inform. Interestingly, this duty to provide information about the preventive measures and the risks of the activity extends not only to other states potentially affected by the hazardous activity,[38] but also directly to the public.[39]

The work of the ILC and the content of the due diligence obligations, including the obligation to prevent harm, is precise. The very general obligation to prevent harm recognized in *Trail Smelter* has thus evolved into a series of performance

[30] *Id.* Article 5 (Implementation) provides: "States concerned shall take the necessary legislative, administrative or other action including the establishment of suitable monitoring mechanisms to implement the provision of the present articles."

[31] *Id.* at 399 (art. 6 and commentary).

[32] Corfu Channel, *supra* note 5, at 22.

[33] 2001 ILC Draft, *supra* note 8, at 402 (art. 7 and commentary); *See also* Rio Declaration, Princ. 17, *supra* note 22.

[34] *Id.* at 406 (art. 8 and commentary).

[35] *Id.* at 412 (art. 10 and commentary).

[36] *Id.* at 410 (art. 9 and commentary); *see also* Robert Kolb, La Bonne Foi en Droit International Public: Contribution à l'Étude Des Principes Généraux du Droit 756 (2000).

[37] 2001 ILC Draft, *supra* note 8. Article 10 requires balancing of, *inter alia*: "(a) The degree of risk of significant transboundary harm and of the availability of means of *preventing* such harm, or minimizing the risk thereof or repairing the harm; (b) The importance of the activity, taking into account its overall advantages of a social economic and technical character for the state of origin in relation to the potential harm for the state likely to be affected; (c) The risk of significant harm to the environment and the availability of means preventing such harm, or minimizing the risk thereof or restoring the environment."

[38] *Id.* at 420 (art. 12 and commentary). *See also* Pierre-Marie Dupuy et Henri Smets, *Pollution transfrontière-Information et Consultation*, 7 Environmental Policy and Law 3 (1981).

[39] 2001 ILC Draft, *supra* note 8, at 422 (art. 13 and commentary).

obligations intended to give full effect to due diligence and cooperation, rather than leaving it up to the goodwill of states.

In the context of international environmental law, general rules of state responsibility apply. This means that in most of the incidents of transboundary environmental pollution it is necessary to establish a link between the state and the private company or enterprise responsible for the operation of the facility emitting the pollution. No clear test has been laid down to determine which entities should be considered to be state-owned and controlled. One can, however, assume that it is not sufficient that the state has initially established the corporate entity.[40] However, the conduct of the company can be considered imputable to the state when the company exercises elements of governmental authority,[41] when the activities are endorsed by the state,[42] or if the private persons were acting under the control of the state.[43]

INTERNATIONAL LAW ON TERRORISM

The phenomenon of terrorist activities covers a wide range of diverse criminal acts. The difficulties encountered by the legal community in trying to find a universally accepted definition of terrorism provide the best reflection of this diversity.[44] Judge Rosalyn Higgins seems particularly unconvinced of the legal value of the notion of terrorism: "'Terrorism' is a term without any legal significance. It is merely a convenient way of alluding to activities, whether of states or of individuals, widely disapproved of and in which either the methods used are unlawful, or the targets protected, or both."[45]

The absence of any universally accepted definition of terrorism and the absence of any possibility of defining it without making political judgments on the moral validity of armed activities, does not, however, prevent a consideration of the consequences under international law for activities that might be classified as terrorism. In fact, many activities of states, or nonstate actors, whether qualified as terrorist acts or not, fall under the general rules of state responsibility.

Before examining the legal consequences of terrorism, it is helpful to recall that the primary obligations imposed on states to prevent terrorism arise out of the same general principle as the obligation of prevention in environmental law – the principle of *sic utere laedas*.

[40] Eastman Kodak Co. v. Iran, 17 Iran-US. Cl. Trib. Rep. 153 (1987).

[41] In the sense of Art. 5 of the Draft Articles. *See* 2001 ILC Draft, *supra* note 8.

[42] United States Diplomatic and Consular Staff in Tehran (U.S. v. Iran) 1980 I.C.J. 3, 34 (May 24).

[43] 2001 ILC Draft, *supra* note 8, at 406 (art. 8 and commentary).

[44] For an excellent evaluation of the different existing definitions, *see* Christian Walter, *Defining Terrorism in National and International Law, in* TERRORISM AS A CHALLENGE TO NATIONAL AND INTERNATIONAL LAW : SECURITY VERSUS LIBERTY? 23 (Christian Walter et al., eds., 2004).

[45] Rosalyn Higgins, *The General International Law of Terrorism, in* TERRORISM AND INTERNATIONAL LAW, 13, 28 (Rosalyn Higgins and Maurice Flory eds., 1997).

Due Diligence in the Context of Terrorism

As far as "terrorism" is concerned, the obligation of due diligence does not arise out of the harm caused to neighboring states, as in *Trail Smelter*, but instead developed out of the context of the protection of aliens against acts of private actors. It was this cross-border aspect that was explored in a considerable number of cases on state responsibility. One of the first appearances of the obligation of due diligence in the context of state responsibility for acts of private actors was the 1925 *British Property in Spanish Morocco* arbitral decision pronounced by Max Huber.[46] Huber concluded that a state could not be held responsible for all acts that occur in its territory. In particular, the state traditionally could not be held responsible for damages caused by civil wars, national uprising, and the activities of rebel groups. Even so, Huber concluded that the state did bear an obligation to exercise due diligence, noting "in general, repression of the offenses is not only a legal obligation of the competent authorities but also an international duty of the State."[47] Moreover, states can be responsible for the acts and omissions of public authorities to remedy the consequences of those events.[48]

Examining the nature and content of the obligation of due diligence, Huber stated that due diligence in Roman law can be applied by analogy to international law; the *diligentia quam in suis* principle conforms with the independence of states in domestic matters, and defines due diligence as the obligation of states to assure a degree of security that can reasonably be expected from aliens.[49] This seems to include punitive measures, such as prosecuting alleged authors of violent acts against aliens, as well as preventive and protective measures.

In the *Corfu Channel* case, the ICJ held, in a first step, that the mere control of a state over its territory and waters was insufficient to establish that the state knew or should have known of any unlawful act perpetrated therein, or of the presence of the authors of such an act.[50] The Court then examined the evidence in support of the contention that the Albanian government knew of the presence of mines in its waters, and concluded, that the Albanian government had the requisite knowledge.[51] The Court then concluded that on the basis of several international obligations, Albania bore international responsibility for not "notifying, for the benefit of shipping in general, the existence of a minefield . . . " The Court recognized in this decision, that this obligation to notify, is based ". . . on certain general and well recognized principles, namely: elementary considerations of humanity . . . ; the principle of freedom of maritime communication; and every

[46] British Property in Spanish Morocco (Spain v. U.K.) 2 R.I.A.A. 615 (1925); Gilbert Guillaume, *Terrorisme et Droit International*, *in* RECUEIL DES COURS DE L'ACADÉMIE DE DROIT INTERNATIONAL, 287, 391 (1989).

[47] Spanish Morocco, *supra* note 46, at 640–647.

[48] *Id.* (translation by the authors).

[49] *Id.* at 644.

[50] Corfu Channel, *supra* note 5, at 18.

[51] *Id.* at 22.

State's obligation not to allow knowingly its territory to be used for acts contrary to the rights of other States."[52]

In the *Teheran Hostages* case, the Court stated that: "The conclusion just reached by the Court that the initiation of the attack . . . cannot be considered itself imputable to the Iranian State does not mean that Iran is, in consequence, free of any responsibility in regard to those attacks; for its own conduct was in conflict with international obligations as a receiving State to take appropriate steps to ensure the protection of the United States Embassy and Consulates. . . . This inaction of the Iranian Government by itself constituted clear and serious violations of Iran's obligations to the United States under the . . . Vienna Convention on Diplomatic Relations."[53]

The Court considered several criteria, as it later did in the *Nicaragua* case, to determine whether Iran had failed to fulfill its duty of due diligence. In concluding that Iran bore responsibility for the acts of individuals, the Court took into account Iran's knowledge that it had an obligation to act, its awareness that action was needed, and the means Iran had at its disposal to ensure respect for its international obligations toward the United States.[54] This last criteria of "sufficient means" could be of eminent importance in the evaluation of a state's fulfillment of its due diligence obligations, and could even lead to an obligation of the state to authorize foreign armed intervention within its territory.[55]

In the context of the *Teheran Hostages* case, and more specifically in the portion of the opinion devoted to establishing the responsibility of the Iranian authorities (i.e., state responsibility), it appears very clearly that the Court considers an inquiry into due diligence obligations to be very similar, if not identical, to a determination about attribution of private acts to the state. The two inquiries appear functionally equivalent with regard to establishing state responsibility for acts of private actors. State responsibility can be established and invoked either on the basis of due diligence, or on the basis of attribution.

The *Nicaragua* case, better known for its "effective control" test of private actors by a foreign state, also dealt with obligations of due diligence. The Court held that Nicaragua's international responsibility was not triggered by its alleged lack of diligence towards stemming the flow of arms to the Salvadorian opposition from its territory. The Court noted: "if the flow of arms is in fact reaching El Salvador without either Honduras, Nicaragua or the United States succeeding in preventing it, it would clearly be unreasonable to demand of the Government of

[52] *Id.*

[53] United States Diplomatic and Consular Staff, *supra* note 42 at 29.

[54] *Id.* at 34 (para.68). Luigi Condorelli, *The Imputability of Acts of International Terrorism*, 19 Israel Yearbook on Human Rights 233 (1990); Dupuy, *supra* note 3, at 11.

[55] Jochen Abr. Frowein, *Der Terrorismus als Herausforderung an das Völkerrecht*, 62 Zeitschrift für Ausländisches Öffentliches Recht und Völkerrecht 879, 884 (2003). *See* Handl in this volume.

Nicaragua a higher degree of diligence than is achieved even by the combined efforts of the other three States."[56]

This duty of due diligence in the context of terrorism has been given further content through treaties, and through the activities of the United Nations in the fight against terrorism.

Beginning with the *Convention for the Prevention and Punishment of Terrorism*[57] concluded under the auspices of the League of Nations in 1937, a considerable number of later Conventions relating to terrorism have included a provision on the obligation of states to prosecute or extradite terrorist suspects.[58] This obligation appears to be a part of the obligation of prevention, and of the obligation of due diligence in general, and is considered to be customary international law.[59]

The General Assembly of the United Nations has adopted several declarations and resolutions to clarify the content of due diligence in the context of terrorism. Although General Assembly Declarations are not legally binding instruments, the obligations mentioned by the Declarations can reflect customary international law.

The General Assembly's consideration of "terrorism" began in the seventies, when the first Resolutions on measures to eliminate international terrorism were issued. Most prominently however, the obligations of states relating to terrorist activities were mentioned in the Friendly Relations Declaration.[60] The starting point of the Declaration was transboundary terrorism: "(. . .) no State shall organize, assist, foment, incite or tolerate subversive terrorist or armed activities directed towards the violent overthrow of the régime of another state, or interfere in civil strife in another state."

This Declaration of the General Assembly seems to suggest that tolerance (which presupposes knowledge) is to be considered as an omission in the context of terrorist activities. In combination with the narrow definition of acts (which must be directed against a "régime" of another state), the obligation of due diligence as laid down in the Declaration of 1970 does little else to clarify the obligation of due diligence with regard to terrorism.

[56] Military and paramilitary activities in and against Nicaragua (Nicaragua v. U.S.) (merits) 1986 I.C.J. 14, 85 (June 27).

[57] Convention for the Prevention and Punishment of Terrorism, Nov. 16, 1937, 19 League of Nations O.J. 23 (1938) (never entered into force).

[58] Guillaume, *supra* note 46, at 354.

[59] François Dubuisson, *Vers un Renforcement des Obligations de Diligence en Matière de Lutte Contre le Terrorisme*, in LE DROIT INTERNATIONAL FACE AU TERRORISME 150 (Karine Bannelier, et al., eds., 2002).

[60] Declaration on Principles of International Law Concerning Friendly Relations and Cooperation Among Member States in Accordance with the Charter of the United Nations, G.A. Res. 2625, U.N. GAOR, 25th Sess., 1883d plen. mtg., Supp. No. 28, at 121, U.N. Doc A/8082 (1970). This Declaration is generally considered as reflecting customary international law. *See Dubuisson, supra* note 59, at 143.

It is different with Resolution 49/60, which was adopted in 1994.[61] The GA therein formulated a catalogue of very concrete obligations arising out of the state duty of due diligence, including the obligation to refrain from "organizing, instigating, facilitating, financing, encouraging or tolerating terrorist activities, to take appropriate practical measures to ensure that their respective territories are not used for terrorist installations or training camps, to ensure the apprehension and prosecution or extradition of perpetrators of terrorist acts, in accordance with the relevant provisions of their national law."[62] Furthermore, this GA Resolution envisions this duty as encompassing an obligation to cooperate by exchanging relevant information to combat and prevent terrorism, by promptly taking all steps necessary to implement the existing international conventions on this subject to which the state is a party.

It seems obvious that the measures mentioned in this resolution considerably strengthen the notion of due diligence. Some of those measures can be considered as binding obligations under general international law.[63]

It can be concluded that the duty of prevention includes in particular: preventing infiltration into or residence in state's territories, giving refuge to terrorists, training, arming, financing or providing any facilities to terrorists. The Security Council's resolutions before and after the attacks of September 11, 2001, requested exactly these measures from the Taliban in connection with Al-Qaeda concerning the territory under their control.[64]

It is, however, difficult to establish a clear line between legal and illegal behavior from the states with regard to their obligation of due diligence. In the judicial decisions and arbitral awards, the criteria of reasonability of the conduct, taking into account the resources of the government, might be decisive in the evaluation of the conduct of states. The obligations included in the notion of due diligence and their fulfillment will have to be examined on a case-by-case basis.

Another scenario is that of state-sponsored terrorism. The obligation of due diligence also covers the accountability of individuals or groups that sponsor acts of terrorism.

[61] Declaration on Measures to Eliminate International Terrorism, G.A. Res. 49/60, U.N. GAOR, 49th Sess., U.N. Doc. A/RES/49/60 (1994).

[62] *Id.*

[63] This is notably the case for the obligation of the states not to give asylum to suspected terrorists (f), because this obligation is a variation of a provision laid down in the Convention Relating to the Status of Refugees, July 28, 1951, 189 U.N.T.S. 137, 176. Article 1(F) provides in relevant part: "The provisions of this Convention shall not apply to any person with respect to whom there are serious reasons for considering that: (b) he has committed a serious non-political crime outside the country of refuge prior to his admission to that country as a refugee; . . ."

[64] See various Chapter VII resolutions including: S.C. Res. 1214, U.N. SCOR, 53d Sess., 3952d mtg., U.N. Doc. S/RES/1214 (1998); S.C. Res. 1267, U.N. SCOR, 54th Sess., 4051st mtg., U.N. Doc. S/RES/1267 (1999); S.C. Res. 1378, U.N. SCOR, 55th Sess., 4415 mtg. para. 2, U.N. Doc. S/RES/1378 (2001).

It is generally known that states that have an interest in the activities of terrorists provide financial support to terrorist groups. In 2001, one day after the attacks of September 11, the General Assembly adopted a resolution calling for international cooperation to prevent and eradicate terrorism and stressed "that those responsible for aiding, supporting or harboring the perpetrators, organizers and *sponsors of such acts* will be held responsible."[65] The legal content of the term of sponsoring activities of terrorists, however, lacks precision. Here again, the legal evaluation of a situation will have to be done with regard to the specific circumstances of the case.

State Terrorism and Issues of Attribution

Another issue, with regard to state responsibility for terrorism is the attribution of acts of private persons to the state (i.e., that in some specific scenarios, acts of private persons such as terrorists are attributable to the state on the basis of state responsibility).[66] The question of imputability is sometimes difficult to distinguish from that of due diligence in the context of terrorism.[67] But, as a general rule, one can assume that a certain degree of failure of compliance with due diligence obligations will lead to the attribution of terrorist acts to the state itself, and will then be treated as "state terrorism" by other states.

The expression *state terrorism*, or *state-sponsored terrorism*, covers a wide range of state involvements in terrorist acts, from direct participation of state's organs to the *ex post facto* approval or endorsement by official authorities.[68]

State Control

A relatively easy case for state responsibility for acts of private persons or entities can be made when the state has transferred the exercise of state functions to private entities. The responsibility is then based on the assumption that the state acts as a unity with regard to international law, regardless of the means, private or public, it uses to fulfill its functions.[69]

A less clear-cut case is when private persons perform public functions under the control or under the direction of state organs. Defining the term of "control" remains difficult. The first time the ICJ had to deal with the definition of "state control" was in the *Nicaragua* case.[70] The Court rejected the claim that every conduct of the *Contras*, including violations of humanitarian law and human rights, was imputable to the United States. But the Court found that the United States was responsible for "planning, direction and support." In general,

65 G.A. Res. 56/1, U.N. GAOR, 56th Sess., 1st mtg., Agenda Item 8, U.N. Doc. A/RES/56/1, which was adopted on September 12th 2001.

66 G. Dahm, J. Delbrück, R. Wolfrum, Völkerrecht 890 (2nd ed. 1989).

67 Condorelli, *supra* note 54.

68 Dupuy, *supra* note 3, at 8.

69 K. Ipsen, Völkerrecht 560 (4th ed 1999).

70 Nicaragua, *supra* note 56.

international jurisprudence tends to establish the level and effectiveness of state control entailing responsibility on a case-by-case basis.[71]

The International Criminal Tribunal for the Former Yugoslavia has adopted a larger definition of the "control" notion.[72] In the *Tadic* case, the Appellate Chamber noted that the issue is whether "armed forces fighting against the central authorities of the same State in which they live (. . .) may be deemed to act on behalf of another State."[73] The Chamber did not find the *Nicaragua* case's "effective control test" to be persuasive,[74] and stated that Yugoslavian control of the Bosnian Serb "military organization" was sufficient to render the conflict international, as it involved "overall control going beyond the mere financing and equipping of such forces and involving (. . .) participation in the planning and supervision of military operations."[75]

The Appellate Chamber did not seem to consider that the notion of "control" under general international law requires that such control should extend "to the issuance of specific orders or instructions relating to single military actions (. . .)"[76]

The ILC had an opportunity, but failed to clarify the ambiguity created by the existence of both the *Nicaragua* and the *Tadic* test. Indeed, Article 8 of the Draft Articles on State Responsibility does not clarify this point, either. Instead, it simply established another test, providing that "The conduct of a person or group of persons shall be considered an act of a State under international law if the person or group of persons is in fact acting on the instructions of, or under the direction or control, of that State in carrying out the conduct."[77]

It is not clear, however, whether there is any legally relevant difference between the notion of "instruction" and "control"[78] and whether these two notions will be interpreted by international practice as constituting two different levels of state responsibility or as equivalent ones. It will be left to the respective international tribunals to interpret these notions with the flexibility required by international relations in general.

[71] Yeager v. Islamic Republic of Iran, 17 IRAN-U.S. CL.TRIB. REP. 92, 103; Loizidou v. Turkey, 1996-VI Eur. Ct. H.R. 2216, 2235-36 (merits) ("[. . .] the Court held, in conformity with the relevant principles of international law governing State responsibility, that the responsibility of a Contracting Party could also arise when as a consequence of military action – whether lawful or unlawful – it exercises effective control of an area outside its national territory. The obligation to secure, in such an area, the rights and freedoms set out in the Convention, derives from the fact of such control whether it be exercised directly, through its armed forces, or through a subordinate local administration.").

[72] Prosecutor v. Tadić, Case No. IT-94-1-T, Appeals Chamber Judgment (ICTY July 15, 1999), *reprinted in* 38 International.Legal.Materials 1518 (1999).

[73] *Id.* at para. 91.

[74] *Id.* at para. 115.

[75] *Id.* at para. 145 (emphasis in original).

[76] *Id.*

[77] 2001 Report 2001, *supra* note 8, at 406 (art. 8 and commentary).

[78] Dupuy, *supra* note 3, at 10.

Of course, state organs directly perpetrating terrorist acts will entail the responsibility of the state wherever such acts have taken place, even if the organs have committed those acts *ultra vires* (acting beyond of their competences but in their official capacity).[79]

Endorsement

State endorsement of acts of private persons is a very different issue to that of the fulfillment of due diligence obligations. The endorsement acts as the equivalent of the state assuming responsibility for the acts of private actors, and thus more likely enters into the concept of imputability rather than that of due diligence. This might have been the nature of the relationship between Canada and the private operator in the *Trail Smelter* arbitration, as Canada assumed responsibility for the smelter's transboundary pollution in the arbitration convention signed by the two states. The role of such state endorsement of private acts also was established by the jurisprudence of the Court in the *Teheran Hostages* case. The Court there considered that Ayatollah Khomeini and other Iranian authorities endorsed this policy and that this endorsement transformed the legal nature of the situation.[80]

The international law of state responsibility, alone, cannot deal with acts of terrorism. In particular, criminal law plays an important part in the fight against terrorism.

International efforts to combat transnational terrorism should focus on any relevant actors, be they States, individuals, or groups. Furthermore, particular attention must be paid to the interaction between criminal accountability and liability of states. From our point of view, public international law can by itself only provide limited instruments in the fight against terrorism.

Terrorism can be combated more effectively through a combination of measures including international and national criminal law, as well as international relations, diplomacy, and, in particular, international cooperation. Conventions alone are unlikely to deter terrorists or their state sponsors.[81]

CONCLUSION

Our analysis of the international legal regimes governing transboundary environmental pollution and international terrorism has shown that, despite the differences between the two issues, they are nevertheless subject to similar general principles of international law. Furthermore, our examination illustrates that whereas the general rules of international responsibility, such as those deriving

79 2001 ILC Draft, *supra* note 8 (art. 7: "Excess of authority or contravention of instructions").

80 United States Diplomatic and Consular Staff, *supra* note 42, at 35.

81 Anthony Aust, *Comment on the Presentation of Volker Röben*, in TERRORISM AS A CHALLENGE FOR NATIONAL AND INTERNATIONAL LAW, *supra* note 44, at 823.

from the obligation of due diligence, might have to be adapted to the specific requirements of differing issues, in the end the general law of state responsibility can effectively cope with potential transboundary harm, irrespective of its nature. The starting point of the strengthening of the primary obligations of due diligence have been respectively, the *Trail Smelter* arbitration and the Corfu Channel case.

International law already has provisions to deal with issues of transboundary harm. Indeed, broad mechanism of attribution of acts of nonstate actors to states, the content of the general obligation of due diligence developed mainly by treaty law, as well as the partial recognition of individuals as "subjects" of international law, are all parts of our existing international legal order. The existence of an appropriate legal regime to deal with such issues, most especially international terrorism, puts into question the many and varied attempts to create new laws to deal with new problems. The fact is that there is an existing legal system, and the onus is on states to implement, promote, and enforce the existing rules, rather than attempting to recreate a wheel that already exists.

19 The Conundrum of Corporate Social Responsibility: Reflections on the Changing Nature of Firms and States

Peer Zumbansen

"We are fiddling, while Rome burns."[1]

INTRODUCTION

Although its value as precedent, paradigm, or standard-setter continues to be disputed,[2] the *Trail Smelter* arbitration[3] plays an important place in our contemporary search for adequate instruments and forms of international environmental regulation. Already the various contexts in which reference is made to *Trail Smelter* communicate its multifaceted messages.[4] Through the eyes of today, the *Trail Smelter* arbitration might seem outdated or skewed, in particular its disputed construction of Canada's responsibility for the transboundary harm that was brought about by a private enterprise.[5] And yet, whereas *Trail Smelter* stands apart from the later development of international law and the doctrine of state

[1] JULIET SCHOR, A SUSTAINABLE ECONOMY FOR THE 21ST CENTURY 16 (1998).

[2] *See* Ellis in this volume.

[3] *See Trail Smelter Arbitral Decision*, 33 AMERICAN JOURNAL OF INTERNATIONAL LAW 182 (1939) [hereinafter "*Trail Smelter* (1939)"]; *Trail Smelter Arbitral Decision*, 35 AMERICAN JOURNAL OF INTERNATIONAL LAW 684 (1941) [hereinafter "*Trail Smelter* (1941)"]. *See* Annex to this volume.

[4] Alfred Rubin, *Pollution by Analogy: The Trail Smelter Arbitration*, 50 OREGON LAW REVIEW 259, 259 (1971) ("Every discussion of the international law of pollution starts, and must end, with a mention of the *Trail Smelter Arbitration* . . . "). Republished in this volume. *See* Alhaji B.M. Marong, *From Rio To Johannesburg: Reflections On The Role Of International Legal Norms In Sustainable Development*, 16 GEORGETOWN INTERNATIONAL ENVIRONMENTAL LAW REVIEW 21, 71 (2003); *see also* Dinah Shelton, *Protecting Human Rights In A Globalized World*, 25 BOSTON COLLEGE INTERNATIONAL & COMPARATIVE LAW REVIEW 273, 307 (2002) (placing *Trail Smelter* in the context of the emerging international law of state responsibility for violations of other states' rights even where these violations originate from private actors.) *But see* Ellis in this volume.

[5] ". . . [U]nder the principles of international law, as well as of the law of the United States, no State has the right to use or permit the use of its territory in such a manner as to cause injury by fumes in or to the territory of another or the properties or persons therein, when the case is of serious consequence and the injury is established by clear and convincing evidence." *See Trail Smelter* (1941), *supra* note 3, at 716.

responsibility,[6] it continues to engage our imagination. *Trail Smelter* continues to resurface as a starting point for thinking about adequate ways to resolve border crossing environmental conflicts, but also other forms of transboundary harm.[7] It does so, precisely, by inspiring ongoing inquiries into the right balance between State-*versus* Market-based strategies of environmental regulation,[8] and by prompting many of the pertinent questions raised by deterritorialized corporate activities, highly diversified regulatory structures, and the limited enforcement competences of traditional political agencies.[9]

In this light, *Trail Smelter* must be read as inviting the following questions: Who bears responsibility for extraterritorial harm caused by transboundary pollution? Should state responsibility for privately induced transboundary harm replace or accompany private responsibility? Does either concept of responsibility respond to the particularly complex challenge posed by a proliferation of decreasingly well-defined environmental harms, dangers, and risks? Shifting, then, our focus away from the state as the exclusive author and enforcer of norms, and instead concentrating on the private actors themselves, leads to the next level of inquiry. What can these private or corporate actors contribute to a comprehensive program of environmental protection? How far can the state legitimately regulate corporate activity without infringing on the corporation's property rights? What is the best mixture of state regulation and corporate self-regulation? Can, and should, there be trade-offs between a "public" environmental protection agenda and the "private" acquisition, sale, and trading of pollution rights?

6 *See* Pierre-Marie Dupuy, *The International Law of State Responsibility: Revolution or Evolution?* 11 Michigan Journal of International Law 105 (1989).

7 *See* Hestermeyer, Anderson, and Miller in this volume. *See also* Edith Brown Weiss, *International Environmental Law: Contemporary Issues and the Emergence of a New World Order*, 81 Georgetown Law Journal 675, 677 (1993) ("... because the *Trail Smelter* arbitration is a rare example of international environmental adjudication [from an] early period, it has acquired an unusually important place in the jurisprudence of international environmental law.")

8 Eric Orts, *Reflexive Environmental Law*, 89 Northwestern University Law Review. 1227, 1241 (1995) ("The inadequacy of command-and-control regulation fuels the hottest growth industry in environmental law ... ")

9 *See* Bruce Ackerman & Richard B. Stewart, *Reforming Environmental Law*, 37 Stanford Law Review 1333 (1985); Bruce Ackerman & Richard B. Stewart, *Reforming Environmental Law: The Democratic Case for Market Incentives*, 13 Columbia Journal of Environmental Law 171 (1988); Benedict Sheehy, *Corporations and Social Costs: The Wal-Mart Case Study*, 24 Journal of Law & Commerce. 1, 8–9 (2004) (drawing attention to the focus on nuisances typical for cases of atmospheric pollution such as in the *Trail Smelter* decisions.) *See also* Michael Anderson, *Transnational Corporations and Environmental Damage: Is Tort Law the Answer?*, 41 Washburn Law Journal 399, 399 (2002 (referring to the *Trail Smelter* decisions as "paradigmatic for international environmental lawyers," at least in part because the case provided a "relatively straightforward" illustration of the issues at stake). For a foundational assessment of the emergence of a risk-society; *see* Ulrich Beck, Risk Society: Towards a New Modernity 1992; *see also* Ulrich Beck, *From Industrial Society to Risk Society: Questions of Survival, Social Structure and Ecological Enlightenment*, 9 Theory, Culture & Society 97 (1992).

This chapter cannot offer satisfying answers to all of these questions. Instead, it will explore the changing role of the state and private actors in environmental regulation. Hence, *Trail Smelter* is taken as starting point for a series of reflections on contemporary struggles over the proper balance between public and private instruments in the field of environmental protection. The assignment of legal responsibility through the 1938 and 1941 *Trail Smelter* arbitral decisions exposes two competing regulatory regimes that, in the post–*Trail Smelter* Era, have undergone dramatic developments and further differentiations. *Trail Smelter's* focus on the responsibility of the state for environmental protection has been strongly questioned. Through the important role that nonstate actors played in the progress of the arbitral deliberations,[10] *Trail Smelter* also implicitly raises various questions about emerging alternative, private regimes of societal self-regulation. Furthermore, *Trail Smelter* invites us to consider the shift away from substantive standards of harm toward the adoption of processes that permit a constant refinement not only of the analytical frame but also of the applied standards and the modes by which environmental goals are pursued.[11]

This chapter, thus, addresses the other side of *Trail Smelter*, attempting to unfold its as-yet untold story of corporate responsibility. This latter story speaks the language of private self-regulation of environmental protection, of corporate self-regulation through codes of conduct, of *soft law*, and of corporate social and environmental responsibility. These narratives emerge when we focus on the business corporation as the primary *locus* for the regulation of environmental harm. The private, self-regulatory challenge of corporate responsibility will thus be discussed in close connection with the changes of the public side of environmental law and the dramatic exhaustion of the state's regulatory capacities.[12] The chapter argues that there is a striking parallel between the state's transformation into a collaborative, contracting, and learning entity that remains dramatically dependent on private knowledge, and the modern business corporation's increasing assumption of public tasks as it grows in size and function, spanning its organization and activities across a seemingly borderless, global arena. Both the state and the firm depend on knowledge to inform their strategic choices in a regulatory environment that has ceased to lend itself to easy consensus, to meaningful

[10] *See* Miller in this volume.

[11] *See* Christopher D. Stone, Where the Law Ends: The Social Control of Corporate Behavior 122–23 (1975) (describing the process-oriented approach taken by the Environmental Protection Act).

[12] See Gunther Teubner, *Regulatory Law: Chronicle of a Death Foretold*, 1 Social & Legal Studies 451 (1992); *see also* Stepan Wood, *The Role of the International Organization for Standardization (ISO) in Governing Environmental Conflict and Corporate Social Responsibility in Developing Countries: Questions for Research*, in Propiedad, Conflicto y Medio Ambiente 15–56 (Beatriz Londoño Toro ed., 2004) (underlining the necessity to look beyond the traditional confines of environmental law and to recognize corporate codes as potentially powerful sources of effective environmental regulation).

deliberation, or to an effective, top-down production and implementation of norms.[13] In this volatile strategic environment, the acquisition and administration of knowledge becomes a challenge for both the "retreating" state[14] and the boundaryless firm.[15] For both, the production of knowledge is characterized by the fragility of intermittently accepted standards, recently taken decisions, and temporarily reached agreements. Fittingly, the "risk society"[16] is the backdrop against which "knowledge management" has emerged, which itself has become the foremost challenge to knowledge-driven entities. This chapter argues that this challenge is put to both the state and to the business corporation as the knowledge society moves the element of risk from the outside of management into its heart. Risks no longer lie outside of norm-production with regard to a norm's real-world consequences. Instead, risks are inherent to the decision-making process itself, as they originate in the social production of knowledge and norms rather than in nature itself.[17]

PUBLIC VERSUS PRIVATE ORDERING: WHAT WE KNOW AND WHAT WE DON'T KNOW ABOUT REGULATORY LAW

The *Trail Smelter* decisions were already contemplating many of these issues. As Russell Miller argues elsewhere in this volume, this contemplative legacy might outweigh the Tribunal's "correctness" in producing a doctrinally sound solution to every problem touched upon in the arbitration.[18] Although the Tribunal addressed the border-crossing dimension of the smelter's pollution, it ultimately took refuge in the construction of state responsibility without offering, in fact, a fully satisfactory justification for this holding.[19] The legacy of this holding is that it has no legacy. International environmental law has not embraced the Tribunal's construction of state responsibility,[20] but has instead embraced a "movement from status to contract."[21] Against the background of ever more, and ever *more detailed*, international environmental law treaties, the role of state responsibility "in addressing global environmental problems" has increasingly been questioned.[22] Meanwhile,

[13] IAN AYRES AND JOHN BRAITHWAITE, RESPONSIVE REGULATION: TRANSCENDING THE DEREGULATION DEBATE (1992).

[14] SUSAN STRANGE, THE RETREAT OF THE STATE (1996).

[15] Robert Boyer and J. Rogers Hollingsworth, *From National Embeddedness to Spatial and Institutional Nestedness*, in CONTEMPORARY CAPITALISM: THE EMBEDDEDNESS OF INSTITUTIONS 433 (Robert Boyer & J. Rogers Hollingsworth eds., 1997).

[16] *See, e.g.*, Michael Power, *From Risk Society to Audit Society*, 3 SOZIALE SYSTEME 3, 5 (1997).

[17] *Id.*

[18] *See* Miller in this volume.

[19] *Trail Smelter* (1941), *supra* note 3, at 716–717.

[20] *See* Ellis in this volume.

[21] For the origin of this idea, *see* HENRY SUMNER MAINE, ANCIENT LAW (1861).

[22] Jutta Brunée, *Of Sense and Sensibility: International Liability Regimes as a Tool for Environmental Protection*, *in* RECONCILING LAW, JUSTICE AND POLITICS IN THE INTERNATIONAL ARENA 110, 113–114 (Canadian Council on International Law 2003); Thomas Gehring & Markus Jachtenfuchs, *Civil Liability for Transboundary Environmental Harm*, 4 EUROPEAN JOURNAL OF INTERNATIONAL LAW 92,

the success of negotiation, contract, and treaty is, itself, curtailed by the fundamental complexities of environmental damages and the resulting challenge of addressing and regulating them.[23] In this light, the law of environmental protection is a case in point for the challenge to regulatory law under conditions of extreme uncertainty.[24] We therefore need to sketch the context of regulatory state law in which environmental law has so long been conceptualized.

Legacies and Legends of *Trail Smelter* in International Environmental Law

The rise and proliferation of international environmental law has shifted the focus from the nation state to international regimes and international, treaty-based conflict resolution.[25] Against this background the failure of the United States to embrace the Kyoto Protocol is a dramatic fall-back.[26] On a more conceptual level, the development of environmental law on the international plane has been propelled prominently by the rise of the precautionary principle that complements the traditional causation-based liability standard with a complex, scientifically open standard.[27] The precautionary principle's most important effect is its improved level of risk assessment with regard to unknown or, at least, difficult-to-assess environmental risks. Despite its disputed function as "precedent" for international environmental law, *Trail Smelter* can be read as addressing the same challenge that is met by the precautionary principle, including the procedural approach that led to the arbitration itself as well as to the bilateral resolution of the conflict.

93 (1993) ("With continuing industrialization and increasing risks of transboundary environmental damage, there is a growing need to establish specific rules that are precise enough to be applicable and that are therefore apt to be 'effective'. However, a derivation of these specific rules in the area of transboundary environmental damage from the general law of state responsibility involves a number of fundamental problems.")

23 Brunnée, *supra* note 22, at 114 ("Important aspects even of central international environmental norms remain opaque. To begin with, the legal status and content of several key norms, such as the precautionary principle, sustainable development, common concern, or common but differentiated responsibilities, remain contested.")

24 *See* Gehring & Jachtenfuchs, *supra* note 22, at 93; *See also*, Bratspies in this volume.

25 *See* Drumbl in this volume.

26 *See* Miller in this volume; *see also* the discussion in W. Bradnee Chambers, *Towards an Improved Understanding of Legal Effectiveness of International Environmental Treaties*, 16 GEORGETOWN INTERNATIONAL ENVIRONMENTAL LAW REVIEW 501 (2004).

27 Jon M. Van Dyke, *The Evolution and International Acceptance of the Precautionary Principle*, *in* BRINGING NEW LAW TO OCEAN WATERS 357 (David D. Caron & Harry N. Schreiber eds., 2004) (arguing that even with remaining skepticism as to its definitional boundaries, the precautionary principle "can no longer be ignored."). *See also* Carolin Hillemanns, *UN Norms on the Responsibilities of Transnational Corporations and Other Business Enterprises with regard to Human Rights*, 4 GERMAN LAW JOURNAL 1065 (2003) (pointing out that the recently published UN Norms on the Responsibilities of Transnational Corporations and Business Enterprises with Regard to Human Rights explicitly command that corporations observe the precautionary principle); *see also* Bratspies in this volume.

In addition, international law makers have viewed the precautionary principle's flexibility as an incentive to engage in international cooperation in environmental risk assessment. Here we can see the valuable enrichment of nation-state based research programs and institutions by an irrevocable trend toward cooperation among states and non-state actors, eventually fueling further the transnational and global growth of norms and standards, promoted by nonstate actors and international organizations.[28]

Meanwhile, this global development in norm-production[29] mirrors the dynamics of a changing regulatory regime within the nation-state. The political economy of the nation-state is most adequately described by the fragmentation of public arenas into specialized discourses and by the emergence of a comprehensive and increasingly deterritorialized knowledge economy.[30] The following section will, albeit briefly, highlight a number of decisive elements of the transformed landscape of the regulatory state at the beginning of the twenty-first century, and will measure the changing face of environmental risk regulation against a backdrop of larger transformations of the political economy of domestic and transnational regulation.

REGULATION AND DISPERSED KNOWLEDGE

The exhaustion of the regulatory (welfare) state on the domestic level,[31] and the much-disputed demise of the state as sole actor on the international legal and institutional plane,[32] constitute the fast-changing context of national, supranational, and transnational regulatory experiments, networks, and political hopes.[33] Environmental protection is a classic case of the "regulatory state under siege,"

28 *See, e.g.*, NILS BRUNSSON & BENGT JACOBSSON, A WORLD OF STANDARDS (2000); Walter Mattli & Tim Büthe, *Setting International Standards: Technological Rationality or Primacy of Power?*, 56 WORLD POLITICS 1 (2003); Peer Zumbansen, *Transnational Law*, *in*: ENCYCLOPEDIA OF COMPARATIVE LAW (Jan Smits ed., 2006).

29 *See* Andreas Fischer-Lescano & Gunther Teubner, *Regime Collisions: The Vain Search for Legal Unity in the Fragmentation of Global Law*, 25 MICHIGAN JOURNAL OF INTERNATIONAL LAW 999 (2004).

30 *See* NICO STEHR, WISSEN UND WIRTSCHAFTEN (2001); MICHAEL S. PIORE & CHARLES SABEL, THE SECOND INDUSTRIAL DIVIDE: POSSIBILITIES FOR PROSPERITY (1984); PETER BURKE, A SOCIAL HISTORY OF KNOWLEDGE: FROM GUTENBERG TO DIDEROT (2000).

31 Jürgen Habermas, *The New Obscurity: The Crisis of the Welfare State and the Exhaustion of Utopian Energies* [1985], *in* THE NEW CONSERVATISM: CULTURAL CRITICISM AND THE HISTORIANS' DEBATE 48 (Shierry Weber Nicholsen trans., 1989).

32 Paul Schiff Berman, *From International Law to Law and Globalization*, 43 COLUMBIA JOURNAL OF TRANSNATIONAL LAW 485, 487 (2005); Peer Zumbansen, *Die Vergangene Zukunft des Völkerrechts*, 34 KRITISCHE JUSTIZ 46, 50–53 (2001).

33 See David Levi-Faur, *The Global Diffusion of Regulatory Capitalism*, 598 ANNALS OF THE AMERICAN ACADEMY OF POLITICAL AND SOCIAL SCIENCE 12 (2005); Oren Perez, *Normative Creativity and Global Legal Pluralism: Reflections on the Democratic Critique of Transnational Law*, 10 INDIANA JOURNAL OF GLOBAL LEGAL STUDIES 25 (2003).

because it constantly faces the multifaceted challenge of identifying the problem, and designing the instruments to combat the problem, while at the same time defining the appropriate scope and direction of the state's response.[34] A clear expression of this normative and institutional challenge is the United States' 1984 *Chevron* decision, in which the U.S. Supreme Court recognized the crucial role of agencies in interpreting statutes.[35] *Chevron* is rightly regarded as a paradigmatic case in transforming American administrative law into a responsive, knowledge-based regulatory regime.[36] Like *Trail Smelter, Chevron* embodies the recognition of the knowledge-based, interpretative role of administrative agencies that is particularly prevalent in the field of environmental law where expert knowledge has long been considered crucial.[37] *Chevron* forms part of a long-term transformation of a legalistic, functionalist understanding of the state into an "expertise-driven" regulatory regime.[38] This transformation is marked by the increased inclusion of private actors in public action, raising not only far-reaching questions as to its democratic accountability and legitimacy[39] but also to the fundamental dilemma of how the state is to gain, produce, and process the necessary regulatory knowledge that is needed.

An essential feature of this contemporary transformation in regulatory theory concerns not only the institutional dimension of the regulatory response to transboundary harm, but its normative quality. Administrative agencies and other regulatory actors increasingly resort to procedural, experimental and, ultimately, *learning* forms of regulation when designing statutes, standards, and sanctions targeted at polluters and other addressees of regulation.[40] The most remarkable feature of these new forms of regulation is the regulator's recognition of its likely failure in designing an ultimately successful and effective regulatory instrument,

34 Erhard Denninger, Verfassungsrechtliche Anforderungen an die Normsetzung im Umwelt- und Technikrecht (1990); Ackerman & Stewart, *Reforming Environmental Law, supra* note 9 .

35 Chevron, Inc. v. Natural Resources Defense Council, Inc., 467 U.S. 837 (1984).

36 See, e.g., E. Donald Elliott, *Chevron Matters: How the Chevron Doctrine Redefined the Roles of Congress, Courts, and Agencies in Environmental Law*, 17 Villanova Environmental Law Journal 1, 3 (2005).

37 *See id.* at 14–15.

38 *See* James M. Landis, the Administrative Process (1938).

39 Jody Freeman, *The Private Role in Public Governance*, 75 New York University Law Review 543, 547 (2000) (referring to the prevailing "hierarchical, agency-centered conception of administrative power").

40 *See* Karl-Heinz Ladeur, *Coping with Uncertainty: Ecological Risks and the Proceduralization of Environmental Law*, in Environmental Law and Ecological Responsibility: the Concept and Practice of Ecological Self-Organization 299 (Gunther Teubner et al. eds., 1994); Karl-Heinz Ladeur, Das Umweltrecht der Wissensgesellschaft (1995); *see also* Liora Salter, *Institutional Learning in Standards Setting, in* Innovation and Social Learning: Institutional Adaptation in an Era of Technological Change 65 (Meric S. Gertler & David A. Wolfe eds., 2002).

in particular, when dealing with an unidentified polluter, a plurality of actors,[41] or, more generally, when having to base a regulatory response or program on constantly changing factual and technical data.[42] Although the *Trail Smelter* Tribunal decidedly faced the last of these factors, it was spared the complex questions of causation and proof implicated when there are multiple or unknown actors.

The postindustrial, "contracting"[43] state no longer finds itself on top of a hierarchical order in which it can effectively set the direction of societal change. Instead, the state has become a label for the political system that forms a mere part of a more comprehensive, encompassing social arrangement that has neither center nor top, but is broken down into a multitude of social systems of autopoietic reproduction.[44] Likewise, the knowledge society knows no central regulator, but is made up of a multitude of decentralized, dramatically fragmented *loci* of knowledge production. This new structure has dramatic consequences for our understanding of law, which hitherto had been described in close association with the entity of the state and its political agencies of norm-creation.[45] In contrast, the law of the knowledge society is decentered from the political system, it forms in communication with different social systems and, consequently, embraces constantly changing conditions of experimental, reflexive norm-production. In this light, the separation of "state and society," the distinction between public and private law, must be seen as historical concepts used to describe law's attachment to particular institutions of norm-creation. The determining characteristic of the "noninterventionist," "postregulatory" law of the knowledge society is its *responsiveness*.[46] Incessantly adapting and changing, this newly responsive vision of law[47] can, at least in theory, assume the regulatory stance most appropriate at any given moment.

Against this background, then, how can we describe the state and the changing nature of the state in supervising these activities? Where the state formerly assumed legislative authority in regulating fields of corporate action with regard to environmental protection, the heavy reliance on scientific evidence and standardization by private entities necessitates a thorough overhaul of the state's

[41] *See* Gunther Teubner, *The Invisible Cupola: From Causal to Collective Attribution in Ecological Liability*, *in* ENVIRONMENTAL LAW AND ECOLOGICAL RESPONSIBILITY, *supra* note 40, at 17.

[42] *See generally* ULRICH BECK, ECOLOGICAL POLITICS IN AN AGE OF RISK (1995). *Trail Smelter* raised the problem of a known polluter but constantly changing technical data.

[43] IAN HARDEN, THE CONTRACTING STATE (1992); Jody Freeman, *The Contracting State*, 28 FLORIDA STATE LAW REVIEW 155 (2000).

[44] See NIKLAS LUHMANN, OBSERVATIONS ON MODERNITY (William Whobrey transl., 1998); Gunther Teubner, *The King's Many Bodies: the Self-Destruction of Law's Hierarchy*, 34 LAW & SOCIETY REVIEW 753 (1997).

[45] This challenge posed by the knowledge society is striking, for example, in constitutional law. *See, e.g.*, Neil Walker, *The Idea of Constitutional Pluralism*, 65 MODERN LAW REVIEW 317 (2002).

[46] *See* AYRES AND BRAITHWAITE, RESPONSIVE REGULATION, *supra* note 13.

[47] Gunther Teubner, *Substantive and Reflexive Elements in Modern Law*, 17 LAW & SOCIETY REVIEW 239 (1983).

prerogative in regulating pollution and other environmental harm. A dense interwoven network of public and private action materializes. The changed nature of the supervision-state (or, the environmental, regulatory, prevention state[48]) is highly volatile, fragile, and dependent on fragmented knowledge, domestically and transnationally.[49] Success depends on the state being able to absorb private knowledge in an optimal manner. The state, therefore, is increasingly expected to engage in innovative cooperation with private parties in initiating, funding, and generating scientific research. Enforcing an effective environmental protection scheme thus requires that state to transform itself from the regulatory interventionist state and to adopt new roles as a moderator of private, commercial self-regulation and public policy interests, on the one hand, and tort law enforcer – both based on national and international law – on the other. Facing constantly growing and differentiating governance problems, the state is pressed to rely on private self-organization in reaching its goals. Although the state must constantly augment and update its working knowledge of these forms of public-private governance, it is necessary to gain a better understanding of the actor on the other, "private" side of this relationship. It is against this background that the following section will attempt to unfold the conundrum of corporate social and environmental responsibility.

THE OTHER SIDE OF *TRAIL SMELTER*: TRACING NARRATIVES OF CORPORATE RESPONSIBILITIES

Whereas *Trail Smelter* clearly speaks to the public, state-oriented dimension of transnational responsibility, its other message is much quieter, less audible, discernible only in its conceptual, theoretical background. It is, in a nutshell, the story of the rise and fall and, eventually, the rebirth of a dramatically transformed regulatory private law.[50] On first impression, *Trail Smelter* seems silent on this private dimension of conflict resolution. However, in the background of the Tribunal's discussion of state responsibility, we can easily discern the potential and the promise of *private* responsibility. Hence, our focus on *Trail Smelter*'s second, untold story shall be on corporate responsibility. It is a story that we find in our contemporary reading of *Trail Smelter*'s central focus on state responsibility. But, our focus will neither be that nor the often-discussed, substantive side of the

48 For these categories, *see* Gralf-Peter Calliess, Prozedurales Recht (1999); Helmut Willke, Supervision des Staates (1997).

49 *See, e.g.*, Anne-Marie Slaughter, A New World Order (2004) (describing worldwide communications between various regulatory agencies, and other state and non-state bodies).

50 *See* Teubner, *supra* note 47; *see also* Peer Zumbansen, *Quod Omnes Tangit: Globalization, Welfare Regimes and Entitlements*, *in* The Welfare State, Globalization, and International Law 135–173 (Eyal Benvenisti & Georg Nolte eds., 2003)

corporation's responsibility.[51] Instead, we are interested in the law-making dimension of corporate social and environmental responsibility (CSR) – in the regulatory context and in the political economy of CSR to be precise. This section will argue that to understand the regulatory dimension of the firm's responsibilities to society at large, we must now, after our brief exploration of the changing dimensions of public, state-centred regulation, look to the corporation itself, to its role and function in a dramatically changing globalized socioeconomic environment.

In studying the political economy of corporate (environmental, social) responsibility, we must place the CSR debate within the context of three connected discourses that indirectly speak to corporate social responsibility. These discourses concern the briefly sketched themes of environmental regulation through "hard" and "soft" law as well as the transformation of the regulatory state into a supervising and moderating state in the knowledge economy. The third discourse that we need to explore concerns the *political economy of the "embedded" corporation*, in other words, the domestic and transnational, regulatory framework and context of corporate activity, but also the norms internal to the organization and governance of the business corporation.[52] Taken together, these discourses inform any assessment of the corporation's larger social, political, and environmental responsibilities. It is the central contention of this chapter, that unfolding the conundrum of corporate social responsibility ("*What is the scope and the content of the corporation's responsibility/ies?*"; "*What are the adequate regulatory mechanisms to implant, consolidate, and enforce these responsibilities?*"; "*Does this lead to an undue (public) intervention into (private) corporate law,* in other words: *Does 'corporate law' extend to the regulation of* CSR?") against the just-described background of surrounding theoretical inquiry can help us to adequately address the wider conditions of any meaningful concept of CSR.

It is here, where we shall turn our focus away from the usual target, the state and its regulatory responses and, instead, *focus on the corporation itself*. Doing so will allow us to gain a better understanding of the social actor that is most often involved in dramatic cases of environmental harm and, increasingly, implicated in the continuing search for an appropriate regulatory response.[53]

[51] This question is captured in the famous debate between Adolf Berle and E. Merrick Dodd. *See* Adolf Berle, *Corporate Powers as Powers in Trust*, 44 HARVARD LAW REVIEW 1049 (1931); E. Merrick, Dodd, *For Whom are Corporate Managers Trustees?* 45 HARVARD LAW REVIEW 1145 (1932); *see also* Lord Wedderburn, *The Legal Development of Corporate Responsibility: For Whom Will Corporate Managers Be Trustees? in* CORPORATE GOVERNANCE AND DIRECTORS' LIABILITIES 3 (Klaus J. Hopt & Gunther Teubner eds., 1985).

[52] For a brilliant exposition on this point, *see* Walter Powell, *The Capitalist Firm in the Twenty-First Century: Emerging Patterns in Western Enterprise, in* THE TWENTY-FIRST-CENTURY FIRM. CHANGING ECONOMIC ORGANIZATION IN INTERNATIONAL PERSPECTIVE 33 (Paul Dimaggio ed., 2001).

[53] *See, e.g.*, Michael Power, *Constructing the Responsible Organization: Accounting and Environmental Representation, in* ENVIRONMENTAL LAW AND ECOLOGICAL RESPONSIBILITY, *supra* note 40, at 369.

What's in a Firm? Unfolding the Conundrum of State Responsibility

The untiring discussion over scope and content, direction and aspirations of "Corporate Social Responsibility" (CSR) is a reminder of what we know and do not know about the very subject and object of our contemporary explorations of the social, environmental, in short: the larger societal, public obligations of the corporation.[54] Whether a corporation is merely a private, profit-oriented undertaking, or whether – perhaps in addition – it bears noncontractual responsibilities to society at large, employees, creditors, the community, and other stakeholders, has not ceased to occupy our minds.[55] This inquiry has certainly contributed to a far-reaching, societal discussion about the function and role of large corporations in society, in both the domestic and the transnational context. It has furthered and instigated a more popular awareness of corporate activities worldwide.[56] But, although case law and literature regarding the corporation's larger role in society abound, little knowledge apparently is advanced about the corporation *itself*. Surely, the corporation, the large publicly held corporation, the multinational enterprise, the embedded corporation, are researched, analyzed, and explored with great scholarly earnestness and policy interest. And, yet, it seems that these many assessments of shareholder primacy or stakeholder theory actually contribute very little to a better understanding of the corporation. And with increasing contestation of the public and political role of corporations,[57] we risk losing sight of the very *locus* of ideological battles.

The "unknown firm" is surely not a promising starting point for our continued discussion of the corporation's societal responsibilities. The corporation remains unknown, because neither the CSR debate nor the larger dispute over convergence or divergence of corporate governance systems[58] deliver a more concrete

54 *See* Joel Bakan, The Corporation: The Pathalogical Pursuit of Profit and Power (2004), which has been made into a multiple award-winning documentary film. *See also* Lawrence Mitchell, Corporate Irresponsibility: America's Newest Export (2001).

55 Simon Deakin, *Squaring the Circle? Shareholder Value and Corporate Social Responsibility in the U.K.*, 70 George Washington Law Review 976 (2002); David Wheeler & David Grayson, *Business and Its Stakeholders*, 32 Journal of Business Ethics 101 (2001).

56 Naomi Klein, No Logo (2000); Saskia Sassen, Globalization and Its Discontents (1998); Peter Cornelius & Bruce Kogut, *Creating the Responsible Firm: In Search for a New Corporate Governance Paradigm*, 4 German Law Journal 45 (2003); Ruth O. Kuras, *Corporate Social Responsibility: A Canada-U.S. Comparative Analysis*, 28 Manitoba Law Journal 303 (2002); Wesley Cragg & Alan Greenbaum, *Reasoning About Responsibilities: Mining Company Managers on What Stakeholders are Owed*, 39 Journal of Business Ethics 319 (2002).

57 The seminal text is Adolf A. Berle, The 20th Century Capitalist Revolution (1954). For a contemporary assessment, *see* Amar Bhatia, *Reading Corporate Law: The Fate of Berle's 20th Century Capitalist Revolution, in* 1 International Studies in Comparative Law and Political Economy (Peer Zumbansen, and John Cioffi eds., *forthcoming* 2006).

58 *See* the provocative contribution to the debate by Henry Hansmann & Reinier Kraakman, *The End of History in Corporate Law*, 89 Georgetown Law Journal 439 (2001); *see also* the *Habilitation* by Hansmann's German student Mathias Siems, Die Konvergenz der Rechtssysteme im Recht der Aktionäre (2005).

description of what actually goes on inside of the corporation, what the corporation *does*, how it *decides*, and how it *adapts to* a dramatically nervous economic and political environment. While an analysis of the socioeconomic and regulatory context and environment in which corporations operate may be necessary to understand the embeddedness of the corporation, our inquiry must extend to the corporation itself. In other words, while we must focus on the "political economy of the corporation,"[59] its socioeconomic, regulatory context in light of national path dependencies and international comparisons,[60] our other target must be the *corporation as a complex organizational entity of social learning*. With a focus on the organizational design of today's corporation, we can begin to understand and to conceptualize the corporation as a complex and innovative institution of social learning in the context of building a sustainable economy. Does *Trail Smelter* offer any guidance or lesson?

The still governing corporate law theory that describes the firm as a nexus of contracts must be reread in light of the changes that affect both the state's and the corporation's activity. Both operate under conditions of an eroding knowledge base and the ensuing demand for better and more adequate risk assessment. The firm becomes, especially as it assumes ever more public tasks in infrastructure provision and public service delivery, a hybrid actor – neither private nor public – at a crossroads of intertwining demands from the "state" and the "market." The theory of the firm can thus be compared to contemporary theoretical enquiries into the theory of the state.

Postheroic Management

The key to understanding the contemporary corporation in the political economy of the deterritorialized knowledge economy is to focus on its capacity to remain innovative.[61] The firm's capacity to engage in innovative production depends on its ability to constantly grow, adapt, and learn. This it can do by letting go of traditional modes of command and control, and, instead, embracing an ironical, distancing, reflecting, and *postheroic* attitude to corporate governance and management.[62] Our urgently sought definition of corporate responsibilities, its public duties and obligations to society at large, especially in an era of scandalous corporate crime,[63] depends fundamentally on our understanding of the firm itself. It is here where we recognize the relevance for our *Trail Smelter* derived theme questioning

[59] The Political Economy of the Company (John Parkinson et al., eds., 2000); John W. Cioffi, *Governing Globalization? The State, Law, and Structural Change in Corporate Governance*, 27 Journal of Law & Society 572 (2000).

[60] Cioffi, *supra* note 59; Mary O'Sullivan, *The Political Economy of Comparative Corporate Governance*, 10 Review of International Political Economy 23–72 (2003).

[61] *See* Mary O'Sullivan, *The Innovative Enterprise and Corporate Governance*, 24 Cambridge Journal of Economics 393 (2000).

[62] Dirk Baecker, Postheroisches Management (1994).

[63] *See* Bakan, *supra* note 54; Klein, *supra* note 56.

the role of the corporation in society, which, in modern parlance, has become the fierce battle between shareholder-value oriented systems of corporate control and those that place a higher emphasis on workers' voice, participation, industrial relations, and a wider consideration of the firm's stakeholders.[64] Whether we lay our emphasis on the shareholder or on the stakeholder dimension of the firm, will have a significant impact on our assignment of duties and obligations of the firm.[65] This is particularly prevalent with regard to disclosure. Where corporate governance reform is predominantly concerned with shareholders, the emphasis is likely to remain placed – at least for the time being – on improvements in the financial auditing schemes. In contrast, were our focus on an improved environmental accountability of the firm, we would indeed direct our initiatives at other areas in corporate organization. Environmental internal auditing, in fact, constitutes a prime example of stakeholder-oriented developments in environmental, corporate self-regulation.[66] Restated thus, the question of the firm's responsibilities cannot be separated from our foundational understanding of the firm.

However, this perspective on the connection between the *political economy of the firm* and the firm's environmental (or wider social) responsibilities, fails to account for our remaining lack of knowledge of the corporation *itself*. Today's large, publicly held and globally operating firms, escape clear definitions, both with regard to their core activities or "competences"[67] and their organizational structure. Increasingly, firms have become unbounded, borderless and virtual, with activities that span multiple areas of industry, manufacture or soft products. Echoing many of the challenges that we identified for the state today in a complex society, the firm constitutes a highly complex organization that operates in a volatile regulatory and competitive environment. Rejecting thus, both overly simplistic categorizations of the firm as either shareholder or stakeholder oriented, the firm of the twenty-first century challenges our learned ways of organizing social behaviour. Shifting the CSR debate away from the control-oriented images of the corporation, is an essential step in beginning to understand the question of the firm's social responsibilities. Instead, the firm must be viewed within a complex web made up of the socioeconomic framework, the embeddedness of the corporation, the internal organization of corporate governance, and the organizational experiments of a constantly evolving, dynamic, multipolar business enterprise. Although the latter two dimensions describe the corporation as a communicative, self-referential being, the first dimension speaks of the embeddedness of the firm,

64 *See* Hansmann & Kraakman, *supra* note 58, on the one hand, and William W. Bratton, *Enron and the Dark Side of Shareholder Value*, 76 Tulane Law Review 1275 (2002), on the other.

65 Simon Deakin, *supra* note 55.

66 See Orts, supra note 8, at 1303–1304; see also Power, From Risk Society to Audit Society, supra note 16.

67 C. K. Prahalad and Gary Hamel, *The Core Competence of the Corporation*, 68 Harvard Business Review 79 (1990).

its socioeconomic and political place in a dramatically changing local and global environment. With the corporation increasingly assuming formerly public functions (welfare, pensions, medical care), *Trail Smelter* reminds us that we must reconsider our understanding of the firm's allegedly exclusively private character. Where it has become increasingly difficult to assign to social activities the label public or private, this certainly extends to our conception of the business corporation. Understanding the firm is the first step toward understanding the challenge of corporate social responsibility.

20 A Pyrrhic Victory: Applying the *Trail Smelter* Principle to State Creation of Refugees

Jennifer Peavey-Joanis

INTRODUCTION

History is repelete with tales of mass expulsions. For centuries, these expulsions have been tolerated – "condemned only sporadically, while repeatedly being undertaken with international acquiescence, indifference, or approval."[1] It was only with the pervasive mass expulsions forced by the Nazi regime in the period leading up to, and during, World War II that international attitudes began to shift. The Nazi-era experiences prompted international consideration of the plight of refugees, and gave birth to a comprehensive and multilateral attempt to clarify international and state responsibilities toward transboundary refugees fleeing state turmoil.

Under the auspices of the United Nations, the international community codified a series of treaties setting out state responsibilities regarding the duty to care for refugees.[2] These agreements start from the proposition that war and internal strife will occur, and will necessarily result in involuntary migrations.[3] The agreements also assume that under conditions of scarcity, such migrations will invariably pose complex problems in the land of refuge.[4] Nonetheless, through these treaty regimes, the international community has largely accepted that the

[1] Jack I. Garvey, *Toward a Reformulation of International Refugee Law*, 26 HARVARD INTERNATIONAL LAW JOURNAL 483, 488 (1985).

[2] Convention Relating to the Status of Refugees, July 28, 1951, 19 U.S.T. 6259, 189 U.N.T.S. 137 [hereinafter "1951 Convention"] and the Protocol Relating to the Status of Refugees, Jan. 31, 1967, 19 U.S.T. 6223, 606 U.N.T.S. 257 [hereinafter "1967 Protocol"].

[3] *See generally* Geneva Convention Relative to the Protection of Civilian Persons in Time of War of August 12, 1949, 75 U.N.T.S. 287, Art 44 [hereinafter "1949 Geneva Convention"] and Protocol Additional to the Geneva Conventions of 12 August 1949, and Relating to the Protection of Victims of International Armed Conflicts (Protocol I), adopted June 8, 1977, 1125 U.N.T.S. 3 [hereinafter "1977 Protocol"].

[4] 1951 Convention, *supra* note 2 at 15 (recognizing, in the Preamble, that "the grant of asylum may place unduly heavy burdens on certain countries, and that a satisfactory solution of a problem of

"unduly heavy burden" of caring for refugees rests primarily on the refuge state and secondarily upon the international community as a whole.[5]

In the latter part of the twentieth century, political upheavals again created large-scale refugee populations. Refuge states began to balk at their increasingly expensive duties towards refugees under these regimes, at least in part because meeting those duties diverted funds from more domestic concerns.[6] This growing resistance to the international agreements defining state duties toward refugees prompted the question of whether a relatively clear liability regime like that laid out in *Trail Smelter* might offer a better approach.

This chapter explores whether *Trail Smelter* offers a desirable alternative approach towards state responsibility and liability for refugees in situations where persecuted civilians have been forced to flee across international borders, thereby causing economic, social, and other burdens to the nonrefouling state.[7] There are two levels to this inquiry. First, on a literal level, can refugees fit into the definition of "harm" to a neighboring state that, according to *Trail Smelter*, states have an obligation to prevent? Second, even if the refugee context fits into the *Trail Smelter* framework, is it useful to conceive of a stream of refugees flowing across a border as akin to a stream of air pollution flowing across that same border?

THE *TRAIL SMELTER* PRINCIPLE

The smelter in Trail, British Columbia was a source of aggravation for property owners in Washington State for decades.[8] Washington farmers initially asserted individual claims that their land, crops, livestock, timber, orchards, town, and business enterprises were damaged by the noxious emissions from the Trail smelter. These claims were eventually pursued by the U.S. government on the farmers' behalf. Canada similarly took up the cause of its corporate citizen, Cominco. Under the Convention establishing the *Trail Smelter* Tribunal, Canada acknowledged a duty to ensure that agencies within its territory utilize that

which the United Nations has recognized the international scope and nature cannot therefore be achieved without international co-operation . . . ")

5 *Id.* at art. 33(1). "No contracting State shall expel or return ("refouler") a refugee in any manner whatsoever to the frontiers of territories where his life or freedom would be threatened on account of his race, religion, nationality, membership of a particular social group or political opinion."

6 Despite the fact that sovereign states entered into the 1951 Convention and 1967 Protocol freely, the UNHCR "detected a distinct trend in an increasing number of States to move away from a law or rights-based approach to refugee protection towards more discretionary and ad hoc arrangements that gave greater primacy to domestic concerns rather than to international responsibilities." 1999 United Nations Yearbook 1127, 1134, U.N. Sales No. E.01.I.4.

7 "Refouler" [French], expulsion or return of a refugee from one state to another. Black's Law Dictionary 1307 (8th ed. 2004).

8 *See* Allum in this volume.

territory in conformity with obligations under international law.[9] The Tribunal's decision awarding damages to the United States was couched in the now-famous statement that:

> no state has the right to use or permit the use of its territory in such a manner as to cause injury by fumes in or to the territory of another or the properties or persons therein, when the case is of serious consequence and the injury is established by clear and convincing evidence.[10]

Thus, damages were awarded largely on a nuisance theory – the famous *Trail Smelter* principle quoted above is basically a restatement of the maxim of *sic utere tuo ut alienum non laedaus* (use your own property so as not to injure that of another) – and were based on the litany of scientifically provable direct damages that the Tribunal detailed in its lengthy arbitral decisions. In addition, the Tribunal answered the third question posed by the parties, "[w]hat measures or régime should be adopted and maintained by the Trail smelter?"[11] by establishing a temporary, and then ultimately, a permanent, operating régime for the smelter. This permanent régime included a series of monitoring, reporting, and capacity-building requirements.[12]

In imposing an operating régime on the smelter, the Tribunal recognized that neither government would accept complete cessation of industrial activity.[13] By setting acceptable emissions levels, the Tribunal and the two nations signaled a willingness to tolerate a degree of transboundary harm. At the same time, however, the Tribunal flatly and summarily rejected as meritless the United States' claim for damages based on a violation of its territorial sovereignty.[14] These last two points, far more than the nuisance theory articulated in the famous *Trail Smelter* principle, have lasting import for how *Trail Smelter* fits, or more accurately does not fit, into any consideration of international duties towards refugees.

9 *Trail Smelter Arbitral Decision*, 33 AMERICAN JOURNAL OF INTERNATIONAL LAW 182, 184 (1939) [hereinafter "*Trail Smelter* (1939)"]. *See* Annex to this volume.

10 *Trail Smelter Arbitral Decision*, 35 AMERICAN JOURNAL OF INTERNATIONAL LAW 684, 716 (1941) [hereinafter "*Trail Smelter* (1941)"]. *See* Annex to this volume.

11 *Trail Smelter* (1939), *supra* note 9, at 183; *Trail Smelter* (1941), *supra* note 10 at 686.

12 *Trail Smelter* (1941), *supra* note 10, at 727–731.

13 John E. Read, *The Trail Smelter Dispute*, 1 CANADIAN YEARBOOK OF INTERNATIONAL LAW 213, 224 (1963) ("The acceptance of the principle of absolute cessation of damage might have shut down the Trail Smelter; but it would also have brought Detroit, Buffalo, and Niagara Falls to an untimely end."). Republished in this volume.

14 *Trail Smelter* (1939), *supra* note 9, at 208 ("The Tribunal is, therefore, of opinion that neither as a separable item of damage nor as an incident to other damages should any award be made for that which the Unites States terms 'violation of sovereignty.'")

REFUGEE LAW AND CURRENT FLOWS ACROSS INTERNATIONAL BORDERS

The first comprehensive instrument protecting refugees, and governing their care and treatment during mass migrations, was created in the direct aftermath of World War II. The 1951 Convention Pertaining to the Status of Refugees[15] derived its philosophical origins from the 1948 Universal Declaration of Human Rights.[16] The treaty defines a refugee as any person who has:

> a well-founded fear of being persecuted for reasons of race, religion, nationality, membership of a particular social group or political opinion, is outside the country of his nationality and is unable or, owing to such fear, is unwilling to avail himself of the protection of that country; or who, not having a nationality and being outside the country of his former habitual residence as a result of such events, is unable or, owing to such fear, is unwilling to return to it.[17]

This convention, which focused narrowly on the plight of European refugees,[18] was expanded through the 1967 Protocol, to encompass refugees throughout the world.[19] Signatories to the 1951 Convention and 1967 Protocol have assumed a series of duties toward refugees. Nonrefouling states commit to providing refugees with: general freedoms (nondiscrimination, religion); juridical status; the ability to engage in gainful employment; and the benefits of welfare (housing, education, and other public relief).[20] The 1951 Convention and the 1967 Protocol do not recognize a state duty to prevent the creation of refugees, nor do they impose a duty on refugee-creating states to reimburse the non-refouling state for the costs of care or other damages. However, the General Assembly has asserted that refugee-creating states bear the twin duties of repatriating "refugees wishing to return to their homes" and compensating "those choosing not to return . . . [including] the

[15] 1951 Convention, *supra* note 2.

[16] "All human beings are born free and equal in dignity and rights. They are endowed with reason and conscience and should act towards one another in a spirit of brotherhood." Universal Declaration of Human Rights art. 1, G.A. Res. 217A, U.N. GAOR, 3d Sess., Supp. No. 3, U.N. Doc. A/810 (1948).

[17] 1951 Convention, *supra* note 2 at art. 1(A)(2), as amended by the 1967 Protocol, *supra* note 2 at art. 1(2).

[18] 1951 Convention, *supra* note 2 at art. 1B(2)(a).

[19] 1967 Protocol, *supra* note 2. The Preamble announced that "[c]onsidering it is desirable that equal status should be enjoyed by all refugees covered by the definition in the [1951 Convention] irrespective of the dateline 1 January 1951 . . . " To that end, the parties agreed to apply the Convention's protection without regard to temporal (art. 1(2)) or "geographic limitation" (art. 1(3)).

[20] 1951 Convention, *supra* note 2. Obligations of contracting parties are limited to the degree that their nationals or other aliens are afforded the same benefits. Noticeably absent is an enforcement mechanism. As a result, there are varying levels of compliance that are highly dependent upon the "overall strength of the rule of law in countries of asylum." Merrill Smith, *Development Aid for Refugees: Leveraging Rights or Missing the Point?*, *in* World Refugee Survey 2005 – Inventory of Refugee Rights 4 (2005).

loss of or damage to property."[21] To date, a total of 145 states are parties to one or both of these instruments.[22]

Within this structure, the General Assembly has charged the United Nations High Commissioner for Refugees (UNHCR) with "supervising . . . the protection of refugees"[23] and with providing "international protection" and assistance to refugees through "permanent solutions."[24] Despite the many subsequent agreements supplementing the 1951 Convention and 1967 Protocol, both the UNHCR Executive Committee (ExCom) and the General Assembly have reaffirmed these early documents as the foundation of the international refugee regime.[25]

According to the UNHCR, the total population "of concern" in 2004 was 19.2 million, with 48%, or 9.2 million, classified as refugees.[26] Disturbing as those numbers are, they grossly understate the scope of the actual refugee problem that exists in the world. For example, excluded from this total are the many thousands who flee social conditions yet cannot demonstrate the "well-founded, individualized fear of persecution" that is the *sine qua non* of the international definition of a refugee.

APPLYING *TRAIL SMELTER* TO CREATION OF REFUGEES

In many areas of international law, it is now generally accepted that "every state is obligated to respect the territorial integrity and rights of other states."[27] To the extent *Trail Smelter* embodied this principle, the arbitration's legacy is far-reaching. However, less than a decade after the *Trail Smelter* decision, the international community determined that where refugees are concerned, the focus of state responsibility and liability would be on the nonrefouling state; a 180-degree turn from the *Trail Smelter* format. Although clearly, people are not "noxious

[21] Luke T. Lee, *The Right to Compensation: Refugees and Countries of Asylum*, 80 AMERICAN JOURNAL OF INTERNATIONAL LAW 532, 535 (1986) (*quoting* G.A. Res. 194 (III), U.N. GAOR, U.N. Doc. A/810 at 21 (1948) (establishing the United Nations Conciliation Commission for Palestine). This principle of return and/or compensation was further endorsed by the Report of the Group of Governmental Experts on International Co-operation to Avert New Flows of Refugees, U.N. Doc. A/41/324 (1986) [hereinafter "Report of the Group"], unanimously adopted by the General Assembly in 1986. G.A. Res. 41/70, U.N. GAOR, 41st Sess., U.N. Doc. A/RES/41/70 (1986). *See* Lee for specific instances where countries have compensated refugees for state-forced expulsion. *But see* Eric Rosand, *The Right to Compensation in Bosnia: An Unfulfilled Promise and a Challenge to International Law*, 33 CORNELL INTERNATIONAL LAW JOURNAL 113 (2000).

[22] A list of parties to the 1951 Convention and the 1967 Protocol is available at http://www.unhcr.ch/cgi-bin/texis/vtx/basics.

[23] 1951 Convention, *supra* note 2, at 15 (Preamble).

[24] G.A. Res. 429(v), U.N. GAOR, 5th Sess., Supp. No. 20, at 48, U.N. Doc.A/1775 (1950).

[25] UNITED NATIONS YEARBOOK, *supra* note 6.

[26] 2004 Global Refugee Trends, Overview of Refugee Populations, New Arrivals, Durable Solutions, Asylum-Seekers, Stateless and Other Persons of Concern to UNHCR (June 17, 2005), *available at*: http://www.unhcr.ch/cgi-bin/texis/vtx/statistics/ opendoc.pdf?tbl=STATISTICS&id=42b283744.

[27] *See* Garvey, *supra* note 1 at 494.

fumes," certain similarities exist between the *Trail Smelter* situation and refugee populations:

> both may cross international boundaries from countries of origin; both such crossings are preventable by the countries of origin; both such crossings are not made with the voluntary consent of the receiving states; and both such crossings may impose economic and social burdens upon the receiving states . . .[28]

In exploring the applicability of a *Trail Smelter*–like model of state responsibility to the creation of refugees, the interesting questions are whether the creation of refugees can truly be attributed to the state; whether alleged harms could satisfy the high evidentiary burdens *Trail Smelter* established; and, if a *Trail Smelter*–like responsibility exists, how damages could be measured? These questions are explored below.

Imputing Acts Creating Refugees to States under *Trail Smelter*

The Universal Declaration of Human Rights, the International Law Commission's (ILC) Draft Articles on State Responsibility, as well as multiple treaties all declare, in varying forms, that a state, through its actors, cannot interfere with the right of its nationals to pursue a life free from fear and persecution. Often, however, in the refugee context, it is not entirely clear who is acting and whether those actors are authorized by the State. The definition of "state" requires: (a) a permanent population; (b) a defined territory; (c) a government; and (d) the capacity to enter into relations with other states.[29] Generally, in situations that create large numbers of refugees, one or more of these criteria have become ambiguous.

Many modern refugee crises are due to internal political power struggles that involve attempts to displace or replace a recognized State entity. When such a conflict creates a refugee problem, it calls into question the "government" element in the definition of a state. A State can only act through individuals and any single actor, acting on behalf of the state, can, at least in theory, be brought to justice under either regional courts of human rights or, now, by the International Criminal Court.[30] But, in order to hold the entire state responsible, rather than individuals acting on behalf of the state, international law looks first to the nexus between the state and the actors who create the refugee problem.[31]

[28] Lee, *supra* note 21 at 554.

[29] 1933 Montevideo Convention on the Rights and Duties of States, Dec. 26, 1933, art. I. 49 Stat. 3097, 165 U.N.T.S. 19; *see also* RESTATEMENT (THIRD) OF THE LAW: THE FOREIGN RELATIONS LAW OF THE UNITED STATES §201, cmt. b, c, d, and e (2003).

[30] Rome Statute of the International Criminal Court, July 17, 1998, 2187 U.N.T.S. 3. The Rome Statute established a permanent institution with the power to exercise its jurisdiction over "persons for the most serious crimes of international concern" including: genocide (art. 6); crimes against humanity (art. 7); and war crimes (art. 8).

[31] ANTONIO CASSESE, INTERNATIONAL LAW 191 (2001).

Only when individuals can be found to have acted on behalf of the State when creating refugees will state responsibility be satisfied, and a state claim that it "did not act willingly" will defeat an imposition of such responsibility under the Draft Articles.

The Trail smelter presented a relatively easy case – its harm-creating activities were conducted under the laws of Canada, with the state's knowledge and encouragement. The case of armed opposition groups that are not acting on behalf of the State and where the State is powerless to stop them is far more complex. Should the State be held responsible for the refugee-creating actions of these groups? Practically and conceptually, imputing an intentional wrong to a state may be difficult when the actors creating population migration are not state actors, but are, rather, insurgent or armed opposition groups. Indeed, international law perennially wrestles with how to make armed opposition groups accountable for abuses committed against civilian populations.[32]

Harm to the Nonrefouling Host State

State responsibility, as recognized by the *Trail Smelter* Tribunal, requires that there be a wrongful act of "serious consequence" and that the injury be established through "clear and convincing evidence." In other words, *Trail Smelter* requires a clear demonstration of harm. Certainly, the impact of refugees on host communities has largely been assumed to be negative.[33] Some commentators express a more optimistic view. "[R]efugees are an opportunity for host governments to "positively transform" their political economy."[34] For example, one aspect that has "largely remained unexplored is the role played by the significant amount of material and financial inputs [to the nonrefouling state] made by UNHCR, its implementing partners (IPs) and donor governments in a refugee populated area."[35] Often, as part of a relief effort, tremendous amounts of money are infused into the local economy in the form of infrastructure, agricultural, and industrial aid. Furthermore, often overlooked are the contributions the individual refugees themselves may be able to make to the nonrefouling state's economy.[36]

32 Liesbeth Zegveld, Accountability of Armed Opposition Groups in International Law 164 (2002).

33 *See Melissa* Phillips, *The Role and Impact of Humanitarian Assets in Refugee-hosting Countries*, UNHCR Working Paper No. 84, March 2003, *available at* http://www.unhcr.ch/cgi-bin/texis/vtx/home/opendoc.pdf?tbl=RESEARCH&id=3e71f- 7fc4&page=research; *See generally* Karen Jacobsen, *Livelihoods in Conflict: The Pursuit of Livelihoods by Refugees and the Impact on the Human Security of Host Communities*, Centre for Development Research Study Working Paper (2002), available at http://www.cdr.dk/ResTHEMES/conflict/migdevfinal.htm.

34 Phillips, *supra* note 33, *citing* Barbara Harrell-Bond, Imposing Aid: Emergency Assistance to Refugees, at 10–11, 331 (1986).

35 Harrell-Bond, *supra* note 34, at 3.

36 Smith, *supra* note 20, at 14, *citing* Jacques Cuénod, *Refugees: Development or Relief? in* Refugees and International Relations 241 (Gil Loescher and Laila Monahan, eds., 1989).

Despite these suggestions that refugee populations may convey some benefit to the nonrefouling state, it seems clear that sudden and massive influxes of refugees have a destabilizing and, at some level, and, for some period, detrimental effect. When the nonrefouling state is already fragile, this burden can be overwhelming. The burden of providing immediate care for refugees, as well as the responsibility to make up for gaps in international refugee–assistance funding falls primarily on the nonrefouling states. Many nonrefouling states already grapple with high infant-mortality rates, low rates of sustainability, low per-capita incomes, and preexisting political instability. Developing countries have voiced their concerns about the imbalanced burden placed on emerging democracies and their struggling neighbors in this regard.

> Much as the rich nations seem to get donor fatigue, likewise those who have been hosting refugees for years with no solution in sight are similarly fatigued . . . the fact that we are required to open our borders to refugees . . . without much support thereafter is very frustrating indeed. Once we admit the thousands and millions we are left alone to cushion both social and economic impacts . . . [37]

Sudden flows of refugees *en masse* may well produce "serious consequence" to the nonrefouling state charged with their care. However, demonstrating harms that satisfy the "clear and convincing" evidentiary standard may prove to be another hurdle altogether.

The Impenetrable Problem of Measuring Damages

Much as the monetary losses associated with sulfur pollution from the Trail smelter were borne by the Washington landowners, the immediate costs of caring for refugees falls on the nonrefouling state. However, where the *Trail Smelter* arbitration established a compensation mechanism to readjust this allocation of financial burdens, the 1951 Convention and 1967 Protocol do not.

In the absence of *Manna* and *quail*[38] the 1951 Convention confers an affirmative duty on nonrefouling states to bear the financial and social burdens of caring for refugees and avoiding a humanitarian crisis. Countries in regions of turmoil must accept refugees as they flee across adjacent borders, whereas more developed states and regions, like North America and Europe, that are typically geographically distant from the turnoil, have the luxury of determining whether to grant admittance. Many entry requests are denied. As a result, "[s]ome nations in Africa and Asia, states with far fewer economic resources than industrialized countries, sometimes host larger numbers of refugees for far longer periods of time."[39]

37 Statement by the Honourable Muhammed Seif Khatib, Minister for Home Affairs of the United Republic of Tanzania, to the 53rd Session of the Executive Committee of the High Commissioner for Refugees (ExCom), Geneva, 1 October 2002, *cited in* Phillips, *supra* note 33.

38 *Exodus* 16:4, 13–16, 31 (New King James Version).

39 UNHCR, Protecting Refugees – Questions & Answers, *available at* http://www.unhcr.ch.

Concerned about the financial costs, domestic social disruption, and political instability, several states have recently refused to allow refugees to cross their borders, in contradiction of clear international duties of nonrefoulement.[40] According to the UNHCR, in 1995, 50,000 Rwandans and Burundis fled to bordering Tanzania seeking refuge.

> What happened next was unprecedented. Rather than welcoming them as the country had regularly done in the past, Dar es Salaam [Tanzania] deployed the army, closed its border and effectively told the refugees to stay away.[41]

Other states are avoiding their responsibilities under international law by creatively reinterpreting definitional concepts like "persecution" and "territory." Even countries with the financial capacity to absorb the costs associated with caring for refugees have refused them admittance. In 1993, the U.S. Supreme Court concluded that the 1951 Convention permitted the federal government to forcibly repatriate Haitian refugees intercepted prior to entering U.S. territorial waters, despite clear indications that the refugees were fleeing feared persecution.[42]

Would a duty for one state to pay another state for the costs of housing and caring for refugees change this troubling trend? In *Trail Smelter*, the state – Canada – assumed responsibility for the wrongful harms caused by Cominco, a private corporation.[43] Imagine how a claim would play out in a refugee scenario. The

[40] Macedonia recently closed its borders twice. Once "in 1997, the border was closed to refugees fleeing the civil unrest in Albania. Two years later, Macedonia closed its borders to keep out those driven from Yugoslavia's Kosovo province by Serbian atrocities and NATO air raids." Roman Boed, *State of Necessity as a Justification for Internationally Wrongful Conduct*, 3 YALE HUMAN RIGHTS & DEVELOPMENT LAW JOURNAL 1, 2 (2000) *citing* Julia Strauss & Robert Fox, *Thousand Try to Flee Growing Chaos in Tirana*, DAILY TELEGRAPH, Mar. 14, 1997, at 15; Michael R. Gordon, *Weather Breaking, and Allies Seek to Intensify Bombing*, N.Y. TIMES, Apr. 5, 1999, at A1, A6. In 1998 Jordan's Crown Prince Hassan explained that Jordan "would have to close [its] borders to asylum-seekers from Iraq: 'We are talking about the possibility of hundreds of thousands of refugees . . . we simply cannot absorb further hundreds of thousands of refugees." Jay Bushinsky, *US Dismisses Latest Iraqi Proposal*, JERUSALEM POST, Feb. 12, 1998, at 1. *See also*,James C. Hathaway and R. Alexander Neve, *Making International Refugee Law Relevant Again: A Proposal for Collectivized and Solution-Oriented Protection*, 10 HARVARD HUMAN RIGHTS JOURNAL 115, 124–24 (1997) (discussing other instances of border closures to prevent a mass influx of asylum-seekers.).

[41] Roman Boed, *supra* note 40, *citing* Augustine Mahiga, *A Change of Direction for Tanzania*, 110 REFUGEES, 14 (1997).

[42] Sale v. Haitian Centers Council, Inc. 509 U.S. 155 (1993). In 1968 the United States had ratified the 1967 Protocol. By becoming a party, pursuant to Art 33.1, the United States pledged nonrefoulment "in any manner whatsoever."

[43] This was strictly a historical accident, in that nuisance claims could not be brought in British Columbia courts since the injured property was not located in the same jurisdiction and Consolidated Mining could not remedy the situation as it had in the past, through "smoke easements." *See generally* JOHN D. WIRTH, SMELTER SMOKE IN AMERICA (2000); *see also* Allum in this volume. Alternatively, Washington did not yet have a long-arm statute permitting jurisdiction over a Canadian company not engaged in any business in that State. BARRY E. CARTER & PHILIP R. TRIMBLE, INTERNATIONAL LAW 1082 (3d ed., 1999).

party seeking reparations is the state obligated to nonrefoulement. The party defending against the action is the state that has just expelled its population *en masse*, either deliberately or as a result of a chaotic political environment. Unlike Canada in the *Trail Smelter* situation, such a state is probably not a stable, solvent government.

There are other, more conceptual hurdles to applying *Trail Smelter* in such a situation. At least one expert on refugee policy has compared a refugee-creating state's duty to Corbin's illustration of a quasi-contractual relationship: "Under compulsion of law. . . . A makes payment of money that it was B's legal duty to pay. In spite of any express refusal, B is under a quasi-contractual duty to reimburse A."[44] According to this line of reasoning, B, as the refugee-creating state would have a quasi-contractual duty to reimburse A, the nonrefouling state. However, under the 1951 Convention, the nonrefouling state has itself assumed the burden of caring for refugees in the same way that its nationals are treated, and to accord refugees no less than the treatment accorded to aliens generally in the same circumstances.[45] Thus, the quasi-contract analogy breaks down at its inception.

Moreover, the 1951 Convention and 1967 Protocol contain no analog to *Trail Smelter's* established régime of permitting a certain concentration of sulfur parts per million depending on the air currents and temperature. The 1951 Convention accepts, rather than prohibits, or even limits, the expulsion of refugees, despite the clearly wrongful nature of the act.

In summary, several unique aspects of the refugee situation make application of the *Trail Smelter* principles inapposite. First, as a contracting party to the 1951 Convention or its 1967 Protocol, the state has given up the right to claim a violation of territorial sovereignty as related to an influx of refugees. A necessary element of the *Trail Smelter* analysis thus may not be present. Second, customary law clearly recognizes a legal duty to care for "stateless" persons crossing borders because of mass expulsions.[46] Thus the putative harm suffered by the nonrefouling state can alternatively be viewed as the fulfillment of a duty that customary international law places on every state. Certainly no such claim can be made for smelter smoke or its toxic effects on land. Even though the General Assembly has characterized refugees as "immensely burdening to the receiving states,"[47] extending the *Trail Smelter* rule to encompass refugee flows is probably far beyond where the General Assembly is itself willing to go. Instead, the international community

44 Lee, *supra* note 21, at 557, *citing* 1 CORBIN ON CONTRACTS § 19, 47–48 (1963).

45 1951 Convention, *supra* note 2.

46 For example, the United States, although not a party to the 1951 Convention, nevertheless assumed responsibility for refugees under its immigration and naturalization laws. *See* Immigration and Nationality Act of 1952, Pub. L. No. 82–414, 66 Stat. 163 (codified as amended at 8 U.S.C.).

47 Garvey, *supra* note 1 at 496, *citing* G.A. Res 35/196, 35 U.N. GAOR, 35th Sess., Supp. No. 48, at 208, U. N. Doc A/35/48 (1980).

and municipal laws have long recognized the duty to respond to a refugee crisis without subsequent compensation, and the Universal Declaration of Human Rights states that "everyone has the right to seek and to enjoy in other countries asylum from persecution."[48]

Unintended Consequences of Applying *Trail Smelter* to the Refugee Context and Potential Solutions

Any attempt to apply *Trail Smelter* in the refugee context also requires careful analysis of the conduct we hope to deter and whether that objective is likely to be attained through these means. The rigidity of the *Trail Smelter* formula makes its use in this context inappropriate. Even relatively stable democracies may be undermined by significant financial stresses. Were *Trail Smelter*–like financial liabilities imposed, states already undergoing the kind of upheaval and instability that generates refugee flows might well suffer further disintegration. In other words, requiring a refugee-creating state to compensate a nonrefouling state might perpetuate the instability that created a refugee crisis in the first place.

Additionally, adherence to the *Trail Smelter* approach might create a perverse incentive for even more serious human rights violations. We certainly want to avoid a system that might encourage states to commit atrocities such as genocide in order to circumvent financial obligations resulting from *Trail Smelter*–like liability for the creation of refugees. After all, those fleeing are likely already considered expendable by the refugee-creating regime. Although genocide is punishable as a crime under international law,[49] after-the-fact prosecution of perpetrators will hardly remedy the situation. The international community must take care not to create incentives for such atrocities in the first place.

As an alternative to monetary reimbursement, there has been a recent emphasis in the international community on proactive methods intended to reduce the costs of refugee care. The UNHCR recently applauded itself for reducing refugee numbers by over one million, attributing the achievement to simultaneous concentration on repatriation.[50] Further efforts have been made to end "warehousing" of refugees and asylum seekers in nonrefouling states.[51] The idea is that if refugees are allowed to participate as productive and contributing members in

48 Universal Declaration of Human Rights art. 14, *supra* note 14.

49 Convention on the Prevention and Punishment of the Crime of Genocide, December 9, 1948, art. 1, 78 U.N.T.S. 277.

50 The UNHCR has begun to focus more on pursuing successful repatriation of refugees. *See* UNHCR Annual Programme Budget 2005, U.N. GAOR, 55th Sess., at 6, U.N. Doc A/AC.96/992 (2004), *available at* http://www.acnur.org.

51 Karen Jacobsen, Local Integration:The Forgotten Solution (October 2003), *available at* www.migrationinformation.org.

society, rather than required to live in camps at a border or refrained from becoming gainfully employed, they are less likely to be a financial strain on the nonrefouling state. Along the same lines, an emphasis on sustainable development can diminish the conflicts generated by a scarcity of resources.[52]

CONCLUSION

Countless international declarations promote the furtherance of the belief that, "[a]ll human beings are born free and equal in dignity and rights. They are endowed with reason and conscience and should act towards one another in a spirit of brotherhood."[53] Yet, the international community has acknowledged that wars will be fought and because of that, threatened populations will flee.[54] When a society is determined to resolve its conflicts through warfare, the world must do more than watch the suffering.

There is no doubt that a state should bear responsible toward its own citizens for harms resulting from refugee status. This will promote a rapid return of the refugee to his/her home country. However, that state responsibility should not extend to claims that, by providing assistance to refugees, a nonrefouling state has suffered compensable harms under international law. The *Trail Smelter* principle that states are internationally responsible for harms occurring outside their borders as a consequence of activities taking place within their jurisdiction, should therefore have only a limited role to play in the context of refugees. If a state were held monetarily liable, à la *Trail Smelter*, for the creation of refugees, it would only further a downward economic and social spiral in the refugee-creating society and likely result in less ability to provide for the immediate care of refugees.

On a more fundamental level, the Law of Refugees rests on a humanitarian premise – a devotion to the welfare of humanity and a foundation of philanthropy are its driving force. Tragic would be the day when the charitable hand was held out after the conflict in order to receive "just payment" for services rendered. Whereas pure application of *Trail Smelter* principle is thus, inappropriate, the spirit of burden-sharing that imbues the *Trail Smelter* decisions is essential in order to achieve state compliance with refugee protection treaties and human

52 The UNCED and the Rio Declaration define "sustainable development" as: "development that meets the needs of the present without compromising the ability of future generations to meet their own needs." The two key concepts are: (1) prioritizing the essential needs of the world's poor; and (2) imposing technological and societal limitations to provide for current and future needs. David A. Wirth, *The Rio Declaration on Environment and Development: Two Steps Forward and One Back, or Vice Versa?*, 29 Georgia Law Review 599, 603–610 (1995).

53 1948 Universal Declaration of Human Rights, *supra* note 16.

54 1949 Geneva Convention, *supra* note 3, at art. 44; 1977 Protocol, *supra* note 3, at art. 73.

rights principles. The duty to care for refugees ought not fall on a nonrefouling state alone.

The international community must do more to prevent refugee crises before they occur. To that end, the international community must refocus on the root causes that produce flows of refugees, rather than simply on the maintenance and care of refugees once created.[55] The United Nations can play a significant role in this process through capacity-building, including: bolstering judicial institutions, strengthening civil society, and stabilizing economic viability. United Nations auspices could also become the framework for comprehensive regional approaches "aimed at strengthening . . . international security, the development of good-neighbourly relations and the creation of an atmosphere of confidence."[56] We have ample evidence that multinational efforts to resolve refugee-creating conflicts can stem the flow of refugees while simultaneously reducing threats to international peace and security. If preventative mechanisms fail, the international community must stand ready to ensure swift adjudication of individual criminal offenders in arenas such as the International Criminal Court or through avenues provided for by regional conventions, such as the Inter-American Court of Human Rights and the European Court of Human Rights.

The international community also should do more to prepare neighboring countries that may feel the greatest impact from a refugee crisis, and to alleviate the burdens once such a crisis does occur. Systems similar to those intended to avert natural disaster crises could ensure a rapid response to a political crisis, including early warning indicators, preventative actions, and community organization.[57] These rapid response systems could help nonrefouling states aid the injured, house the homeless, feed the hungry, and heal the suffering.

King Pyrrhus of Epirus defeated the Romans in 280 B.C. at Heraclea and 279 B.C. at Asculum.[58] In winning, he expended vast amounts from the treasury and lost one-third of his army.[59] When congratulated on the victory, he purportedly replied, "One more such victory . . . and we are utterly undone."[60] The Pyrrhic War, 280–275 B.C., has retained its linguistic place in history through the phrase "Pyrrhic Victory" – a victory gained at too great a cost. As a global community

[55] Luke T. Lee, *The Preventative Approach to the Refugee Problem*, 28 WILLAMETTE LAW REVIEW 821 (1992). "The international community has elected to treat the symptoms [care, maintenance, and repatriation] but not the root causes of the disease." *See* Maria Stavropoulou, *The Right Not to be Displaced*, 9 AMERICAN UNIVERSITY JOURNAL OF INTERNATIONAL LAW AND POLICY 687, 698 (1994) (concluding that any one cause cannot be treated "independent of other causes.").

[56] Report of the Group, *supra* note 21, at 2–25, ¶ 64.

[57] *Id.* at ¶ 66(g).

[58] For an account of Pyrrhus's exploits, *see* 9 PLUTARCH'S LIVES 411–417 (Bernadotte Perrin trans., 1998)

[59] *Id.*

[60] *See* 12 THE OXFORD ENGLISH DICTIONARY 945 (2d ed. 1989) ("Pyrrhic Victory: a victory gained at too great a cost.").

we should not seek compensation for an infringement on territorial sovereignty particularly where doing so would cause undue burden to an already fragile nation. Doing otherwise would be a "Pyrrhic Victory," creating a monetary gain for the nonrefouling state at the cost of the potential demise or further destabilization of the refugee-creating state – a cost far greater to the world as a whole, and to future generations than the small victory for that nonrefouling state.

21 Transboundary Harm: Internet Torts

Holger P. Hestermeyer

INTRODUCTION

No presentation of today's approaches to transboundary harm would be complete without a glance at Internet law. Of particular interest is the question of how to deal with tortious acts committed online where the act is committed in one country and the harm occurs in another. The difference between the solution that has been adopted for transboundary harm in that area of law and the approach to transboundary (environmental) harm chosen in the *Trail Smelter* arbitration[1] could hardly be more striking. In the *Trail Smelter* case, private international law failed to provide a solution for the United States farmers whose fields were damaged as a result of fumigations connected with smelting activities in Canada. Rather than pursuing their claims in national courts, the farmers nudged the United States into taking up their case, and the countries involved found the solution for the situation on the international plane in the form of an *ad hoc* arbitral tribunal.[2] The well-known public international law rule, "that no State has the right to use or permit the use of its territory in such a manner as to cause injury by fumes in or to the territory of another or the properties or persons therein (. . .),"[3] salvaged the situation and was the key to the parties' obtaining indemnification for the harm they had suffered.

The days when "cyberspace" pioneers promoted a vision in which the Internet was a space apart that would require a similar international tribunal, or, as most wished, no lawyers at all, are long gone. When a tort is commited online the harmed party routinely turns to national courts for help, raising significant

[1] *Trail Smelter Arbitral Decision*, 33 AMERICAN JOURNAL OF INTERNATIONAL LAW 182 (1939) [hereinafter "*Trail Smelter* (1939)"]; *Trail Smelter Arbitral Decision*, 35 AMERICAN JOURNAL OF INTERNATIONAL LAW 684 (1941) [hereinafter "*Trail Smelter* (1941)"]. *See* Annex to this volume.

[2] Convention on Damages Resulting From Operation of Smelter at Trail, British Columbia, June 5, 1935, U.S.-Canada, 49 Stat. 3245. *See* Annex to this volume.

[3] *Trail Smelter* (1941), *supra* note 1, at 716.

concerns about personal jurisdiction.[4] Where private international law failed in the *Trail Smelter* arbitration, it succeeded in Internet cases.

Nevertheless, today's law on personal jurisdiction on the Internet is far from satisfactory. The recent *Yahoo!* case has rekindled fears that posting information online exposes someone to the risk of being sued for a tort, such as defamation, trademark infringement, copyright infringement, or unfair competition, in *any* foreign jurisdiction. Insecurity about jurisdiction is heightened by the fact that national courts often apply widely varying domestic laws to Internet content, instilling Web site hosts with the fear of being sued for content that is protected as free speech in the country in which the website is based. I will describe the current approaches to personal jurisdiction over Internet actors in two major jurisdictions, the United States and Germany, and venture a proposal on how the shortcomings of the current approach could be remedied. I will conclude by illustrating the differences between the situation in the *Trail Smelter* arbitration and Internet law in order to highlight the causes of the failure of private international law in the *Trail Smelter* case.

THE *YAHOO!* CASE

The discussion on Internet jurisdiction has gained enormous prominence with the decision in *UEJF & LICRA v. Yahoo!*,[5] in which Article R 645–1 of the French *Code Pénal* prohibiting the exhibition of Nazi propaganda was pitted against the protection of free speech in the United States.

French associations combating racism discovered in April 2000 that they could access anti-Semitic material in France *via* auction sites hosted by Yahoo!, which is based in the United States. In the resulting legal proceedings, the French courts exercised jurisdiction and found violations of Art. R 645–1 of the French *Code Pénal*. Yahoo! was ordered to take all necessary measures to dissuade and make impossible any French access on yahoo.com to auction sites for Nazi memorabilia and revisionist Web sites, for example, by using blocking software that identifies the locales from which Web sites are accessed.[6] Yahoo! France

[4] For another example of jurisdictional issues arising in the context of disputes over transboundary harm, *see* Vennemann in this volume.

[5] UEJF & LICRA v. Yahoo!, Inc. & Yahoo! France, Tribunal de Grand Instance Paris (T.G.I. Paris), May 22, 2000, No. 00/05308, *available at* http://www.meldpunt.nl/juris/ yahoo1.html; UEJF & LICRA v. Yahoo!, Inc. & Yahoo! France, T.G.I. Paris, November 20, 2000, No. 00/05308 (affirming the May 22 order), *available at* www.juriscom.net/txt/ juristr/cti/tgiparis20001120.pdf.

[6] There is some confusion in the French judgment concerning its scope. Taken literally, the court orders Yahoo! to make all access to the offensive sites impossible. This is the interpretation on which the United States court seemed to rely, thus dodging the question of whether the First Amendment is affected by blocking access for French users only. Given the attention paid by the French Court to the technical question of blocking access from France only, the overly broad holding seems to be a drafting error.

also was ordered to warn all users that following the hyperlink to continue their research on yahoo.com might put them in violation of French law.[7]

Yahoo!'s defense to the allegations, primarily that compliance with the order was impossible, was significantly undermined by the evidence that Yahoo! knew it was addressing French Internet users: the yahoo.com Web site displayed French-language commercials if accessed from France. Given that, in order to target commercials to French users, Yahoo! was already using the identifying technology that would permit it to block access to these sites from France, it was concluded that compliance with the order imposed no extravagant burden.[8]

Yahoo! largely complied with the French order. It undertook the relevant changes to the yahoo.fr Web site and amended its auction policy to prohibit items promoting, glorifying, or directly associated with groups such as the Nazis or which violated Yahoo!'s hate-speech policy. However, some items remained available.[9]

In spite of its compliance with the French order, and rather than pursuing an appeal in France, Yahoo! filed a complaint in the United States Federal District Court for the Northern District of California seeking a declaratory judgment that the French court orders were not enforceable in the United States. The French claimants filed a motion to dismiss for lack of *in personam* jurisdiction. The District Court rejected the motion, reasoning that the defendants had purposefully availed themselves of the benefits of California by sending a cease-and-desist letter to Yahoo!, using the U.S. Marshals Service to serve process and by requesting a French court to order that Yahoo! perform acts in California.[10] The District Court then granted Yahoo!'s motion for summary judgment holding that enforcement of the French orders in the United States would violate the First Amendment.[11]

In August 2004, the Court of Appeals for the Ninth Circuit reversed the District Court,[12] holding that UEJF & LICRA had not purposefully availed themselves of the jurisdiction of California. The Circuit Court reasoned that the French complainants had not yet asked a United States District Court to enforce the French orders and that enforcing their legal rights under French law did not constitute wrongful conduct targeted at Yahoo!. The case is set to continue with an *en banc* rehearing in the Ninth Circuit.[13]

7 UEJF & LICRA v. Yahoo!, Inc. & Yahoo! France, May 22, 2000, No. 00/05308, *available at* http://www.meldpunt.nl/juris/ yahoo1.html.

8 UEJF & LICRA v. Yahoo!, Inc. & Yahoo! France, T.G.I. Paris, November 20, 2000, No. 00/05308 (affirming the May 22 order), *available at* www.juriscom.net/txt/juristr/ cti/tgiparis20001120.pdf.

9 For the background of the case, *see also* Yahoo! v. LICRA, 145 F.Supp.2d 1168 (N.D.Cal. 2001); Yahoo! v. LICRA, 169 F.Supp.2d 1181, 1189 (N.D.Cal. 2001); Yahoo! v. LICRA, 379 F.3d 1120 (9th Cir. 2004).

10 Yahoo! v. LICRA, 145 F.Supp.2d 1168 (N.D.Cal. 2001).

11 Yahoo! v. LICRA, 169 F.Supp.2d 1181, 1189 (N.D.Cal. 2001).

12 Yahoo!, 379 F.3d at 1127 (Circuit Judge Brunetti dissenting).

13 Yahoo! Inc. v. LICRA, 399 F.3d 1010 (9th Cir. 2005).

INTERNET JURISDICTION IN THE UNITED STATES

Since *Hanson v. Denckla*,[14] American courts have required that the defendant "purposefully avail" herself of the privilege of conducting activities within the forum state in order to fulfill the constitutional "minimum contacts" requirement for establishing a court's *in personam* jurisdiction. For the Internet this test poses more questions than answers. A Web site content provider knows that the page is accessible worldwide. Does that mean that the content provider means to purposefully avail herself of the "privilege of conducting activities" in every state from which the Web site is accessible? In an early set of cases courts, tended to answer this question in the affirmative and thus applied what could be called an accessibility test.[15]

This broad approach changed in 1997, when the Federal District Court of the Western District of Pennsylvania decided *Zippo Manufacturing v. Zippo Dot Com*[16] and pronounced a new standard for purposeful availment in Internet cases. Zippo Dot Com, an Internet news service provider based in California, and with Pennsylvanian subscribers, was sued by Zippo Manufacturing, the Pennsylvanian maker of lighters, for trademark infringement and dilution.

The District Court applied what would become known as the *Zippo* sliding scale. At one end of the scale (providing for jurisdiction) are those cases in which the defendant clearly does business over the Internet, entering into contracts with residents of the jurisdiction that involve the knowing and repeated transfer of files. Maintaining such an *active Web site*, the Court explained, allows for the assertion of jurisdiction. At the other end of the scale (not providing jurisdiction) are situations in which the defendant merely posted information to a Web site. The Court reasoned that these *passive Web sites* do not allow the assertion of jurisdiction. The middle ground is occupied by "interactive Web sites where a user can exchange information with the host computer. In these cases, the exercise of jurisdiction is determined by examining the level of interactivity and commercial nature of the exchange of information that occurs on the Web site."[17] The Court held that Zippo Dot Com was on the active website end of the spectrum, and that it had purposefully availed itself of jurisdiction in Pennsylvania.[18] Other courts soon adopted the *Zippo* test.[19]

[14] 357 U.S. 235, 253 (1958). *See* World-Wide Volkswagen v. Woodson, 444 U.S. 286 (1980).

[15] Inset Systems v. Instruction Set, 937 F. Supp. 160, 165 (D. Conn. 1996); Mariz v. Cybergold, 947 F. Supp. 1328 (E.D.Mo. 1996). *See* Michael A. Geist, *Is There a There There? Toward Greater Certainty for Internet Jurisdiction*, 16 BERKELEY TECHNOLOGY LAW JOURNAL 1345, 1361 (2001).

[16] 952 F. Supp. 1119 (W.D. Pa. 1997).

[17] Id. at 1124.

[18] Id. at 1126.

[19] Cybersell v. Cybersell, 130 F.3d. 414 (9th Cir. 1997); Mink v. AAAA Development, 190 F.3d 333, 336 (5th Cir. 1999); Soma Medical Int'l v. Standard Chartered Bank, 196 F.3d 1292, 1296 (10th Cir. 1999). Markus Rau, *"Minimum Contacts" und "Personal Jurisdiction" über auswärtige Gesellschaften im Cyberspace*, RECHT DER INTERNATIONALEN WIRTSCHAFT [RIW] 761 (2000).

In the face of strident scholarly criticism of the *Zippo* sliding scale test, largely focusing on the contingency of the technical question of a Web site's interactivity, courts have implicitly[20] and, at times, explicitly[21] started to move away from the test. Some commentators have heralded this turn away from *Zippo* as a return to the constitutional standards of jurisdiction.[22] This is somewhat incorrect, given that *Zippo* was an attempt to further define, and not abandon, the constitutional standard of purposeful availment. In the absence of the *Zippo* test, a different refinement of that standard will be needed. The precedent of *Calder v. Jones* has proven to be particularly attractive after *Zippo* fell into disfavor.[23] In *Calder*, an entertainer residing in California brought a libel suit in California against the writer and editor of a Florida-based tabloid because of a story impugning the entertainer's professionalism. The tabloid sold numerous copies in California and the Supreme Court upheld the jurisdiction of the California courts, reasoning that a court can assert jurisdiction over a nonresident defendant when the defendant has expressly aimed his intentional and allegedly tortious actions at the forum state, knowing that that conduct would cause harm in that state. This effects test, alongside the *Zippo* test, has become one of the most commonly used jurisdictional tests for online torts.[24]

A third standard advocated by Professor Geist is the "targeting test." It tries to establish the intent of the defendant by examining the steps taken to enter or avoid a jurisdiction and asserts jurisdiction only if the forum was targeted.[25] The factors to be taken into account in determining the intended target jurisdictions can be a matter of much discussion, but, given the wide variety of possible factual situations, it is questionable whether an exhaustive list of factors is desirable. An examination of the totality of circumstances to determine which jurisdictions were targeted seems most reasonable, perhaps informed by an open-ended, illustrative list of criteria to mitigate the vagueness of the standard. Courts have started to take targeting into account, often in a rather complex mix in which different tests are blended into one or applied side by side. A good example of this approach is the 2003 case *Toys "R" Us v. Step Two*.[26] Step Two, a Spanish corporation, owns the *Imaginarium* trademark in Spain and franchises toy stores under that trademark in Spain and other countries, but not in the United States. It also operates interactive Spanish-language Web sites under the domain name "imagniariumnet.com" and

[20] Panavision Int'l v. Toeppen, 141 F.3d 1316, 1322 (9th Cir. 1998).

[21] GTE New Media Services v. BellSouth, 199 F.3d 1343 (D.C. Cir. 2000).

[22] *Note, D.C. Circuit Rejects Sliding Scale Approach to Finding Personal Jurisdiction Based on Internet Contacts*, 113 HARVARD LAW REVIEW 2128 (2000).

[23] 465 U.S. 783 (1984).

[24] *See* Blakey v. Continental Airlines, 751 A.2d 538, 543 (N.J. 2000); Nissan Motor Co. v. Nissan Computer, 89 F. Supp. 2d 1154, 1160 (C.D. Cal. 2000); Millennium Enterprises v. Millennium Music, 33 F. Supp. 2d 907, 921–922 (D. Or. 1999); Carefirst of Maryland v. Carefirst Pregnancy Centers, 334 F. 3d 390 (4th Cir. 2003).

[25] Geist, *supra* note 15, at 1380–1386.

[26] 318 F.3d 446 (3d Cir. 2003).

similar names, on which shoppers can buy goods by entering their credit card number, delivery address, and phone number. The Web sites list the product prices in Euro and provide a contact phone number in Spain. Toys "R" Us, which owns similar "Imaginarium" stores and the *Imaginarium* trademark in the United States, sued for trademark infringement. Toys "R" Us adduced evidence of two United States sales with United States credit cards. Both sales were initiated by Toys "R" Us, with the purchased goods shipped to a Toys "R" Us employee in Madrid and then forwarded to the United States. Even though the Court cited *Zippo* as the seminal authority, it went on to state that a commercially interactive Web site in itself is not sufficient to establish jurisdiction. There must be evidence, the Court explained, that the defendant purposefully availed herself "by directly targeting its web site to the state, knowingly interacting with the residents of the forum state via its Web site, or through sufficient other related contacts," which may be non-Internet related. The Court concluded that there was no purposeful availment and confirmed the result by applying the *Calder* effects test.[27]

U.S. courts have come a long way from simply asserting jurisdiction on the basis of the accessibility of the Web site within the forum. Nevertheless, there still is no certainty as to the approach United States courts will take towards Internet jurisdiction. Three approaches seem to compete: The *Zippo* test, the *Calder* effects test, and Professor Geist's targeting test. This breeds uncertainty for Web site hosts as to when they can expect U.S. courts to assert jurisdiction.

INTERNET JURISDICTION FOR TORTS IN GERMANY

German courts approach the question of personal jurisdiction from a different angle, examining whether the dispute falls within their competence *ratione loci*, their "local jurisdiction." Generally, the German rules on local jurisdiction govern both the question of local jurisdiction within Germany as well as the question of international jurisdiction.[28] European integration has brought an additional set of rules. In deciding on their local jurisdiction in international cases, German courts now first have to consider the applicable norms of Council Regulation (EC) No. 44/2001[29] or of the Lugano Convention on Jurisdiction and the Enforcement of Judgments in Civil and Commercial Matters.[30] As far as is relevant for the present discussion, these rules provide that a tortfeasor who resides in an European Union Member State (or, for the Lugano Convention, an EFTA Member State) can

27 *See Symposium: Personal Jurisdiction in the Internet Age*, 98 Northwestern University Law Review 409 (2003–2004).

28 Othmar Jauernig, Zivilprozessrecht 21 (28th ed. 2003).

29 Council Regulation on Jurisdiction and the Recognition and Enforcement of Judgments in Civil and Commercial Matters, *see* Council Regulation 44/2001, 2001 O.J. (L 12) 1.

30 Lugano Convention on Jurisdiction and the Enforcement of Judgments in Civil and Commercial Matters, Sept. 16, 1988, 1988 O.J. (L 319) 9, *reprinted in* 28 International Legal Materials 620 (1989).

be sued in the courts of any of three fora: the state in which he resides,[31] the place where the tortious act was committed, or where the injury was caused.[32] Attempts to introduce special jurisdictional regulations for the Internet, including the adoption of a "country of origin" standard, according to which a Web site would only be subject to the laws and jurisdiction of the country of origin of the Web site, have failed.[33]

When the tortfeasor does not reside within the European Union or the territory covered by the Lugano Convention, the German national rules for local competence apply. I will focus on these rules. Section 32 of the *Zivilprozessordnung* (ZPO – Code of Civil Procedure) provides that "[f]or tort actions the court, in the district of which the tort was committed, is competent."[34] The same jurisdictional rule is provided for in competition law by the *Gesetz gegen den unlauteren Wettbewerb* (UWG – Law against Unfair Competition).[35] Courts have clarified the provision by stating that the tort is committed both where the tortious act took place and where the relevant protected legal interest was harmed.[36]

For Internet cases the risk of a broad interpretation is evident: although the location of the tortious act can be either the location of the Web server[37] or of the uploading of the relevant data by the tortfeasor,[38] the place where the legal interest is harmed could theoretically be anywhere, as information on the Internet can be accessed anywhere. Hence, German courts would have jurisdiction for any tortious claim connected with Internet content accessible in Germany.

Courts have had to face a similar problem in press defamation cases. Arguably, in those cases the harm is done wherever the defamatory statement is read,

[31] Council Regulation 44/2001, art. 2, 2001 O.J. (L 12) 1.

[32] Council Regulation 44/2001, art. 5 no. 3, 2001 O.J. (L 12) 1. *See* Dietmar Czernich, *in* DIETMAR CZERNICH & STEFAN TIEFENTHALER & GERHARD KODEK, KURZKOMMENTAR EUROPÄISCHES GERICHTSSTANDS- UND VOLLSTRECKUNGSRECHT, Art. 5 para. 81 (2nd ed. 2003).

[33] The Directive of the European Parliament and of the Council from 8 June 2000 on certain legal aspects of information society services, in particular electronic commerce, in the Internal Market ("Directive on electronic commerce"), does not regulate jurisdiction, *see* Parliament and Council Directive 2000/31/EC, art. 1 para. 4, 2000 OJ (L 178) 1. *See* Graham Smith, *Directing and Targeting – The Answer to the Internet's Jurisdiction Problems?*, 5 COMPUTER LAW REVIEW INTERNATIONAL/COMPUTER UND RECHT INTERNATIONAL [CRi] 145, 148 (2004).

[34] Translation by author.

[35] Gesetz gegen den unlauteren Wettbewerb, from July 7, 2004, §14 para. 2 (BGBl. I, p. 1414) (formerly §24).

[36] Rufus Pichler, *in* HANDBUCH MULTIMEDIA-RECHT. RECHTSFRAGEN DES ELEKTRONISCHEN GESCHÄFTSVERKEHRS, §31 para. 54 (Thomas Hoeren & Ulrich Sieber eds., loose-leaf, last updated 2005); Torsten Bettinger & Dorothee Thum, *Territorial Trademark Rights in the Global Village – International Jurisdiction, Choice of Law and Substantive Law for Trademark Disputes on the Internet. Part One*, 31 INTERNATIONAL REVIEW OF INDUSTRIAL PROPERTY AND COPYRIGHT LAW [IIC] 162, 170–172 (2000); for the UWG, *see* Wolfgang Büscher, *in* LAUTERKEITSRECHT – KOMMENTAR ZUM GESETZ GEGEN DEN UNLAUTEREN WETTBEWERB §14 para. 23 (Karl-Heinz Fezer ed., 2005).

[37] THOMAS HOEREN, RECHTSFRAGEN IM INTERNET 381 (2003).

[38] *Contra* Pichler, *supra* note 36, at §31, para. 123.

exposing any journal to liability in Germany simply because the journal happens to be read there. Germany's highest civil court, the *Bundesgerichtshof* (BGH – Federal Court of Justice), recognized the apparent danger of such loose attribution of jurisdiction. In 1977 it had to decide a defamation case in which the plaintiff sued an Austrian journal allegedly containing defamatory statements.[39] The plaintiff had obtained an issue of the journal in Germany by mere happenstance and had then procured further issues *via* a German bookseller. The *Bundesgerichtshof* decided that these circumstances were not sufficient to assert jurisdiction. The Court concluded that jurisdiction only lies where the journal purposefully is distributed in Germany. This standard resembles the "purposeful availment" standard of the Supreme Court's *Hanson* decision. Transposed to Internet cases, a court could only assert jurisdiction where the person posting the Web site purposely made it accessible for the relevant jurisdiction.[40]

The provision on jurisdiction for unfair competition cases (§ 14 para. 2 UWG) has also been limited by the *Bundesgerichtshof*. In those cases, a harm to a protected legal interest can only take place in markets where the competition interests of the competitors actually clash.[41]

In the first cases in which the issue of Internet jurisdiction was raised, several German courts were willing to assert jurisdiction arguing merely that the relevant Web site was accessible within their jurisdiction. In the mid-1990s the *Landgericht München I* (Regional Court of Munich) and the *Landgericht Nürnberg-Fürth* (Regional Court Nürnberg-Fürth) asserted local jurisdiction under the old equivalent of §14 para. 2 UWG for violations of the Competition Act solely on the basis of the accessibility of the offensive text in the district of the court.[42] Even though some courts referred to the "purposeful accessibility" standard stemming from press law, the criterion has not necessarily been applied to limit jurisdiction. In 1997 the *Kammergericht* (Higher Regional Court Berlin) relied on that criterion to assert jurisdiction under §32 ZPO in a dispute about the domain names "concert-concept.com" and "concert-concept.de," a dispute between a German and a United States company with a representation in Berlin.[43] The *Kammergericht* merely stated that the domain name could purposely be accessed in Germany and failed to engage in any further discussion of the point. It seems that its argument was that Internet content providers are well aware that they make content available worldwide and hence distribute it purposely in the whole

39 Bundesgerichtshof, NEUE JURISTISCHES WOCHENSCHRIFT [NJW], 30 (1977) 1590.

40 Peter Mankowski, *Anmerkung*, 18 COMPUTER UND RECHT [CR] 450 (2002).

41 *See* BGHZ 35, 329 (334); Büscher, *supra* note 36, at para. 23.

42 LG München I, COMPUTER UND RECHT [CR], 13 (1997) 155, 156; LG Nürnberg-Fürth, COMPUTER UND RECHT [CR], 13 (1997) 415, 416.

43 KG, COMPUTER UND RECHT [CR], 13 (1997) 685. *See* LG Düsseldorf, NEUE JURISTISCHE WOCHENSCHRIFT-RECHTSPRECHUNG REPORT [NJW-RR], 51 (1998) 980. The claim was based on the "right to the name" under §12 BGB.

world.[44] Other appellate courts have adopted a similar line of reasoning. In November 2001, the *Oberlandesgericht München* (Higher Regional Court Munich) asserted jurisdiction in a dispute concerning the domain name "literaturhaus.com," reasoning solely that the domain names were accessible in Germany.[45]

Many commentators voiced their concern over such a broad assertion of local jurisdiction and discussed how a limitation could reasonably be achieved.[46] The most thorough treatment remains Pichler's extensive commentary on court competence in Internet matters published in 2000.[47] He advocates an objective test to evaluate whether a Web site has been made "purposely accessible" in Germany, examining whether Germany was targeted by taking account of such factors as the content of a Web site, its language, and the markets the content provider hopes to reach.[48]

In competition cases, a growing body of case law has since adopted such a limiting approach, commonly basing it on the requirement of a clash of competition interests in the relevant forum. The apparently first clear case in the direction was handed down by the *Oberlandesgericht Frankfurt* (Higher Regional Court Frankfurt) at the end of 1998.[49] At issue was the advertisement and offer for sale of pens on the Internet by the defendant. The defendant had also offered the pens at two trade fairs in Frankfurt. The Court held that the *locus* of an online tort is any place at which the homepage can purposely be accessed and results in a clash of interests, which is the place at which the clash of competitive interests occurs. Germany was a location of the tort in the case because the advertisement described the main export markets as "worldwide," necessarily making Germany a target for sales as well. In reaching that conclusion, the Court also considered the language of the homepage (English), the kind of goods offered, and other circumstances of the case.

The *Oberlandesgericht Bremen* (Higher Regional Court Bremen) expanded on that holding in a case of wrongful advertising online in 2000.[50] The plaintiff wanted the Court to assert jurisdiction on the basis that online advertisements are accessible anywhere and hence also in Bremen. The Court, however, ruled that it had no jurisdiction on the basis of the old equivalent of §14 para. 2 UWG.

44 Birgit Bachmann, *Der Gerichtsstand der Unerlaubten Handlung im Internet*, 18 PRAXIS DES INTERNATIONALEN PRIVAT- UND VERFAHRENSRECHTS [IPRAX] 179, 185 (1998).

45 OLG München, COMPUTER UND RECHT [CR], 18 (2002) 449, 450.

46 *See e.g.*, Christopher Kuner, *Internationale Zuständigkeitskonflikte im Internet*, 12 COMPUTER UND RECHT [CR] 453 (1996).

47 Pichler, *supra* note 36.

48 Note that the determination has to take account of the specifics of the course of action in question. *See* Christopher Koch, *Internationale Gerichtszuständigkeit und Internet*, 15 COMPUTER UND RECHT [CR] 124 (1999).

49 OLG Frankfurt, KOMMUNIKATION & RECHT [K&R], 2 (1999) 138.

50 OLG Bremen, COMPUTER UND RECHT [CR], 16 (2000) 770.

Importing the "purposeful accessibility" criterion as well as the requirement of a clash of competition interests in the relevant forum it argued that the global accessibility of a Web site is a mere technical characteristic of that particular medium. The Court reasoned that the decision regarding the markets in which the competitors "meet" has to be made on a case-by-case basis. For online competition cases, the Court concluded that courts have to determine which recipients are targeted by the competitor. In making this determination, courts have to take account of such factors as the content of a homepage, its design, and the places where the relevant goods are offered. The Court also suggested that the Web site could contain a disclaimer excluding certain markets.

Similar analyses of whether the forum was targeted by the defendant have now become commonplace in Internet competition cases. Thus, the *Oberlandesgericht Hamburg* (Higher Regional Court Hamburg) held that online advertising of a washing machine by a defendant located in Coburg, some 300 miles from Hamburg, fell into its jurisdiction. Customers in Hamburg were targeted, the Court explained, because they could have made use of the offer by having the product delivered, despite the costs involved in doing so. The Court indicated that the case would have come out differently for the advertising of local services, such as the repair of washing machines.[51]

Thus, whereas German courts now apply a targeting approach in competition cases, they are still likely to rely on a mere accessibility test for other tort cases. The different treatment is a result of the requirement of a clash of competition interests in the forum in competition cases. This additional requirement has limited jurisdiction. There is, however, no proper reason to treat other tort cases differently. In those cases, too, it would be improper to assert jurisdiction if the forum was not targeted by the defendant. Courts could import a targeting test under the criterion of purposeful accessibility. However, to this day it remains uncertain whether they will do so, thus exposing a broad range of Internet actors to German jurisdiction.

THE SOLUTION: AN INTERNATIONAL CONVENTION ON INTERNET JURISDICTION

Both in the United States and in Germany there is still considerable uncertainty about Internet jurisdiction for tort actions. At times, courts assert jurisdiction on the sole basis of accessibility of the Web site within the forum, sometimes disguising this approach as an application of a different standard. Such a mere effects test, without any targeting component, potentially exposes Web site hosts to liability anywhere in the world.

[51] OLG Hamburg, MULTIMEDIA UND RECHT [MMR], 6 (2003) 538. *See also* OLG Brandenburg, MULTIMEDIA UND RECHT [MMR], 5 (2002) 463.

A targeting test would be a proper answer to the challenge of Internet jurisdiction. Targeting is flexible enough to answer to the particularities of each case and it is technology-neutral, thus it is open to the future development of the Internet. Targeting also enables Web site hosts to minimize exposure to foreign jurisdictions by inserting disclaimers that a Web site is not meant to target certain jurisdictions or by applying geofiltering technology.[52] Of course, where the Web site host engages in repeated sales with a jurisdiction that is purportedly not targeted, courts could still assert jurisdiction under the test. Legal security for the Web site host could further be enhanced by providing for a nonexhaustive list of factors to be taken into account and weighed for deciding on jurisdiction, such as the language of the Web site, the services or goods advertised, the markets in which the Web site host is active, the manner and content of the presentation,[53] modalities of payment, the use of geolocation technology, disclaimers, and the top-level domain used.[54] However, it must be pointed out that the French courts' assertion of jurisdiction in the *Yahoo!* decision, which have caused such a sensation, are in strict conformity with a targeting approach. After all, Yahoo! placed French-language advertising on its yahoo.com Web site, strong evidence that it was targeting France.

Of course, Web site hosts would only be secure if all states agreed to adopt the same test, a prospect imaginable only in the form of an international convention. An obvious vehicle for achieving this international solution were the negotiations on the Hague Convention on Jurisdiction and Foreign Judgements in Civil and Commercial Matters with the ambitious goal of a comprehensive convention on international jurisdiction and the enforcement of foreign judgments, which commenced in 1992.[55] During the negotiations, the United States aimed at facilitating the enforcement of U.S. judgments abroad, whereas the European Union hoped to achieve a harmonization of jurisdictional rules and limitations to U.S. jurisdiction. The first draft convention,[56] adopted in 1999, was not adapted to the needs of e-commerce. This led to the announcement that the Internet ramifications of the convention would be examined by a group of specialists.[57] The draft convention's provision on torts provided that a plaintiff may bring an action in tort or delict in the courts of the state in which the act occurred or "in which the injury arose, unless the defendant establishes that the person claimed to be responsible could not reasonably have foreseen that the act or omission could result in an injury of the same nature in that State."[58] Where the action is brought in the

[52] There should not be a requirement to use such technology, as that would run the risk of destroying the specific advantages of the Internet.

[53] Smith, *supra* note 33, at 150.

[54] Büscher, *supra* note 36, at para 26.

[55] NIKLAS GANSSAUGE, INTERNATIONALE ZUSTÄNDIGKEIT UND ANWENDBARES RECHT BEI VERBRAUCHERVERTRÄGEN IM INTERNET 167 (2004).

[56] Preliminary Draft Convention on Jurisdiction and Foreign Judgments in Civil and Commercial Matters, available at http://www.hcch.net/upload/wop/jdgm_drafte.pdf [hereinafter "Draft"].

[57] Footnote 1 of the Draft.

[58] Art. 10 para. 1 of the Draft.

courts of a state only on the basis that the injury arose there, those courts "shall have jurisdiction only in respect of the injury that occurred (. . .) in that State, unless the injured person has his or her habitual residence in that State."[59] As regards the Internet, the provision suffers from the same shortcomings as the traditional jurisdictional tests in the United States and Germany: the defendant can reasonably foresee that content posted to the Internet can be read anywhere in the world, so that the purported limitation of jurisdiction contained in the Draft convention might prove hollow. This was pointedly remarked by the group of experts that convened in Ottawa in 2000.[60] Sadly, the group fell short of proposing a solution: national case law was regarded as too unsettled to make any clear, universal, or harmonizing statements. Even the attempt to allow a defendant to avoid certain jurisdictions by taking "reasonable steps" to avoid acting in or toward those states failed, as the notion of what would constitute "reasonable" steps was still in flux.[61] Differences between the negotiating states in many areas proved too great and the objective of the project was later scaled down to a convention on choice of court agreements in business-to-business cases,[62] thus excluding the question of online torts.

Despite the collapse of the original Hague Judgments Project, it is advisable to push for an international adoption of the targeting approach to Internet tort jurisdiction. It must be conceded that any such approach would necessarily give deference to courts and their evaluation of the case at hand. It is also foreseeable that member states' courts' will diverge on such questions as what constitutes "reasonable" steps to exclude a foreign jurisdiction. However, even the vague targeting approach would represent a considerable improvement over the current state of the law, in which Web site hosts are potentially exposed to liability in every jurisdiction on the mere basis of the world-wide accessibility of their Web site.

WHY PRIVATE INTERNATIONAL LAW FAILED IN *TRAIL SMELTER*

Even though *Trail Smelter* has become one of the best-known and most cherished cases in international environmental law, the facts of the case made it an unlikely candidate for an international tribunal: private parties suffered harm from the acts of another private party. The facts seemed to point to just another private international law case. John Knox, in his contribution to this volume, judges harshly the illogic of the forum chosen in the *Trail Smelter* dispute, declaring that it was the wrong tribunal for resolving what was essentially a private matter.[63]

59 Art. 10 para. 4 of the Draft.

60 Catherine Kessedjian, *Electronic Commerce and International Jurisdiction. Ottawa, 28 February to 1 March 2000*, PREL. DOC. NO. 12., 8 (2000).

61 Avril Haines, *The Impact of the Internet on the Judgments Project: Thoughts for the Future*, PREL. DOC. NO. 17, paras. 17, 18 (2002).

62 Masato Dogauchi & Trevor Hartley, *Preliminary Draft Convention on Exclusive Choice of Court Agreements*, PREL. DOC. NO. 26, paras. 4 et seq. (2004).

63 *See* Knox in this volume.

The main reason for the breakdown of private international law, which prevented national courts from resolving the matter, was a historical anomaly. The British Columbia Courts, at the time, allowed suits for injury to land only where the land was located and thus would have refused to grant the remedy sought. A solution on the American side of the border was prevented by the inability of the smelter to acquire "smoke easements" in Washington state, as it had done to resolve equivalent claims brought by Canadian farmers whose land had been damaged. Washington smoke easements were out of the question because the law of that state prohibited selling real property to foreigners.[64]

Other factors certainly helped the farmers bring their case up to an international tribunal. International civil litigation was not yet as common and fears that courts abroad would discriminate against foreigners abounded. Cases such as *Trail Smelter* were rare enough to justify the enormous administrative hassle involved in setting up an international tribunal. The dispute in *Trail Smelter* also had the public international law facet of Canadian infringement on United States territorial sovereignty (*via* air pollution). Finally, the Canadian interest in the welfare of the polluting smelter was so immense as to make the challenges to its industrial practices a matter of national interest.

Most of these factors are not present in Internet tort jurisprudence. A similar international arbitral tribunal for Internet torts would be inconceivable for one reason alone: there are simply far too many cases being litigated in a whole number of national courts to imagine a centralized approach to the matter.

The (unjust) criticism voiced by United States commentators against the French court's assertion of jurisdiction in the *Yahoo!* case illustrates, however, that an acute skepticism of the fairness of foreign tribunals survives to this day, one of the forces that thwarted private law resolution of the *Trail Smelter* dispute. Harmonized private international law rules on personal jurisdiction for Internet torts would help quell this fear, or at least keep it within reasonable limits. In any event, the suspicions or frustration presently confounding the field of Internet torts are highly unlikely to lead to a total failure of private international law as occurred in the *Trail Smelter* dispute.

64 John E. Read, *The Trail Smelter Dispute*, 1 The Canadian Yearbook of International Law 213, 222 (1963). Republished in this volume.

22 International Drug Pollution? Reflections on *Trail Smelter* and Latin American Drug Trafficking

Judith Wise and Eric L. Jensen

INTRODUCTION

The illicit Latin American drug trade causes almost incalculable harm in both the producing and consuming states, while generating huge economic rewards for the traffickers. Since the "War on Drugs" was declared twenty years ago, vast sums have been spent to eliminate the flow of illegal drugs into the United States, but these efforts have failed to reduce this international drug trade substantially.[1] Would a reframing of the problem bring a more definite resolution? Perhaps other transboundary, or cross-border legal regimes could offer fruitful insight. Might principles derived from the *Trail Smelter* arbitration be useful as applied to Latin American drug trafficking? We conclude that neither the *Trail Smelter* "polluter pays" principle of state accountability for transboundary environmental harms nor the related state obligation to regulate against future continuing harm can be usefully extended to Latin American drug trafficking. The reasons for this conclusion illuminate challenges in applying the "polluter pays" principal in even its original international environmental law context.

LEGAL THEORY AND THE *TRAIL SMELTER* ARBITRATION

For purposes of this discussion, we presume the reader's familiarity with the basic events and holdings of the *Trail Smelter* arbitration.[2] Before leaving the *Trail Smelter* arbitration for the drug trafficking world, certain key attributes of the *Trail Smelter* arbitration fact pattern should be noted. The Trail smelter's domicile and ownership were parallel (i.e., the same state was both host and home to the smelter). Canada, as host and home of the polluter, possessed effective regulatory

[1] *See* Jurg Gerber and Eric L. Jensen, Drug War, American Style: The Internationalization of Failed Policy and Its Alternatives (2001).

[2] For a detailed discussion of the *Trail Smelter* dispute, *see generally* John D. Wirth, Smelter Smoke in North America (2000); *see also* Allum in this volume.

capacity. That is, following the *Trail Smelter* arbitration, Canadian regulations did in fact restrict the smelter's harmful emissions. Those harmed by the Trail smelter possessed political efficacy. The affected Washingtonians, failing in efforts to obtain relief directly from the Trail smelter, organized and complained to the U.S. federal government, which undertook to espouse their claims in the pursuit of a remedy at international law.

To consider whether and how to export the *Trail Smelter* principles,[3] it is helpful to paint the *Trail Smelter* decision with broader theoretical strokes. *Trail Smelter* articulated the attachment of (i) state liability for *privately* caused harms and (ii) affirmative regulatory duties. Traditional legal and economic theories applied to the *Trail Smelter* facts dictate the same outcome: reduced transboundary pollution. Beginning with traditional legal theory, from the perspective of the common law, the Trail smelter's transboundary pollution was a nuisance involving particulate matter encroaching on private property. Due to strong doctrinal preferences for protecting private property rights, the law typically grants relief for nuisances of this kind through an injuction or money damages. The harm was conceptualized as such within Canada, and Cominco, the smelter owner, purchased smoke easements from affected Canadian farmers. Then-existing restrictions on foreign land ownership precluded Cominco from applying the same solution in the United States.[4] Because the Trail smelter's pollution crossed an international border, framing the dispute solely as a common law private nuisance is inadequate to capture fully the harms involved. Transboundary pollution also encroaches on sovereign territory and the sovereign's right to control the resources therein.[5] Because sovereignty is protected by states acting among states under customary international law (augmented, in this case, by international treaty law), the remedy is against the polluter host *state*, which owes a duty to sister states to control the conduct of the enterprise it hosts to avoid further internationally offensive conduct. The availability of a remedy *between* states reflects their mutual respect for state sovereignty under international law.

From an economic perspective, the remedy available in *Trail Smelter* causes the polluting party to bear (more of) the real costs of its activities, rather than externalizing them, giving the polluter an incentive to improve its practices. The polluter is also the party best positioned to research and implement pollution control technologies. With the *Trail Smelter* remedy, the smelter will only continue its activities if it can take cost-effective measures to render the process less environmentally harmful or where smelting is so remunerative that it remains

3 We refer to the *Trail Smelter* principles to mean both the polluter pays principle and the state duty to regulate activities to prevent transboundary harms.

4 *See* WIRTH, *supra* note 2.

5 See William Snape III, et al., *Protecting Ecosystems Under the Endangered Species Act: The Sonoran Desert Example*, 41 WASHBURN LAW JOURNAL 14, n.143 (2001); Nicholas A. Robinson, *Befogged Vision: International Environmental Governance a Decade After Rio*, 27 WILLIAM & MARY ENVIRONMENTAL LAW & POLICY REVIEW 299, n. 82 (2002).

worthwhile to the smelter notwithstanding the liability exposure. Even without the state regulation component of the remedy, shifting liability for some or all of the externalized costs of smelting to the polluter's host state gives that state an incentive to regulate to avoid such costs of the activities it consents to host. With these analytical tools in hand, we now turn from smelting to drug trafficking and from the relationship of the United States with its northern to its southern neighbors.

LATIN AMERICAN DRUG TRAFFICKING – FRAMING AND LEGAL THEORY

We specify what we mean by the term "Latin American drug trafficking" so we may analyze how *Trail Smelter* might apply to its harms. Latin American drug trafficking includes both market activities and security activities. Market activities include: crop cultivation, the import of the raw materials for drug production, production of drugs for export, and the export and distribution of final drug products. Security activities include: crop destruction, production site interdiction, border policing, other import controls, and point-of-sale policing. States that are primarily producers in the drug trade (in any combination of cultivating crops, processing raw plants or chemicals into marketable drugs, or coordinating and conducting their export and distribution) are referred to here as producer states. States that are primarily consumers of drugs are, in turn, consumer states.[6]

State liability, *à la Trail Smelter*, for drug trafficking harms to private individuals of other states might be configured in multiple ways, four of which will be considered here. Producer states could be liable for introducing harmful substances into consumer state markets. Consumer states could be liable for drug enforcement externalities (i.e., for harms from its export of security and eradication activities). Consumer states could also be liable for being the source of the demand[7] that motivates producer state supply activities.[8] Or, from the perspective of decriminalization proponents, consumer states could be liable for their restrictive regulations to the extent these, by raising the market price, shift the production calculus in producer states relative to substitute crops and exports (e.g., food and textiles).

6 We avoid the terms import and export, which apply to drugs *and* security in inverse directions.

7 Demand for drug imports is not static. As a result of criminalization of possession, consumption statistics are hard to ascertain. The United States Office of National Drug Control Policy details the methodological difficulties in describing actual consumption, *available at* http://www.whitehousedrugpolicy.gov/publications/drugfact/drug_avail/chpt5.pdf. Interdiction and arrest records are typically substituted. For trafficking trends, *see generally* United States Drug Enforcement Administration, BACKGROUND PAPER ON DRUG TRAFFICKING *available at* http://www.usdoj.gov/dea/concern/drug_trafficking.html.

8 Because cash is fungible, it is not the money itself but the condition of its exchange for drugs that renders the flow of funds harmful. Purchase money is a proxy for demand.

Defining the universe of "Latin American drug trafficking" requires identifying which Latin American countries are involved, in what roles, and to what extent. Any such description is a temporal array, not a fixed map.[9] Finished cocaine, for example, is a regional product.[10] Until 1996, Peru was the world's largest coca cultivator; now Colombia is. Peru and Bolivia contend for second position, with Peru's cultivation declining and Bolivia's rising.[11] Most Peruvian cocaine base is routed to Colombia for production of finished cocaine for export, 90% of which is destined for the U.S. market. Bolivian base is routed through or to Brazil, Argentina, and Chile.[12] Drug production in Latin America is mainly for export to the United States, but a portion travels to Europe and Africa and another remains for domestic consumption. Significant quantities of heroin, marijuana, methamphetamine, and ecstasy[13] consumed in the United States also originate in Latin America.[14] With drug production and distribution comes corruption, money laundering, and organized crime.[15] Legal trade with the United States is significant for most Latin American producer states, but these states do not collectively represent a large portion of U.S. legal trade.[16]

COSTS OF LATIN AMERICAN DRUG TRAFFICKING

The interrelation of drug production and distribution markets with enforcement and eradication programs renders any typology of drug trafficking costs or harms an oversimplification. Nevertheless, to compose coherent verses to this sad song and to track the suggested liability propositions, we bifurcate costs roughly into: (a) costs in health, social and economic functioning, safety and public order, and criminal justice borne within consumer states; and (b) costs in producer states in their capacities as drug producers and exporters and in their capacities as importers

9 Since 1986, the President of the United States has performed part of this task in an annual, statutorily mandated drug certification ritual. *See* Eugene. E. Bouley Jr., *The Drug War in Latin America: Ten Years in a Quagmire*, *in* GERBER AND JENSEN, *supra* note 1, at 69–195.

10 For specific information about various producer states, *see* CIA WORLD FACTBOOK (2005), *available at* www.cia.gov/cia/publications/factbook/index.html.

11 Eradication and alternative crop programs in Bolivia have not kept pace with recent increases in cultivation. *See* CIA WORLD FACTBOOK (Bolivia), *supra* note 10.

12 *Id.*

13 3, 4-methylenedioxymethamphetamine (MDMA).

14 CIA WORLD FACTBOOK, *supra* note 10 (Mexico and Colombia).

15 A majority share of the proceeds of narcotics sales in the United States is laundered or invested in Colombia through the black market Colombian peso exchange. CIA WORLD FACTBOOK, *supra* note 10 (Colombia). Mexican drug trafficking is controlled by drug syndicates, *Id.*, (Mexico); *see also* MIKE GRAY, DRUG CRAZY: HOW WE GOT INTO THIS MESS AND HOW WE CAN GET OUT 33 (2000) (describing drug gang rivalry that plagues border cities).

16 For specific details about trade flows between Latin American countries and the United States, *see* The Economic Commission for Latin America and the Caribbean, *Principales Destinos de Exportaciones*, available at http://www.eclac.cl/cgi-bin/getprod.asp?xml=/comercio/noticias/paginas/2/21312/P21312.xml&xsl=/comercio/tpl/p18f.xsl&base=/comercio/tpl/top-bottom.xsl.

of security from consumer states. Social perspectives may tilt whether particular effects are viewed as costs or benefits. We have erred on the side of inclusiveness, thereby reflecting inconsistent social views.

Costs in Consumer States

Consumer states endure the health costs of drug trafficking.[17] These include public and private health care costs for drug treatment and related conditions, suffering due to acute and chronic physical illnesses, addiction/compulsive use, effects of maternal use, increased STD and other disease transmission including HIV/AIDS, poor substance quality control, deterrence of voluntary drug treatment, and restricted medicinal uses.[18] Social and economic functioning impairments include reductions in workplace and academic productivity, degraded parenting, peer-user influence, reductions in user self-esteem and reputation (including employability), criminal experience accrued in drug markets, acquaintance with criminal networks, elevated market price of controlled substances, and infringement on personal liberty.

Drug trafficking also raises safety and public order costs in consumer states. Among the costs attributable directly or indirectly, in whole or in part, to drug trafficking are: drug-related accidents (work, roadway, etc.); property/acquisitive crime; violence (at points of sale, enforcement, and consumption); disorder and perceptions of pubic disorder; fear in public and private spaces (with related restrictions on and of mobility); property value degradation near drug markets; and observably widespread violation of law. Elevated safety and public order costs imply (without wholly explaining) rising criminal justice costs. These include policing costs, court costs, incarceration costs, use of scarce jail/prison space, court congestion and delay, police and other (e.g., educational institutional) invasion of privacy, corruption and demoralization of legal authorities, violation of law, dilution of respect for the rule of law through widespread violation, devaluation of arrest as a moral sanction, fines, lost time and income (in court, prison), legal expenses, stigma and other effects of criminal and prison records (including political disenfranchisement), and fear of apprehension.

Not all costs related to drug trafficking are cross-border harms, yet there can be no meaningful separation between domestic and international harms. Municipally endogenous costs (harms felt within the state in which the activity originates) have international implications. Costs borne domestically by producer states, for example, limit such states' regulatory efficacy and interdiction efforts, thereby

17 Robert J. MacCoun and Peter Reuter, Drug War Heresies: Learning from Other Vices, Times, and Places 106–107 (2001) (listing a taxonomy of drug harms); Eric L. Jensen et al., *Social Consequences of the War on Drugs: The Legacy of Failed Policy*, 15 Criminal Justice Policy Review 100 (2004).

18 *See* Gonzales v. Raich, 121 S. Ct. 2195 (2005).

permitting export of drugs. Likewise, costs borne in consumer states motivate policies of security export.

Costs in Producer States

The apparent benefits of the drug trade (e.g., use of otherwise unproductive land, material enrichment), have the pernicious effect of supporting the drug trafficking system that brings with it a host of social costs far in excess of those benefits. Costs in producer states include: distortion or reinforcement (as the case may be) of traditional social hierarchies, violence by and against drug enforcement and interdiction efforts, violence by drug traffickers in local communities as they force planting and expropriate profits from the rural poor, expanded access to weaponry, the expenses associated with state interdiction projects and their attendant opportunity costs in health and welfare spending, increased drug use, corruption of the church, and inflation (the infusion of "narcodollars" in local markets raises living costs for all, rendering participation more compelling). The drug trafficking system becomes entrenched through use of otherwise unviable land and provision of a remunerative export crop relative to substitutes. As multinational agribusiness displaces traditional agriculture, the rural poor have at once an increased need for cash and less access to fertile areas where they might cultivate market crops. This supports the turn to coca production, as coca thrives in the jungle and in less fertile areas, adding destruction of jungle and old-growth forests to the list of producer state harms.

The cost list continues, including: the supplanting and overpowering of the state by drug traffickers, contributions by drug traffickers to their communities that garner loyalty and dependency[19] by funding needs the state cannot or will not provide (thus weakening the civil state and removing state incentives to fund social needs by other means) and, in some cases, poor communities valorizing individual drug traffickers as romantic rebels or nationalist heroes. Harm is mitigated by the remoteness of the points of retail sale and by the less restrictive use and possession regulations in most producer states.

Producer states are also importers of security products, programs, and services from consumer states. Costs of eradication and enforcement efforts abound. For example, zero coca tolerance policies of producer states infringe on traditional use practices in the Andes, where coca has been used traditionally and recently to stave off hunger. U.S. eradication and enforcement programs in Latin American states

[19] *See* Bouley, *supra* note 9. *See also* Fernando Garcia Argañarás, *Harm Reduction at the Supply Side of the Drug War: The Case of Bolivia, in* HARM REDUCTION: A NEW DIRECTION FOR DRUG POLICIES AND PROGRAMS 99–115 (Patricia G. Erickson, et al. eds., 1997). That Latin American states consent to or invite these activities means, on a formal level, their sovereignty has not been compromised. The relative wealth and differential trade and aid dependency between the parties calls into question the voluntary character of consent in this context. *See Principales Destinos*, *supra* note 16.

may infringe on the sovereignty of those states through direct military involvement and through political pressure. These programs militarize Latin American policing, which hinders the development of nascent democracies. With an eternally debatable degree of awareness, the United States has supported paramilitaries and death squads operating under the banner of antidrug trafficking.[20] Producer states have used weapons, aircraft, and arguably consumer state military personnel earmarked for counter-trafficking efforts in their own counterinsurgency efforts.[21] Typical crop eradications methods (e.g., aerial pesticide spraying) tend toward those most environmentally harmful and toxic to local populations.[22]

As neat as the listing here may appear, harm vectors are not so conveniently separable as to lend themselves to a calculation of net harm for which a producer or consumer state could pay the other money damages. Because drug trafficking is illegal, harms stemming from the activity itself are interwoven with harms attributable to the illegality of the activity, and/or the enforcement context. Activity harms in producer states might include rising social violence, political degradation, development opportunity costs, economic distortions through producer state dependency on drug rather than substitute export markets (as food exporters become food importers), imports infected by money laundering, and environmental degradation. How can these be extricated from enforcement harms, such as the arming of paramilitary militias, U.S. military involvement in producer state affairs, and issues of the extraterritorial reach of U.S. law?[23]

TRAIL SMELTER CLAIMS UNLIKELY

Even if harms could be conceptually bound and netted, it is doubtful that a claim could or would be brought employing *Trail Smelter* principles. Jurisdiction presents a significant hurdle, as do states' perceptions of their respective interests. Respect for state sovereignty under international law requires that states

[20] *See* Bouley, *supra* note 9, at 182.

[21] Colombia has for some time engaged in active opposition to drug trafficking. Colombia's aerial eradication program employs planes provided by the United States. The United States also provides support for Colombia's military and paramilitary opposition to drug trafficking. See MARK BOWDEN, KILLING PABLO: THE HUNT FOR THE WORLD'S GREATEST OUTLAW 61, 68 (2001). through "Plan Colombia," $7.5 billion in aid has been committed by the United States, major national donors (mainly E.U. members), the World Bank, the Inter-American Development Bank and other international lending institutions. The United States has provided Green Beret troops to train Colombian officers and enlisted personnel, Blackhawk troop transport helicopters, and Huey II helicopters. CIA WORLD FACTBOOK, *supra* note 10 (Colombia). Plan Colombia was conceived and pitched as an effort against drug trafficking, but the Colombian government's efforts are increasingly directed against its armed leftist rebels. This redirection has in part resulted from the inseparable interrelations between drug trafficking and the armed groups in Colombia, but it also suggests a discursive response to world events.

[22] Bouley, *supra* note 9, at 185–186.

[23] *See, e.g.*, Verdugo-Urquidez v. United States, 513 U.S. 1114 (1995).

consent to the legal regimes under which they might incur liability. Before a *Trail Smelter*–like claim for state liability or obligation to regulate could attach, states would have to enter into one or more agreements providing for such a remedy (unless they were deemed to have done so under existing instruments or under customary international law). None of the three UN conventions applicable to Latin American drug trafficking provides such a remedy.[24] Although consumer and producer states have undertaken a mutual obligation to regulate, we are not aware of an instrument or body through which the producer or consumer states have consented to liability for drug trafficking harms.[25] Consumer and producer states are not likely to perceive an interest in consenting to such a regime, as both types of states would be unlikely to employ it.

Consumer states would be unlikely to bring *Trail Smelter* claims because producer states are effectively judgment-proof. Recognizing the poverty of producer states, consumer states are unlikely to pursue money damages on behalf of their nationals or as leverage to persuade producer states to develop more effective harm-reduction regimes. Instead, consumer states condition aid on cooperation, as the United States does by certifying or decertifying a state in relation to its antitrafficking efforts.[26]

[24] These are the 1961 Single Convention on Narcotic Drugs (as amended in 1972) *reprinted in* 11 INTERNATIONAL LEGAL MATERIALS 804 (1972) (hereinafter the "1961 Convention"), the 1971 Convention on Psychotropic Substances, *reprinted in* 10 INTERNATIONAL LEGAL MATERIALS 261 (1971) (hereinafter the "1971 Convention") and the 1988 Convention Against the Illicit Traffic in Narcotic Drugs and Psychotropic Substances, *reprinted in* 28 INTERNATIONAL LEGAL MATERIALS 493 (1989) (hereinafter the "Vienna Convention"). Hans-Joerg Albrecht, *The International System of Drug Control: Developments and Trends, in* GERBER AND JENSEN, *supra* note 1 at 49–60. The principle states involved in Latin American drug trafficking are all parties to all three conventions. United Nations Office of Drugs and Crime, http://www.unodc.org/unodc/en/ drug_and_crime_conventions.html. The Vienna Convention obliges states parties to take measures (and report these measures to the International Narcotics Control Board established under the 1961 Convention) to harmonize drug enforcement regulations, including criminalizing cultivation, production, distribution, possession, and money laundering (Art. 3). Additional provisions address forfeiture (Art. 5), extradition (Art. 6), mutual assistance (Art. 7), and measures to reduce cultivation and demand (Art. 14). Disputes under the Vienna Convention that cannot be resolved by peaceful consultation are referred under Article 32 to the International Court of Justice. Although Article 32 of the Vienna Convention states parties have consented to IJC jurisdiction, that jurisdictional reach would not encompass inefficacy of regulations, harms *due to* regulations, or harms from trafficking generally.

[25] The exception would be a claim against a state for extraterritorial measures taken *without* the consent of the state on whose territorial integrity the conduct infringed (e.g., military intervention by a consumer state in the name of drug control after 1988). *See* Vienna Convention, art. 2, *supra* note 24.

[26] The Foreign Assistance Act, directs the President of the United States to compile a list of countries involved in drug trafficking or production, and to identify those countries that have "failed demonstrably" to cooperate with U.S. antidrug efforts. Uncooperative countries are decertified and ineligible for most U.S. aid. In addition, the United States will vote against that country receiving loans from international lending institutions. *See generally* 22 U.S.C. § 2291 et seq. *See also* Bouley, *supra* note 9.

Producer states would be equally unlikely to bring *Trail Smelter* claims. Many producer states have relatively weak economies, high levels of corruption, political sectors opposed to antitrafficking (including drug traffickers, cultivators, and insurgent opposition groups) and, in certain cases, tolerance of drug trafficking. Producer states are unlikely to espouse the claims of their harmed nationals, particularly those most afflicted by drug trafficking: their politically marginalized rural poor populations.[27] Rather than an adversarial relationship, a producer state opposing drug trafficking seeks aid from, or mutual assistance with, other states in the relevant market. A state that chooses otherwise risks being (if it is not already) outgunned by the drug traffickers who compete for the state's constituents. Drug traffickers have presented very real threats to states' capacity to govern.[28] Weak antitrafficking producer states have a powerful incentive to ally with stronger states for military support. Also, inequality between weak producer states and strong consumer states, as well as aid dependency of the former on the latter means that, rather than antagonize wealthy, powerful, consumer states through litigation, producer states tend to prefer formal cooperation against a perceived common enemy.

Even if these hurdles could be overcome, applying *Trail Smelter* principles to international drug trafficking would be unlikely to accomplish the goals of either diminishing harms or strengthening antitrafficking regulation. From a traditional legal perspective, a *Trail Smelter* remedy would make sense only if discrete harms could be identified that were not counterbalanced by opposite harm vectors.[29] As discussed earlier, boundary problems and the interwoven complexity of drug trafficking harms make discerning such a situation difficult. From an economic perspective, applying the *Trail Smelter* principles to this context would be desirable only if they reduced net systemic harm; but if the principles prompted more litigation without net systemic harm reduction, applying *Trail Smelter* could make a bad situation worse.

Neither analytical approach supports application of *Trail Smelter* to Latin American drug trafficking for its potential effect on regulation. State liability and an expanded[30] duty to regulate are unlikely to induce higher levels of countertrafficking regulation in either consumer or producer states. The United States, the archetypal consumer state, with its overwhelmed courts and exploding prison population, does not need state liability to induce higher levels of

[27] Only states have standing to bring claims against other states in international fora.

[28] *See* Bowden, *supra* note 21, at 55.

[29] *Trail Smelter* itself nearly went down that garden path when Canada threatened to counterclaim that industry in Detroit caused parallel harm to Ontario. *See generally* Wirth, *supra* note 2. That parallel harm, however, was caused by private actors in different industries, operating in different markets and affecting different populations. In drug trafficking, as discussed later, the harms are up and down the distribution chain of a single market.

[30] *Trail Smelter* is not needed to induce states' recognition of an obligation to regulate; that was obtained by each state's signature or ratification of the Vienna Convention.

regulation than it already has.[31] Colombia, the archetypal producer state, with its extensive trafficking-funded guerilla insurgency[32] does not need state liability to induce its own opposition to drug trafficking. Besieged by a barrage of bombings, kidnappings, and assassinations of politicians, candidates, judges, and antitrafficking personnel, Colombia did not need any further incentive for its own antitrafficking regulation.[33] At the times Colombia has been terrorized by drug trafficking, concern to avoid liability by not permitting drug cultivation and exports could not have noticeably increased its level of antitrafficking vigilance. Colombia lacked regulatory capacity, not motivation. The grim reality of Colombian drug traffickers' war on the Colombian government for its cooperation with the U.S. "war on drugs" only underscores that only regulatory efficacy not regulatory will or formal regulation can reduce harm. State liability has little hope of providing a regulatory incentive, because it obtains exactly when states are least capable of effective domestic regulation. Moreover, states themselves have various interests in perpetuating drug trafficking and may well enact appropriate regulations without effectively enforcing them.

Distribution Chain Causality

Even if imposing *Trail Smelter* liability offered advantages under a traditional or economic theory, and even if states consented to such claims and defied expectations by bringing them, the complexity of proving causation would make prevailing on such a theory both expensive and unlikely. In *Trail Smelter*, causation, although disputed, was relatively direct, with particulate smelter emissions

31 One wonders, though, whether consumer state liability alone would induce not necessarily more but perhaps more effective regulation. *See* GRAY, *supra* note 15, at 155 (2000); Martin Killias, *Switzerland's Drug Policy as an Alternative to the American War on Drugs?, in* GERBER AND JENSEN *supra* note 1, at 241–259.

32 Colombia has the largest producer state guerrilla insurgency. Peru's movement has re-emerged. *See* May 11, 2005, Testimony before the House Committee on International Relations by Jonathan D. Farrar, Deputy Assistant Secretary of State, Bureau for International Narcotics and Law Enforcement Affairs, *available at* http://www.state.gov/g/inl/rls/rm/46214.htm (eradication efforts in Peru hampered by recent events in coca growing regions indicating activity by remnants of Shining Path movement in supporting drug trafficking). Bolivia has significant popular opposition to state eradication policies seen in wide domestic support for Bolivian cultivators' opposition to the Bolivian state's coca suppression efforts. *See* Shawn Blore, *Despite the Calm, Reform Pressures Challenge Bolivia*, TORONTO GLOBE AND MAIL, June 28, 2005, at A12 (Evo Morales, head of coca growers association, leader of largest opposition party, close second in 2002 presidential election); *Data in UNODC Reports Show Andean Coca Cultivation on a Strong Rebound*, LATIN AMERICAN WEEKLY REPORT, June 21, 2005, at *1 (Bolivian protest movement led by coca growers association head and largest opposition party leader Evo Morales prompted the resignation of President Carlos Mesa); Juan Forero, *Bolivia's Newly Elected Leader Maps His Socialist Agenda*, THE NEW YORK TIMES December 20, 2005 (Evo Morales decisively elected on December 18, 2005). The issue is effectively framed in Bolivia as sovereignty versus imperialism, as in the case of popular opposition to the gas pipeline export contracts.

33 *See* BOWDEN, *supra* note 21, at 52–64; GRAY, *supra* note 15, at 121–124.

traveling in visible clouds to reach obviously damaged crops and forests.[34] In the context of drug trafficking, the claim would have to be for distribution *chain* harm – a dynamic, developing concept in both municipal and international law. These distribution chain claims could allege harms caused indirectly along the chain (market conduct) or that the allegedly liable entity caused or fostered the existence of the chain (market creation). Under either theory, proving causation would be difficult.[35] Distribution chain causality problems plague litigation in many other contexts, including notably international trade, guns, tobacco, and diamonds. Complexity does not, of itself preclude legal action, but it does reduce the hope of attaching liability.[36] Antitrust and antidumping experiences remind that for distribution chain liability, causality is problematic even for legal market activities.

For market *creation* claims, there is the additional wrinkle of cross-causality. Effective enforcement and eradication efforts in one market may create opportunities for alternative markets to develop elsewhere. Should Colombia have a cause of action against Peru for cooperating with U.S. enforcement and eradication efforts in the early 1990s if these efforts fostered expanded coca cultivation in Colombia as a substitute for imported cocaine base? Should Belize and Jamaica have a cause of action against Mexico for cooperating with the U.S. enforcement and eradication efforts in the 1970s, if these efforts fostered cannabis cultivation in those countries?

IMPLICATIONS

Considering the applicability of the *Trail Smelter* principles to Latin American drug trafficking suggests a similar inquiry for other cross-border legal and illegal markets.[37] Amidst concern about a global regulatory race to the bottom, with multinational enterprises of increasing size and importance, forum shopping for least-cost locations, there has been an understandable turn to international law as a mechanism to regulate cross-border effects of domestic activities. Thinking about how to hold cross-border actors accountable for harms caused by their chosen market activities, whether legal or illegal, points to the importance of questioning the presumptions of formal state equality and state regulatory efficacy. A theory of

[34] *See Trail Smelter Arbitral Decision*, 33 American Journal of International Law 182 (1939) [hereinafter "*Trail Smelter* (1939)"]; *Trail Smelter Arbitral Decision*, 35 American Journal of International Law 684 (1941) [hereinafter "*Trail Smelter* (1941)"]. *See* Annex to this volume.

[35] Complexity does not necessarily deter litigation involving harmful market activities, including product liability in torts, conspiracy and organized crime in criminal law, and unfair trade practices in international trade.

[36] For litigation alleging distribution chain theories, *see* The Brady Center to Prevent Gun Violence Legal Action Project Docket, *available at* http://www.gunlawsuits.org/docket/docket.php.

[37] The *Trail Smelter* arbitration is rarely invoked even in environmental contexts. States prefer prospective emissions treaties. *See generally* Peter Sand, *Lessons Learned in Global Environmental Governance*, 18 Boston College Environmental Affairs Law Review 213 (1991); *see also* Knox in this volume.

state liability as a mechanism to reach the conduct of private actors, as apparently attractive as it may be, begs two important questions: liability to whom; and which state(s)?

State Inequalities

The *Trail Smelter* arbitration as a dispute proceeding under international law presumed formal state equality. There is, however, no positive state equality of power or economic or military resources. Given this reality, in the drug trafficking context, states negotiate (whether fairly or not) for mutual assistance in combating drug trafficking rather than engage in international litigation to allocate liability.[38] Aside from extreme differentials in power and wealth, the complex web of trade and other interdependencies between producer and consumer states in the drug trafficking context makes resort to international dispute resolution proceeding improbable. States vary widely in their respective and relative regulatory efficacy. Poor and weak states are less well positioned to deter the cross-border externalities of the hosted enterprises they need to attract and retain. The worth of a state's legal obligation to regulate hosted enterprises is only as good as its regulatory capacity.

If international dispute resolution proceedings (which are the minority of state actions taken under the rubric of international law) cannot induce states to minimize the cross-border harms of the enterprises they host – whether because the proceedings are unlikely or because states lack regulatory efficacy – the theory of liability itself should nevertheless be memorialized in multilateral treaties with informal regulatory mechanisms. The *Trail Smelter* principles could thereby represent not a means for recovery but a reframing of the relationship between states and the activities they host.[39]

Home/Host Variations

It is a truism of globalization that the unity present in the *Trail Smelter* case between home and host cannot be presumed.[40] In contending with harms caused

[38] A heavily negotiated multidimensional and multilateral treaty, such as the Agreement Establishing the World Trade Organization, conceivably could clarify responsibilities for drug trafficking harms and describe a dispute resolution mechanism to address these.

[39] International legal institutions host ostensibly equal negotiations among unequal actors dominated by the interests of the powerful, but these same institutions, operating under a knowingly false presumption of formal equality, have also fostered international cooperative development and redistributive programs. The state differentials writ large in the context of Latin American drug trafficking are more or less ubiquitous and have been so throughout the dual reign of international legal discourse and the nation state. Clearly normative equality trumps positive inequality under international law as we know it. Indeed, international personality and formal equality are and have been the birthright of nation states, one of the brass rings of nation-state formation.

[40] Had a Canadian company not purchased the Trail smelter in 1906 from its then owners in the United States, *see Trail Smelter* (1939 and 1941), *supra* note 34, Canadian regulation would not have been at issue.

by multinational enterprises, international lawyers contend with allocating regulatory responsibility between home and host states. Multinational enterprises by definition operate in multiple regulatory jurisdictions. As thorny as the jurisdictional issues in suing private multinational actors may be, they pale in comparison with the quagmire of allocating *Trail Smelter*-like accountability[41] among regulators in multiple home/host states.

Any application of *Trail Smelter* would have to account for the variable combinations of home and host domicile of harmful enterprises. Permutations abound, including local ownership of a wholly local enterprise (local home, local host), foreign ownership of a wholly local enterprise (foreign home, local host), combined local and foreign ownership of a wholly local enterprise (blended home, local host), foreign ownership of a not wholly local enterprise (foreign home, multiple hosts), and so on. Examples show how home and host domicile matter. Imagine a two-state case with enterprise "Parent" domiciled in State A (its home) that operates through "Subsidiary," domiciled in State B (its host). Would State A residents harmed in State A by Subsidiary's activities in State B be more likely to press State A to bring a *Trail Smelter* claim against State B or to raise claims in State A courts against Parent on a veil-piercing theory or against Subsidiary, perhaps on an "effects" theory of extraterritorial jurisdiction.

Adding a third state increases the complexity geometrically. Imagine now a State A domiciliary or its subsidiary that acts in State B and harms residents of State C. There is an unresolved issue in apportioning liability between States A and B for the harms and for their (effective) failure (whether mutual or not) to regulate the enterprise, particularly where regulations in State A and State B are not harmonized or governed by international agreements. An even more likely challenge to the operational utility of *Trail Smelter* is the multistate case of a multinational enterprise that operates in various states, navigating inconsistent and uneven regulation as it allocates its activities according to the least-cost location for each of its risky activities. Beyond the problems of allocating liability and regulatory responsibility to the states involved, causal issues relating to control, decision making, and structural limitations on liability further complicate the question of which home or host states would be appropriate defendants in a *Trail Smelter* action on these facts.

CONCLUSION

The *Trail Smelter* principles do not apply and should not be applied to Latin American drug trafficking. *Trail Smelter* provided for state accountability on a nuisance theory; harms from drug trafficking do not fit into the *Trail Smelter* construct in a way that would make that the basis of an appropriate remedy. Drug trafficking

[41] One commentator relies on *Trail Smelter* to support her proposition that home *and* host states must regulate multinationals. *See* Dinah Shelton, *Protecting Human Rights in a Globalized World*, 25 Boston College International & Comparative Law Review 273, 307 (2002).

harm, as used here, refers to the externalities of two related exports: drugs and security. These harms are more complex and diffuse than those at issue in *Trail Smelter* and causation is far more attenuated. Moreover, the states involved are not of equivalent power vis-à-vis one another nor vis-à-vis their own populations. As a result, the *Trail Smelter* mechanisms of inducing domestic enforcement or regulation to avoid state accountability are unlikely to be effective. It is equally unclear how to apportion state liability for harms caused by *cooperative* transnational law enforcement efforts. These limits to the utility of the *Trail Smelter* principles to Latin American drug trafficking illuminate factors central to the regulation of cross-border markets generally, presenting challenges to relying on international litigation to correct for market failure in light of variations in domestic regulatory efficacy.

23 Application of International Human Rights Conventions to Transboundary State Acts

Nicola Vennemann

INTRODUCTION

Trail Smelter is the first decision in which state responsibility for transboundary environmental harm was recognized.[1] Long before scholars started talking about globalization,[2] the facts of *Trail Smelter* demonstrated how inextricably linked people are despite the national borders separating them. The *Trail Smelter* arbitration is a prominent part of international law's response to transboundary environmental harm, but how does international law respond to transboundary or extraterritorial human rights violations?[3]

The international human rights conventions' treaty organs have developed a complex body of jurisprudence on this question, which seeks to respond to the main problem in cases of transboundary or extraterritorial human rights violations, namely, that state obligations arising from international human rights conventions are limited to securing the convention rights to persons under the states' jurisdiction. Article 1 of the European Convention for the Protection of Human Rights and Fundamental Freedoms (ECHR), for example, states that "the High Contracting Parties shall secure to everyone *within their jurisdiction* the rights and freedoms defined in Section I of this Convention."[4] The (extra)territorial scope of states' obligations under human rights conventions, thus, depends on whether a member state was exercising jurisdiction over the victim when the violation took place.

[1] *Trail Smelter Arbitral Decision*, 33 AMERICAN JOURNAL OF INTERNATIONAL LAW 182 (1939) [hereinafter "*Trail Smelter* (1939)"]; *Trail Smelter Arbitral Decision*, 35 AMERICAN JOURNAL OF INTERNATIONAL LAW 684 (1941) [hereinafter "*Trail Smelter* (1941)"].

[2] *See* THE GLOBALIZATION OF WORLD POLITICS 14 (John Baylis and Steve Smith eds., 1997).

[3] Transboundary/extraterritorial state acts are understood here as acts taken by state authorities outside of their territory as well as acts taken on a state's territory but producing effects outside of the territory. The implications of the creation of international organizations by states for the protection of human rights will not be dealt with.

[4] European Convention for the Protection of Human Rights and Fundamental Freedoms, Nov. 4, 1950, 213 UNTS 221 [hereinafter "ECHR"].

For at least two reasons, *Trail Smelter*, although involving a transboundary harm, can have only limited value in efforts to interpret the requirement of state jurisdiction in the context of international human rights law. First, *Trail Smelter* only dealt with those sets of cases in which activities on the state's territory have consequences for persons on foreign soil. It did not face the question of the circumstances under which state responsibility might attach to actions taken outside a state's sovereign territory. This question, however, is highly relevant in the context of international human rights law as states increasingly take actions outside their own territory that adversely affect the human rights of individuals on foreign soil.

Second, as *Trail Smelter* involved Canada's voluntary assumption of responsibility for the harm caused by the private smelter, it does not provide any guidance for deciding cases in which private actors violate the human rights of persons abroad. In the context of human rights law, states are not directly responsible for the acts of private persons; they only incur liability under the conventions if they fail to fulfill their positive obligation to secure the convention rights against infringements by private actors. In cases in which private actors cause the human rights violations of persons abroad, it is particularly difficult – as will be seen later in this analysis – to establish that the state exercised jurisdiction over the persons whose human rights have been affected or violated.

If *Trail Smelter's* engagement of a state's responsibility over actions occurring within its territory, including actions undertaken by private actors, promises little as regards the problem of transboundary or extraterritorial human rights violations, what is the solution? As noted earlier, a states' human rights responsibility for extraterritorial acts will hinge on critical definitions of state jurisdiction. This chapter investigates the meaning and content of the term "jurisdiction" in the context of international human rights. The European Court of Human Rights (ECourtHR) recently confronted this definitional question in *Bankovic*.[5] Tracing the Court's reasoning in that case provides an interesting exploration of the contours and limits of this term and highlights the fundamental distinctions operating in the approaches taken by international human rights law and international environmental law to establishing responsibility for transboundary or extraterritorial harm. Thus, *Bankovic* and the cases that follow illuminate the limits on *Trail Smelter's* applicability to international human rights questions.

JURISDICTION UNDER THE ECHR

According to Article 1 of the ECHR, "jurisdiction" is necessary to establish member states' responsibility for transboundary acts. In interpreting Article 1, the European Commission of Human Rights (ECommHR) and the ECourtHR have found

[5] Bankovic et al. v. Belgium et al., 2001 – XII Eur. Ct. H.R. (2001).

that, at least in principle, member states can be held responsible under the ECHR even when acting outside their territory. The decisions of these two organs have recognized four sets of circumstances when this may be the case: the acts of consular or diplomatic authorities of a state on foreign soil; law enforcement or adjudication on foreign soil; the arrest of persons on foreign soil; and military occupation of foreign territory. The richest body of case law on the issue involved the last of these categories, in particular the ECommHR's practice and the ECourtHR's case law concerning Turkey's responsibility for human rights violations in the Turkish-occupied northern Cyprus.

Two tracks of reasoning can be distinguished. On the one hand, the ECommHR has articulated a relational concept of jurisdiction. In grounding states' responsibility for extraterritorial acts on the exercise of control over the person on foreign soil, it conceives the concept of jurisdiction as designating a particular relationship between the person concerned and the state. It held, for example, that, as authorized agents of Turkey, Turkish armed forces in northern Cyprus "bring any other person in Cyprus within the jurisdiction of Turkey [...] to the extent that they exercise control over such persons. Therefore, in so far as these armed forces, by their acts or omissions, affect such persons' rights or freedoms under the Convention, the responsibility of Turkey is engaged."[6] At the same time, the ECommHR as well as the ECourtHR have also suggested that a state's responsibility for extraterritorial acts may arise as a result of the exercise of control over foreign territory. Based on the object and purpose of the Convention, the ECommHR has held that "Article 1 might also cover areas outside the national territory provided that such areas were under the effective control of the Government concerned."[7] Similarly, the ECourtHR has held that the responsibility of a contracting state can arise when "as a consequence of military action – whether lawful or unlawful – it exercises effective control of an area outside its national territory. The obligation to secure in such an area, the rights and freedoms set out in the Convention derives from the fact of such control, whether it is exercised directly, through its armed forces, or through a subordinate local administration."[8] The ECourtHR has further pointed out that, if a state has effective overall control over a given territory, its responsibility cannot be confined to the acts of its own soldiers or officials, but also extends to the acts of the local administration which survives by virtue and at the pleasure of the occupying state's military.[9]

6 Chrysostomos and Papachrysostomou v. Turkey, App. Nos. 15299/89 and 15300/89, 86-A Eur. Comm'n H.R. Dec. & Rep. 4, 26 (1991).

7 Cyprus v. Turkey, App. No. 25781/94, 13 Eur. Comm'n H.R. Dec. & Rep. 85, 151 (1996).

8 Loizidou v. Turkey (Preliminary Objections), 310 Eur. Ct. H.R. (ser. A) para. 62 (1995); Loizidou v. Turkey (Merits and Just Satisfaction), 1996–VI Eur. Ct. H.R. para. 52. *See also* Ilascu and Others v. Moldova and Russia, App. No. 48787/99 Eur. Ct. H.R., section I, 2 b (2004), *available at* http://cmiskp.echr.coe.int/tkp197/portal.asp?sessionId=3345367&skin=hudoc-en&action=request.

9 Cyprus v. Turkey, 2001–IV Eur. Ct. H.R. para. 77 (2001).

At first glance, these two approaches seem to diverge. The first approach bases the state's jurisdiction, and thus, the limiting force of the ECHR on that state, on the exercise of control over the persons on foreign territory. The second approach bases the state's jurisdiction on the exercise of control over the foreign territory itself. However, even if the latter approach seems to define jurisdiction as a geographically limited area, ultimately it is still based on the state's control and authority over persons and therefore in conformity with the relational concept of jurisdiction: as a consequence of the military occupation of the territory, the state exercises control over all persons and property in that territory. As in its own territory, the state will be presumed to exercise jurisdiction over all persons in the occupied foreign territory and it will have to secure, as a consequence, the entire range of Convention rights to those persons.

The Bankovic Case

In the *Bankovic* case the Grand Chamber of the ECourtHR demonstrated, in a decision of principle,[10] the limited extent to which it was willing to apply the ECHR standards to extraterritorial state acts.[11] For the first time, it set out in an abstract manner the limits of states' human rights obligations for trans-boundary state acts in a way that seemed to depart from the Court's previous case law.

From March to June 1999, NATO operated air strikes against the Federal Republic of Yugoslavia.[12] On 23 April 1999, one of the Radio Televizije Srbije (RTS) buildings was hit by a missile launched from a NATO forces' aircraft.[13] Two of the four floors of the building collapsed, sixteen persons were killed and sixteen seriously injured.[14] The applicants, victims of the attack or relatives of deceased victims, claimed that NATO Member states had violated, *inter alia*, their right to life as guaranteed by the ECHR. Without dealing with the question of the liability of states for acts of international organizations, the Court's Grand Chamber declared the application inadmissible because it considered that the applicants were not "under the jurisdiction" of the respondent states for the purposes of article 1 of the ECHR.[15]

The main question the Court had to answer was whether the applicants fell within the jurisdiction of the respondent states "as a result" of the impugned

[10] G. Cohen-Jonathan, *La territorialisation de la juridiction de la Cour européenne des droits de l'homme*, 13 REVUE TRIMESTRIELLE DES DROITS DE L'HOMME 1069, 1074 (2002). In this sense, see Georg Ress, *State Responsibility for Extraterritorial Human Rights Violations – The Case of Bankovic*, 6 ZEITSCHRIFT FÜR EUROPARECHTLICHE STUDIEN 73, 74 (2003).

[11] Bankovic, supra note 5. For a summary and a short appraisal, see Frank Schorkopf, *Grand Chamber of the European Court of Human Rights Finds Yugoslavian Bombing Victims' Application Against NATO Member States Inadmissible*, 3 GERMAN LAW JOURNAL NO. 2 (2002), at http://www.germanlawjournal.com/article.php?id=133.

[12] Bankovic, *supra* note 5, at para. 8.

[13] *Id.* at paras. 9, 10.

[14] *Id.* at para. 11.

[15] *Id.* at para. 82.

act, namely the bombardment of the RTS building. This question had to be decided in accordance with the interpretation guidelines of Article 31 of the Vienna Convention on the Law of Treaties.[16] As to the ordinary meaning of the term "jurisdiction," the ECourtHR stated that, from the standpoint of public international law, the jurisdictional competence of a state was primarily territorial, other bases of jurisdiction being exceptional and requiring special justification in the particular circumstances of each case.[17] The ECourtHR concluded that, state practice, particularly the fact that no state had ever made a derogation pursuant to article 15 of the ECHR for extra-territorial military missions, confirmed this territorial-based interpretation of the term "jurisdiction."[18]

The ECourtHR then gave an overview of the cases in which it had, *exceptionally*, accepted that acts of the contracting states performed or producing effects outside their territories constituted the exercise of jurisdiction within the meaning of Article 1.[19] These include cases in which "the respondent state, through the effective control of the relevant territory and its inhabitants abroad as a consequence of military occupation or through the consent, invitation or acquiescence of the Government of that territory, exercises all or some of the public powers normally to be exercised by that Government."[20] The ECourtHR also noted that in cases involving the activities of a state's diplomatic or consular agents abroad and on board craft and vessels registered in, or flying the flag of that state, "customary international law and treaty provisions have recognized the extra-territorial exercise of jurisdiction by the relevant state."[21]

The ECourtHR concluded that such *exceptional* circumstances, amounting to the extraterritorial exercise of jurisdiction by the respondent states, were not present in the *Bankovic* case.[22]

The ECourtHR notably rejected the applicants' contention that Article 1's reference to jurisdiction includes the positive obligation to secure the ECHR rights in a manner proportionate to the level of control exercised in any given extraterritorial situation. This was, in the ECourtHR's view, "tantamount to arguing that anyone adversely affected by an act imputable to a Contracting state, wherever in the world that act may have been committed or its consequences felt, is thereby brought within the jurisdiction of that state for the purpose of Article 1 of the [ECHR]."[23] The ECourtHR found this interpretation to be inconsistent with the wording of article 1 and noted that it would even render the words "within their jurisdiction" meaningless.[24] Furthermore, the ECourtHR held that this interpretation would lead to the equation of the question, whether a person was within the jurisdiction of a state, with the question, whether the person was a victim of a violation of ECHR rights.[25]

[16] *Id.* at para. 55.
[17] *Id.* at para. 61.
[18] *Id.* at para. 63.
[19] *Id.* at para. 67.
[20] *Id.* at para. 71.
[21] *Id.* at para. 73.
[22] *Id.* at para. 75.
[23] *Id.*
[24] *Id.*
[25] *Id.*

In the light of these considerations, the Court, in a unanimous decision, concluded that the applicants did not come under the jurisdiction of the respondent states on the account of the extraterritorial act in question.

The *Bankovic* decision has been criticized by many as a step backward in the protection of human rights from extraterritorial state acts. From this perspective, both the ECourtHR's resort to a jurisdictional mechanism to avoid the case *and* the approach the ECourtHR adopted towards the concept of jurisdiction are highly questionable. Interestingly, in its most recent decisions concerning extraterritorial state acts, the ECourtHR has moved away from *Bankovic's* narrow approach to jurisdiction and confirmed that it does not intend to give up the relational approach to jurisdiction developed in its previous jurisprudence. In the *Öcalan* case,[26] which concerned the arrest of a Turkish citizen on foreign soil, and in the *Issa* case,[27] which concerned the alleged assassination of shepherds by Turkish armed forces in Iraq, the ECourtHR confirmed the validity of the "authority and control test" as grounds for establishing article 1 jurisdiction. In light of this recent case law, it seems likely that *Bankovic* was a result-oriented judgment in a politically sensitive case.

REDEFINING JURISDICTION

Reflecting on the ECHR's unsatisfactory conceptualization of jurisdiction in *Bankovic*, this section outlines a comprehensive definition of "jurisdiction" and offers a series of general criteria that ought to inform a conclusion regarding a member state's article 1 jurisdiction and the attending applicability of the ECHR protections. The criteria rely on the practice of the ECommHR and much of the case law of the ECourtHR, but deviate notably from the ECourtHR's findings in *Bankovic*.

Extraterritorial Acts and the Abuse of Rights

Even assuming the narrowest interpretation of Article 1 jurisdiction, that is, limiting states' obligations to their own sovereign territory, certain extraterritorial acts would nonetheless engage the state's responsibility under the ECHR.

If a state intentionally places a situation that is normally governed by the ECHR outside of the scope of the ECHR, for example, by bringing persons to foreign countries in order to commit convention violations there, and by this means attempts to circumvent the ECHR's provisions, it cannot be claimed that the

[26] Öcalan v. Turkey (Merits), App. No. 46221/99, Eur. Ct. H.R. (2003), *available at* http://cmiskp.echr.coe.int/tkp197/portal.asp?sessionId=3345367&skin=hudoc-en&action =request.

[27] Issa and Others v. Turkey, App. No. 31821/96, Eur. Ct. H.R. (2004), *available at* http://cmiskp.echr.coe.int/tkp197/portal.asp?sessionId=3345367&skin=hudoc-en&action=request.

state is fulfilling its obligation to apply the ECHR standards in good faith as required by Article 17 of the ECHR and more generally by Article 26 of the Vienna Convention on the Law of Treaties.[28] As a consequence of this abuse of rights, the state will be regarded as having violated the substantive provisions of the ECHR.[29]

The Legality of the State's Extraterritorial Act under General International Law

In *Bankovic*, the ECourtHR found that the ordinary meaning of jurisdiction in international law is essentially territorial, and that the exercise of extraterritorial jurisdiction requires special justification. But the meaning given to a term in the context of general international law does not necessarily have to correspond to the meaning of the same term in the context of international human rights law.[30] Article 27 of the Vienna Convention on the Law of Treaties provides that a treaty be interpreted "in accordance with the ordinary meaning to be given to the terms of the treaty in their context and in the light of its object and purpose."[31] The ordinary meaning of a term cannot be evaluated without reference to the context in which it is used; the different elements of Article 27 must be applied as an integrated or interdependent whole.[32]

International law and its definition of jurisdiction as "the lawful power of a state to define and enforce the rights and duties, and control the conduct, of natural and juridical persons,"[33] invoke a conception of jurisdiction that is strongly influenced by the objective of avoiding conflicts between states' extraterritorial

[28] Vienna Convention on the Law of Treaties, May 23, 1969, art. 26, 1155 UNTS 331 [hereinafter "VCLT"].

[29] *See also* Issa, *supra* note 27, at para. 71; Boudellaa v. B&H and FB&H, App. Nos. No. CH/02/8679, CH/02/8689, CH/02/8690 and CH/02/8691, Human Rights Chamber for Bosnia and Herzegovina para. 192 (2002), *available at* http://www.hrc.ba/database/ decisions/CH02-8679%20BOUDELLAA%20et%20al.%20Admissibility%20and% 20Merits%20E.pdf ("A measure of the national authorities, which has as its sole object the evasion of an obligation, is equivalent to a violation of that provision. This is also implicit from the rationale underlying Article 17 of the Convention [. . .]").

[30] The criticism expressed here is not that the Court wrongly assessed the notion of jurisdiction in international law, but that an understanding of jurisdiction differing from that of general international law is necessary in the context of human rights conventions. *See* Ress, *supra* note 10, at 83.

[31] VCLT, *supra* note 28, at art. 27.

[32] T.O. Elias, The Modern Law of Treaties 74 (1974); J. Cerone, *Minding the Gap: Outlining KFOR Accountability in Post-Conflict Kosovo*, 12 European Journal of International Law 469, 480 (2001).

[33] B. H. Oxman, *Jurisdiction of States*, *in* III Encyclopedia of Public International Law 55 (1997).

jurisdiction and the territorial jurisdiction of other states.[34] That is, limits on jurisdiction have to be established in order to protect the independence and sovereign equality of states by balancing each state's interest in exercising jurisdiction in order to advance its own policies with each state's interest in avoiding interference with its policies resulting from the exercise of jurisdiction by foreign states.[35]

In the context of human rights, however, the function of the term jurisdiction is different. It does not serve to delimit different states' exercise of sovereign power, but to protect the individual affected by acts of a state.[36] In view of the object and purpose of the human rights conventions, the legality of the acts cannot be relevant to the question whether the state exercises jurisdiction.[37] Therefore, the nationality of the person or the existence of any other recognized basis of extraterritorial jurisdiction *vis-à-vis* the person concerned is irrelevant.[38] Whether a state has Article 1 jurisdiction has to be determined without reference to the legality of the act in general international law.[39] This has been confirmed recently by the ECourtHR in the above-mentioned post-*Bankovic* case of *Issa*.[40]

Authority and Control

As used in Article 1, the term "jurisdiction" does not describe a geographically defined area, but rather a particular relationship between the respondent state

34 *See* B. Schäfer, *Der Fall Bankovic oder Wie eine Lücke geschaffen wird*, 7 MENSCHENRECHTS-MAGAZIN 149, 154 (2002).

35 Oxman, *supra* note 33, at 56.

36 In the words of *Cohen-Jonathan*, the Court denies the particularity of the ECHR in comparison with general international law, *see* Cohen-Jonathan, *supra* note 10, at 1080.

37 J. A. Carrillo-Salcedo, *Article 1*, *in* LA CONVENTION EUROPÉENNE DES DROITS DE L'HOMME – COMMENTAIRE ARTICLE PAR ARTICLE 135, 136 (L.-E. Pettiti et al., eds., 2nd ed. 1999); J. Frowein, *Artikel 1 (Geltungsbereich)*, in EUROPÄISCHE MENSCHENRECHTSKONVENTION, 19 (J. Frowein and W. Peukert eds., 2nd ed. 1996); J. VELU AND R. ERGEC, LA CONVENTION EUROPÉENNE DES DROITS DE L'HOMME 68 (1990).

38 J. L. CHARRIER, 2003–2004 CODE DE LA CONVENTION EUROPÉENNE DES DROITS DE L'HOMME NOS. 10, 14 (2002); C. PAPPA, DAS INDIVIDUALBESCHWERDEVERFAHREN DES FAKULTATIVPROTOKOLLS ZUM INTERNATIONALEN PAKT ÜBER BÜRGERLICHE UND POLITISCHE RECHTE 159 (1996); VELU AND ERGEC, *supra* note 37, at 68.

39 *See* Issa, *supra* note 27, at para. 71. *See also* Thomas Burgenthal, *To Respect and to Ensure: State Obligations and Permissible Derogations*, *in* THE INTERNATIONAL BILL OF RIGHTS: THE COVENANT ON CIVIL AND POLITICAL RIGHTS 72, 77 (L. Henkin ed., 1981); J. Cerone, *Minding the Gap: Outlining KFOR Accountability in Post-Conflict Kosovo*, 12 EUROPEAN JOURNAL OF INTERNATIONAL LAW 469, 478 (2001); T. Meron, *Extraterritoriality of Human Rights Treaties*, 89 AMERICAN JOURNAL OF INTERNATIONAL LAW 79, 80 (1995); M. Nowak, U.N. COVENANT ON CIVIL AND POLITICAL RIGHTS: CCPR COMMENTARY Art. 2 CCPR, NO. 28 (1993); PAPPA, *supra* note 38, at 158; Schäfer, *supra* note 34, at 156, 160; VELU AND ERGEC, *supra* note 37, at 68.

40 *See* Issa, *supra* note 27, at para. 71.

and the individual alleging a convention violation.[41] It is defined by legal scholars as well as by the different treaty organs of various human rights regimes as the exercise of effective control or authority over persons or objects. The requisite authority and control does not have to be exercised legally; actual or *de facto* control is sufficient.

Presumption of Control over the Persons and Objects in Occupied Territory

For the application of the ECHR, it is essential that *the person* alleging a convention violation is subject to the state's jurisdiction. Hence, it must be established that the state has authority and control *over persons*. However, control over persons can be so widespread and strong, as a result of the occupation of foreign territory for example, that the responsibility of the state can be considered as extending to the occupied territory. The authority and control over the persons on that territory will be presumed with the result that the state has to secure the whole range of ECHR rights to the persons on that territory. There is no need to examine in each case whether the state exercised control over the person alleging a convention violation.

Total control over foreign territory can be the result of military occupation, as in the case of the Turkish occupation of northern Cyprus. Multidimensional peacekeeping operations that significantly involve international organizations or states in the administration of the territory may also serve to establish jurisdiction over the territory in question.[42] In general, it is sufficient that the state exercises some of the attributes of the sovereign, it is not necessary that it exercises the whole range of jurisdictional rights.[43] But total control over a territory also can be based on the explicit or implicit acquiescence of a foreign state, for example as the result of an international treaty concluded with that state. The United States, for example, occupies the territory at Guantanamo Bay, Cuba, and exercises its jurisdiction there in application of a lease agreement between the United States and the Republic of Cuba. According to this agreement, the United States "shall exercise complete jurisdiction and control over and within said areas."[44] The

[41] CHARRIER, *supra* note 38, at no. 10; Cohen-Jonathan, *supra* note 10, at 1075; PAPPA, supra note 38, at 158.

[42] H. Krieger, *Die Verantwortlichkeit Deutschlands nach der EMRK für seine Streitkräfte im Auslandseinsatz*, 62 ZEITSCHRIFT FÜR AUSLANDISCHES ÖFFENTLICHES RECHT UND VOLKERRECHT 669, 674 (2002); S. Williams and S. Shah, *Case Analysis* – Bankovic and Others v. Belgium and 16 Other Contracting States, 7 EUROPEAN HUMAN RIGHTS LAW REVIEW 775, 781 (2002).

[43] Krieger, *supra* note 42, at 675.

[44] Agreement Between the United States of America and the Republic of Cuba for the Lease (Subject to Terms to be Agreed upon by the Two Governments) to the United States of Lands in Cuba for Coaling and Naval Stations, Feb. 16–23, 1903, 26 U.S.T. 418. *See* J. Fitzpatrick, *Sovereignty, Territoriality, and the Rule of Law*, 25 HASTINGS INTERNATIONAL AND COMPARATIVE LAW REVIEW 303, 309 (2002).

prisoners detained at Guantanamo Bay, therefore, are under the jurisdiction of the United States and entitled to all rights guaranteed by the international human rights conventions to which the United States is a party.[45]

Control over Persons or Objects

The control of the state *vis-à-vis* the individual alleging a violation of his/her rights establishes the existence of jurisdiction. The state's control over persons does not have to extend to all aspects of the person's life. It is sufficient if the state controls one aspect. An example would consist of the consular agents of a state refusing to issue a passport to one of the state's citizens. In such a case, the state only exercises control concerning the person's freedom of movement between states. However, this is sufficient to engage the state's responsibility under the conventions.

The convention rights the state is obliged to respect are commensurate with the control exercised by the state. The state does not have to secure the whole range of convention rights to the person. This would be impossible as the only means for doing so is the instrument of diplomatic protection. Only insofar as the state has control over the person does it have to secure the convention guarantees.

In addition, the state's control over persons does not have to be permanent or durable. The state act allegedly violating the ECHR itself may establish states' jurisdiction. If, for example, a state agent in his official functions kills a person abroad, the act of killing establishes the state's jurisdiction over the person and at the same time constitutes the act to be examined as to its conformity with the convention.

Control through Intentional State Acts that Directly Affect Persons

The problem with the ECHR jurisprudence regarding jurisdiction is that it defines jurisdiction as the exercise of control without developing a complementary definition of what "control" means.

There is general consensus that the state has sufficient control if it exercises sovereign rights on foreign territory. The state also has sufficient control if a person is physically at the hand of state agents, for example, if a person is detained by

45 Fitzpatrick, *supra* note 44, at 337; J. A. Frowein, *Der Terrorismus als Herausforderung für das Völkerrecht*, 62 ZEITSCHRIFT FÜR AUSLANDISCHES ÖFFENTLICHES RECHT UND VOLKERRECHT 879, 903 (2002); Jordan J. Paust, *Antiterrorism Military Commissions: Courting Illegality, Michigan Journal of International Law*, 23 MICHIGAN JOURNAL OF INTERNATIONAL LAW 1, 24 (2001). The Inter-American Commission on Human Rights considers the Guantánamo detainees to be under United States authority and control, the American Declaration of the Rights and Duties of Man therefore being applicable. *See Observations on Precautionary Measures requested by the Center for Constitutional Rights, the Human Rights Clinic at Columbia Law School and the Center for Justice and International Law with respect to the detainees in Guantanamo Bay*, 41 INTERNATIONAL LEGAL MATERIALS 532 (2002).

state agents on foreign soil. Under this reasoning, a state also must be regarded as having the requisite control if its agents, while acting on foreign territory, to destroy a person's belongings or kill a person. This principle ought to apply not only when state agents are physically present on foreign territory but also when they use modern precision weapons from their own territory. A good example of such "remote control" involves the 2002 incident during which a missile fired from an unmanned U.S. CIA surveillance aircraft over Yemen killed six suspected Al Quaeda terrorists.[46] Control is thus exercised by all state acts that have an effect on persons.

Some might object that this definition would entail a boundless state responsibility. However, in the notion of authority and control, a limit on the state's responsibility for its extraterritorial acts is inherent in the implied subjective facet of control. The state should be deemed to have exercised control (and thus, jurisdiction) via acts that affect persons on foreign soil only if it acted intentionally. The state must act with the aim of achieving the effect or the state has to know that its acts will affect persons.

The state can only be said to control persons that are directly affected by its acts. If persons are only affected through the intermediate action of another state, then the state does not have control over them. The limiting role that the notion of direct control can play in defining state responsibility is exemplified by examining the situation of a state that exports torture equipment to another state knowing the importing state will use the equipment to torture its citizens. The export of this equipment certainly is a state act that affects persons, and the state knows that these persons will be affected. However, a separate subject of international law comes between the exporting state and the individual concerned. The exporting state does not directly *control* (for the purpose of torturing) the individual in this example.

One could be tempted to compare this case with *Soering*,[47] in which the extradition of a person was held to violate Article 3 ECHR because it was foreseeable that the applicant would be subjected to conditions violative of Article 3 ECHR in the receiving state. Arguably, if the *Soering* test is applied to the export of torture equipment, the exporting state could be responsible under the human rights conventions because it was foreseeable that the other state would use this equipment to that end.[48] But there is a major difference between the two cases. In *Soering*, the United Kingdom unambiguously had jurisdiction over Soering at the time of the decision regarding extradition; he was detained in the United Kingdom. Thus, Article 1 jurisdiction was satisfied. If a state has this degree of control over a person, it will be liable for all foreseeable consequences even if another subject of

46 See David Kretzmer, *Targeted Killing of Suspected Terrorists: Extra-Judicial Executions of Legitimate Means of Defence?*, 16 European Journal of International Law 171 (2005).

47 Soering v. United Kingdom, 195 Eur. Ct. H.R. (ser. A) (1989).

48 *See* Ress, *supra* note 10, at 75 and 85.

international law intervenes. Foreseeability of harm alone, without the requisite control, however, will not establish Article 1 jurisdiction.

The notion of direct control also limits states' responsibility for the acts of private persons affecting persons on foreign soil. Whereas states are not directly responsible for the acts of private persons under the ECHR, their responsibility may generally be engaged for having omitted to protect the convention rights against violations by private actors. In that case states violate so-called positive obligations inherent in the ECHR. State omissions alone, however, are not sufficient to establish jurisdiction over persons on foreign soil. If a person on foreign soil suffers a violation of convention rights through the acts of private persons on the state's territory, the state itself does not have authority and control over that person. The fact that the state has control over the persons committing the convention violations does not suffice to establish the state's human rights responsibility. This constitutes a significant point of departure from the rule articulated in *Trail Smelter* in the context of international environmental law where Canada was deemed responsible for the acts of its citizens because those acts occurred within Canadian territory.

Considering that control is exercised by all intentional state acts that directly affect persons, it is important to underline the need of a *state* act. That is, it must be possible to impute that action to the state in accordance with the terms of general international law. Hence, the agent has to be under the control of the state and this agent, whose acts are imputable to the state, must exercise control over persons by intentionally taking acts that directly affect persons.

Summarizing the New Criteria for Jurisdiction

The elements of this comprehensive definition of jurisdiction, according to which extraterritorial responsibility for human rights violations would be established would be:

1. The person acting must be a state agent.
2. The act of the state agent must have an effect on people.
3. The state agent must intend to have the resulting effect or at least know about this effect.

CONCLUSION

Article 1 of the ECHR contains a limitation to the convention's personal scope. Only those persons who are under the jurisdiction of the state benefit from the rights contained therein. As a consequence of this personal restriction, the territorial scope of the convention is restricted. Whereas the state regularly exercises jurisdiction over all persons on its territory, the exercise of jurisdiction will be much more difficult to establish with regard to persons on foreign soil. Decisions of the ECommHR and the ECourtHR reveal that both bodies recognize that jurisdiction can be exercised extraterritorially if the state exercises authority and control over persons abroad. According to the concept of jurisdiction articulated here, all state acts that directly affect persons abroad constitute the exercise of state jurisdiction and make the ECHR applicable. However, the prerequisite that a state exercise authority and control serves to exclude from a state's human rights responsibility for those acts of private actors that affect persons on foreign soil. In these cases, the state does not have authority and control over the persons on foreign territory and therefore does not incur responsibility under the convention.

Herein lies the main difference with the principles developed by the Tribunal in the *Trail Smelter* arbitration which contain no analogous limitation of the persons protected by its rule of state responsibility. That the state has authority and control over the persons causing the harm is considered to be sufficient to engage state responsibility. If *Trail Smelter* were examined under a human rights perspective before the ECourtHR or the ECommHR, Canada's responsibility could only be engaged if it could be shown that Canada exercised authority and control over those persons in the United States affected by the fumes of the smelter. As Canada itself did not run the smelter, Canada did not exercise jurisdiction over the persons abroad so that its human rights responsibility would not be engaged. Only if Canada itself had operated the smelter and knew that persons in the United States would be affected, could it be said, on the basis of the concept of jurisdiction developed here, that Canada exercised jurisdiction and that its responsibility under the human rights conventions therefore would be engaged.

In light of the increasingly transboundary or extraterritorial range of state action, a primary concern of international human rights law is to develop a concept of jurisdiction that makes states liable for these acts. The reluctance to apply the international law of human rights to extraterritorial state acts is no longer tolerable. The human rights conventions are living instruments that have to be interpreted in a dynamic manner so as to render the rights contained therein effective. In our globalized world, it is vital to recognize state responsibility for transboundary human rights violations. At least in this respect, if not as regards the finer elements of the jurisprudence, the *Trail Smelter* arbitration, which recognized the state's responsibility for transboundary environmental harm for the first time, might serve as a source of inspiration.

ANNEX A

Convention Between the United States of America and the Dominion of Canada Relative to the Establishment of a Tribunal to Decide Questions of Indemnity and Future Regime Arising from the Operation of Smelter at Trail, British Columbia.

49 Stat. 3245

Signed, April 15, 1935

Senate advice and consent to ratification June 5, 1935

Ratified by the President of the United States June 12, 1935

Ratified by the United Kingdom, in respect of Canada, July 20, 1935

Ratifications exchanged at Ottawa August 3, 1935

Proclaimed by the President of the United States August 7, 1935

The President of the United States of America, and His Majesty the King of Great Britain, Ireland and the British dominions beyond the Seas, Emperor of India, in respect of the Dominion of Canada,

Considering that the Government of the United States has complained to the Government of Canada that fumes discharged from the smelter of the Consolidated Mining and Smelting Company at Trail, British Columbia, have been causing damage in the State of Washington, and

Considering further that the International Joint Commission, established pursuant to the Boundary Waters Treaty of 1909, investigated problems arising from the operation of the smelter at Trail and rendered a report and recommendations thereon, dated February 28, 1931, and

Recognizing the desirability and necessity of effecting a permanent settlement,

Have decided to conclude a Convention for the purposes aforesaid, and to that end have named as their respective plenipotentiaries:

The President of the United States of America:

Pierre De L. Boal, Chargé d'Affaires ad interim of the United States of America at Ottawa;

His Majesty the King of Great Britain, Ireland and the British dominions beyond the Seas, Emperor of India, for the Dominion of Canada:

The Right Honourable RICHARD BEDFORD BENNETT, Prime Minister, President of the Privy Council and Secretary of State for External Affairs;

Who, after having communicated to each other their full powers, found in good and due form, have agreed upon the following Articles:

ARTICLE I

The Government of Canada will cause to be paid to the Secretary of State of the United States, to be deposited in the United States Treasury, within three months after ratifications of this Convention have been exchanged, the sum of three hundred and fifty thousand dollars, United States currency, in payment of all damage which occurred in the United States, prior to the first day of January, 1932, as a result of the operation of the Trail Smelter.

ARTICLE II

The Governments of the United States and of Canada, hereinafter referred to as "the Governments," mutually agree to constitute a tribunal hereinafter referred to as "the Tribunal," for the purpose of deciding the Questions referred to it under the provisions of Article III. The Tribunal shall consist of a chairman and two national members.

The chairman shall be a jurist of repute who is neither a British subject nor a citizen of the United States. He shall be chosen by the Governments, or, in the event of failure to reach agreement within nine months after the exchange of ratifications of this Convention, by the President of the Permanent Administrative Council of the Permanent Court of Arbitration at The Hague described in Article 49 of the Convention for the Pacific Settlement of Inter-national Disputes concluded at The Hague on October 18, 1907.

The two national members shall be jurists of repute, who have not been associated directly or indirectly, in the present controversy. One member shall be chosen by each of the Governments.

The Governments may each designate a scientist to assist the Tribunal.

ARTICLE III

The Tribunal shall finally decide the questions, hereinafter referred to as "the Questions", set forth hereunder, namely: –

(1) Whether damage caused by the Trail Smelter in the State of Washington has occurred since the first day of January, 1932, and, if so, what indemnity should be paid therefor?

(2) In the event of the answer to the first part of the preceding Question being in the affirmative, whether the Trail Smelter should be required to refrain from causing damage in the State of Washington in the future and, if so, to what extent?

(3) In the light of the answer to the preceding Question, what measures or regime, if any, should be adopted or maintained by the Trail Smelter?

(4) What indemnity or compensation, if any, should be paid on account of any decision or decisions rendered by the Tribunal pursuant to the next two preceding Questions?

ARTICLE IV

The Tribunal shall apply the law and practice followed in dealing with cognate questions in the United States of America as well as International Law and Practice, and shall give consideration to the desire of the High Contracting Parties to reach a solution just to all parties concerned.

ARTICLE V

The procedure in this adjudication shall be as follows:

1. Within nine months from the date of the exchange of ratifications of this agreement, the Agent for the Government of the United States shall present to the Agent for the Government of Canada a statement of the facts, together with the supporting evidence, on which the Government of the United States rests its complaint and petition.

2. Within a like period of nine months from the date on which this agreement becomes effective, as aforesaid, the Agent for the Government of Canada shall present to the Agent for the Government of the United States a statement of the facts, together with the supporting evidence, relied upon by the Government of Canada.

3. Within six months from the date on which the exchange of statements and evidence provided for in paragraphs 1 and 2 of this Article has been completed, each Agent shall present in the manner prescribed by paragraphs 1 and 2 an answer to the statement of the other with any additional evidence and such argument as he may desire to submit.

ARTICLE VI

When the development of the record is completed in accordance with Article V hereof the Governments shall forthwith cause to be forwarded to each member of the Tribunal a complete set of the statements, answers, evidence and arguments presented by their respective Agents to each other.

ARTICLE VII

After the delivery of the record to the members of the Tribunal in accordance with Article VI the Tribunal shall convene at a time and place to be agreed upon by the two Governments for the purpose of deciding upon such further procedure as it may be deemed necessary to take. In determining upon such further procedure and arranging subsequent meetings, the Tribunal will consider the individual or joint requests of the Agents of the two Governments.

ARTICLE VIII

The Tribunal shall hear such representations and shall receive and consider such evidence, oral or documentary, as may be presented by the Governments or by interested parties, and for that purpose shall have power to administer oaths. The Tribunal shall have authority to make such investigations as it may deem necessary and expedient, consistent with other provisions of this Convention.

ARTICLE IX

The Chairman shall preside at all hearings and other meetings of the Tribunal, and shall rule upon all questions of evidence and procedure. In reaching a final determination of each or any of the Questions, the Chairman and the two members shall each have one vote, and, in the event of difference, the opinion of the majority shall prevail, and the dissent of the Chairman or member, as the case may be, shall be recorded. In the event that no two members of the Tribunal agree on a question, the Chairman shall make the decision.

ARTICLE X

The Tribunal, in determining the first question and in deciding upon the indemnity, if any, which should be paid in respect to the years 1932 and 1933, shall give due regard to the results of investigations and inquiries made in subsequent years.

Investigators, whether appointed by or on behalf of the Governments, either jointly or severally, or the Tribunal, shall be permitted at all reasonable times to enter and view and carry on investigations upon any of the properties upon which damage is claimed to have occurred or to be occurring, and their reports may, either jointly or severally, be submitted to and received by the Tribunal for the purpose of enabling the Tribunal to decide upon any of the Questions.

ARTICLE XI

The Tribunal shall report to the Governments its final decisions, together with the reasons on which they are based, as soon as it has reached its conclusions in respect to the Questions, and within a period of three months after the conclusion of proceedings. Proceedings shall be deemed to have been concluded when the Agents of the two Governments jointly inform the Tribunal that they have nothing additional to present. Such period may be extended by agreement of the two Governments.

Upon receiving such report, the Governments may make arrangements for the disposition of claims for indemnity for damage, if any, which may occur subsequently to the period of time covered by such report.

ARTICLE XII

The Governments undertake to take such action as may be necessary in order to ensure due performance of the obligations undertaken hereunder, in compliance with the decision of the Tribunal.

ARTICLE XIII

Each Government shall pay the expenses of the presentation and conduct of its case before the Tribunal and the expenses of its national member and scientific assistant.

All other expenses, which by their nature are a charge on both Governments, including the honorarium of the neutral member of the Tribunal, shall be borne by the two Governments in equal moieties.

ARTICLE XIV

This agreement shall be ratified in accordance with the constitutional forms of the Contracting Parties and shall take effect immediately upon the exchange of ratifications, which shall take place at Ottawa as soon as possible.

IN WITNESS WHEREOF, the respective Plenipotentiaries have signed this Convention and have hereunto affixed their seals.

Done in duplicate at Ottawa this fifteenth day of April, in the year of our Lord, one thousand, nine hundred and thirty-five.

PIERRE DE L. BOAL [SEAL]
R. B. BENNETT [SEAL]

ANNEX B

Trail Smelter Arbitral Tribunal Decision, April 16, 1938

This Tribunal is constituted under, and its powers are derived from and limited by, the Convention between the United States of America and the Dominion of *Canada* signed at Ottawa, April 15, 1935, duly ratified by the two parties, and ratification exchanged at Ottawa, August 3, 1935 (hereinafter termed "the Convention").

[The Tribunal here described the composition of the panel]

The duty imposed upon the Tribunal by the Convention was to "finally decide" the following questions:

(1) Whether damage caused by the Trail Smelter in the State of Washington has occurred since the first day of January, 1932, and, if so, what indemnity should be paid therefor?
(2) In the event of the answer to the first part of the preceding question being in the affirmative, whether the Trail Smelter should be required to refrain from causing damage in the State of Washington in the future and, if so, to what extent?
(3) In the light of the answer to the preceding question, what measures or regime, if any, should be adopted or maintained by the Trail Smelter?
(4) What indemnity or compensation, if any, should be paid on account of any decision or decisions rendered by the Tribunal pursuant to the next two preceding questions?

[description of Tribunal evidence gathering process omitted]

The Tribunal is prepared now to decide finally Question No. 1, propounded to it in Article III of the Convention; and it hereby reports its final decision on Question No. 1, its temporary decision on Questions No. 2 and No. 3, and provides for a temporary regime thereunder and for a final decision on these questions and on Question No. 4, within three months from October 1, 1940.

Wherever, in this decision, the Tribunal has referred to decisions of American courts or has followed American law, it has acted pursuant to Article IV as follows: "The Tribunal shall apply the law and practice followed in dealing with cognate questions in the United States of America . . ."

In all the consideration which the Tribunal has given to the problems presented to it, and in all the conclusions which it has reached, it has been guided by that primary purpose of the Convention expressed in the words of Article IV, that the Tribunal "shall give consideration to the desire of the high contracting parties to reach a solution just to all parties concerned," and further expressed in the opening paragraph of the Convention as to the "desirability and necessity of effecting a permanent settlement" of the controversy.

The controversy is between two Governments involving damage occurring in the territory of one of them (the United States of America) and alleged to be due to an agency situated in the territory of the other (the Dominion of Canada), for which damage the latter has assumed by the Convention an international responsibility. In this controversy, the Tribunal is not sitting to pass upon claims presented by individuals or on behalf of one or more individuals by their Government, although individuals may come within the meaning of "parties concerned," in Article IV and of "interested parties," in Article VIII of the Convention and although the damage suffered by individuals may, in part, "afford a convenient scale for the calculation of the reparation due to the State" (see Judgment No. 13. Permanent Court of International Justice, Series A, No. 17, pp. 27, 28).

[The Tribunal here extensively describes the terrain surrounding the Trail Smelter and gave an introduction to the economic development of the area which included extensive population data]

About the year 1896, there was established in Northport a business which has been termed the "Breen Copper Smelter," operated by the LeRoi Mining and Smelting Company, and later carried on by the Northport Smelting and Refining Company which was chartered in 1901. This business employed at times from five hundred to seven hundred men, although, as compared with a modern smelter like the Trail Smelter, the extent of its operations was small. The principal value of the ores smelted by it was in copper, and the ores had a high sulphur content. For some years, the somewhat primitive method of "heap roasting" was employed which consisted of roasting the ore in open piles over wood fires, frequently called in mining parlance, "stink piles." Later, this process was changed. About seventy tons of sulphur were released per day. This Northport Smelting and Refining Company intermittently continued operations until 1908. From 1908 until 1915, its smelter lay idle. In March, 1916, during the Great War, operation was resumed for the purpose of smelting lead ore, and continued until March 5, 1921, when it ceased business and its plant was dismantled. About 30 tons of sulphur per day were emitted during this time. There is no doubt that damage was caused to some extent over a more or less restricted area by the operation of this smelter plant.

The record and evidence placed before the Tribunal does not disclose in detail claims for damage on account of fumigations which were made between 1896 and 1908, but it does appear that there was considerable litigation in Stevens County courts based on such claims. It also appears in evidence that prior to 1908, the company had purchased smoke easements from sixteen owners of land in the vicinity covering 2,330 acres. It further appears that from 1916 to 1921, claims for damages were made and suits were brought in the courts, and additional smoke easements were purchased from thirty-four owners of land covering 5,556.7 acres. These various smoke easements extended to lands lying four or five miles north and three miles south and three miles east of Northport and on both sides of the river, and they extended as far as the boundary line.

In addition to the smelting business, there have been intermittent mining operations of lead and zinc in this locality, but they have not been a large factor in adding to the population.

[The Tribunal here gave a brief description of the timber industry and extensively described the condition of local farmland and agriculture in the region]

In early days, it was believed that, owing to soil and climatic conditions, this locality was destined to become a fruit-growing region, and a few orchards were planted. For several reasons, of which it is claimed that fumigation is one, orchards have not thrived. . . .

In 1896, a smelter was started under American auspices near the locality known as Trail. In 1906, the Consolidated Mining and Smelting Company of Canada, Limited, obtained a charter of incorporation from the Canadian authorities, and that company acquired the smelter plant at Trail as it then existed. Since that time, the Canadian Company, without interruption, has operated the Smelter, and from time to time has greatly added to the plant until it has become one of the best and largest equipped smelting plants on this continent. In 1925 and 1927, two stacks of the plant were erected to 409 feet in height and the Smelter greatly increased its daily smelting of zinc and lead ores. This increased product resulted in more sulphur dioxide fumes and higher concentrations being emitted into the air; and it is claimed by one Government (though denied by the other) that the added height of the stacks increased the area of damage in the United States. In 1916, about 5,000 tons of sulphur per month were emitted; in 1924, about 4,700 tons; in 1926, about 9,000 tons – an amount which rose near to 10,000 tons per month in 1930. In other words, about 300-350 tons of sulphur were being emitted daily in 1930. (It is to be noted that one ton of sulphur is substantially the equivalent of two tons of sulphur dioxide or SO_2.)

From 1925, at least, to the end of 1931, damage occurred in the State of Washington, resulting from the sulphur dioxide emitted from the Trail Smelter. As early as 1925 (and there is some evidence earlier) suggestions were made to the Trail Smelter that damage was being done to property in the northern part of Stevens

County. The first formal complaint was made, in 1926, by one J. H. Stroh, whose farm . . . was located a few miles south of the boundary line. He was followed by others, and the Smelter Company took the matter up seriously and made a more or less thorough and complete investigation. This investigation convinced the Trail Smelter that damage had been and was being done, and it proceeded to negotiate with the property owners who had made complaints or claims with a view to settlement. Settlements were made with a number of farmers by the payment to them of different amounts. This condition of affairs seems to have lasted during a period of about two years. In June, 1928, the County Commissioners of Stevens County adopted a resolution relative to the fumigations; and on August 25, 1928, there was brought into existence an association known as the "Citizens' Protective Association." Due to the creation of this association or to other causes, no settlements were made thereafter between the Trail Smelter and individual claimants, as the articles of association contained a provision that "no member herein shall make any settlement for damages sought to be secured herein, unless the written consent of the majority of the Board of Directors shall have been first obtained."

It has been contended that either by virtue of the Constitution of the State of Washington or of a statute of that State, the Trail Smelter (a Canadian corporation) was unable to acquire ownership or smoke easements over real estate, in the State of Washington, in any manner. In regard to this statement, either as to the fact or as to the law, the Tribunal expresses no opinion and makes no ruling.

The subject of fumigations and damage claimed to result from them was first taken up officially by the Government of the United States in June, 1927, in a communication from the Consul General of the United States at Ottawa, addressed to the Government of the Dominion of Canada.

In December, 1927, the United States Government proposed to the Canadian Government that problems growing out of the operation of the Smelter at Trail should be referred to the International Joint Commission, United States and Canada, for investigation and report, pursuant to Article IX of the Convention of January 11, 1909, between the United States and Great Britain. Following an extensive correspondence between the two Governments, they joined in a reference of the matter to that Commission under date of August 7, 1928. It may be noted that Article IX of the Convention of January 11, 1909, provides that the high contracting parties might agree that "any other question or matters of difference arising between them involving the rights, obligations or interests of either in relation to the other, or to the inhabitants of the other, along the common frontier between the United States and the Dominion of Canada shall be referred from time to time to the International Joint Commission for examination and report. . . . Such reports shall not be regarded as decisions of the question or matters so submitted either on the facts or on the law, and shall not, in any way, have the character of an arbitral award."

The questions referred to the International Joint Commission were five in number, the first two of which may be noted: First, the extent to which property in the State of Washington has been damaged by fumes from Smelter at Trail, B.C.; second, the amount of indemnity which would compensate United States interests in the State of Washington for past damages.

[The Tribunal here described the proceedings before the International Joint Commmission]

On February 28, 1931, the Report of the Commission was signed and delivered to the proper authorities. The report was unanimous and need not be considered in detail. Paragraph 2 of the report, in part, reads as follows:

> In view of the anticipated reduction in sulphur fumes discharged from the Smelter at Trail during the present year, as hereinafter referred to, the Commission therefore has deemed it advisable to determine the amount of indemnity that will compensate United States interests in respect of such fumes, up to and including the first day of January, 1932. The Commission finds and determines that all past damages and all damages up to and including the first day of January next, is the sum of $350,000. Said sum, however, shall not include any damage occurring after January 1, 1932.

In paragraph 4 of the report, the Commission recommended a method of indemnifying persons in Washington State for damage which might be caused by operations of the Trail Smelter after the first of January, 1932, as follows:

> Upon the complaint of any persons claiming to have suffered damage by the operations of the company after the first of January, 1932, it is recommended by the Commission that in the event of any such claim not being adjusted by the company within a reasonable time, the Governments of the United States and Canada shall determine the amount of such damage, if any, and the amount so fixed shall be paid by the company forthwith.

This recommendation, apparently, did not commend itself to the interested parties. In any event, it does not appear that any claims were made after the first of January, 1932, as contemplated in paragraph 4 of the report.

In paragraph 5 of the report, the Commission recommended that the Consolidated Mining and Smelting Company of Canada, Limited, should proceed to erect and put in operation certain sulphuric acid units for the purpose of reducing the amount of sulphur discharged from the stacks.

[The Tribunal here described these installations]

Two years after the signing of the International Joint Commission's Report of February 28, 1931, the United States Government on February 17, 1933, made representations to the Canadian Government that existing conditions were entirely unsatisfactory and that damage was still occurring, and diplomatic negotiations were renewed. Correspondence was exchanged between the two countries, and although that correspondence has its importance, it is sufficient here to say, that it resulted in the signing of the present Convention.

* * *

PART TWO

The first question under Article III of the Convention which the Tribunal is required to decide is as follows:

> (1) Whether damage caused by the Trail Smelter in the State of Washington has occurred since the first day of January, 1932, and, if so, what indemnity should be paid therefor.

In the determination of the first part of this question, the Tribunal has been obliged to consider three points, *viz.*, the existence of injury, the cause of the injury, and the damage due to the injury.

The Tribunal has interpreted the word "occurred" as applicable to damage caused prior to January 1, 1932, in so far as the effect of the injury made itself felt after that date. The words "Trail Smelter" are interpreted as meaning the Consolidated Mining and Smelting Company of Canada, Limited, its successors and assigns.

In considering the second part of the question as to indemnity, the Tribunal has been mindful at all times of the principle of law which is set forth by the United States courts in dealing with cognate questions, particularly by the United States Supreme Court in Story Parchment Company v. Paterson Parchment Paper Company (1931), 282 U.S. 555 as follows: "Where the tort itself is of such a nature as to preclude the ascertainment of the amount of damages with certainty, it would be a perversion of fundamental principles of justice to deny all relief to the injured person, and thereby relieve the wrongdoer from making any amend for his acts. In such case, while the damages may not be determined by mere speculation or guess, it will be enough if the evidence show the extent of the damages as a matter of just and reasonable inference, although the result be only approximate." *[further citations omitted]*

The Tribunal has first considered the items of indemnity claimed by the United States . . . "on account of damage occurring since January 1, 1932, covering: (a) Damages in respect of cleared land and improvements thereon; (b) Damages in respect of uncleared land and improvements thereon; (c) Damages in respect of livestock; (d) Damages in respect of property in the town of Northport; (g) Damages in respect of business enterprises."

With respect to Item (a) and to Item (b) . . . the Tribunal has reached the conclusion that damage due to fumigations has been proved to have occurred since January 1, 1932, and to the extent set forth hereafter.

Since the Tribunal has concluded that, on all the evidence, the existence of injury has been proved, it becomes necessary to consider next the cause of injury. This question resolves itself into two parts – first, the actual causing factor, and second, the manner in which the causing factor has operated.

[The Tribunal here offered some preliminary comments on causation]

On the basis of the evidence, the United States contended that damage had been caused by the emission of sulphur dioxide fumes at the Trail Smelter in British Columbia, which fumes, proceeding down the valley of the Columbia River and otherwise, entered the United States. The Dominion of Canada contended that even if such fumes had entered the United States, they had caused no damage after January 1, 1932. The witnesses for both Governments appeared to be definitely of the opinion that the gas was carried from the Smelter by means of surface winds, and they based their views on this theory of the mechanism of gas distribution. The Tribunal finds itself unable to accept this theory. It has, therefore, looked for a more probable theory, and has adopted the following as permitting a more adequate correlation and interpretation of the facts which have been placed before it.

It appears from a careful study and comparison of recorder data furnished by the two Governments, that on numerous occasions fumigations occur practically simultaneously at points down the valley many miles apart this being especially the fact during the growing season from April to October. It also appears from the data furnished by the different recorders, that the rate of gas attenuation down the river does not show a constant trend, but is more rapid in the first few miles below the boundary and more gradual further down the river. The Tribunal finds it impossible satisfactorily to account for the above conditions, on the basis of the theory presented to it. The Tribunal finds it further difficult to explain the times and durations of the fumigations on the basis of any probable surface-wind conditions. The Tribunal is of opinion that the gases emerging from the stacks of the Trail. Smelter find their way into the upper air currents, and are carried by these currents in a fairly continuous stream down the valley so long as the prevailing wind at that level is in that direction.

[The Tribunal here explored technical details about wind patterns]

The conclusions above, together with a detailed study of the intensity of the fumigations at the various stations from Columbia Gardens down the valley, have led to deductions in regard to the rate of attenuation of concentration of sulphur dioxide with increasing distance from the Smelter which seem to be in accord both with the known facts and the present theory. The conclusion of the Tribunal on this phase of the question is that the concentration of sulphur dioxide falls off very rapidly from Trail to a point about 16 miles downstream from the Smelter, or 6 miles from the boundary line, measured by the general course of the river; and that at distances beyond this point, the concentration of sulphur dioxide is lower and falls off more gradually and less rapidly.

* * *

The above conclusions have a bearing both upon the cause and upon the degree of damage as well as upon the area of probable damage.

The Tribunal will now proceed to consider the different classes of damage to cleared and to uncleared land.

(1) With regard to cleared land used for crops, the Tribunal has found that damage through reduction in crop yield due to fumigation has occurred in varying degrees during each of the years, 1932 to 1936; and it has found no proof of damage in the year 1937.

[The Tribunal here described the affected property]

The Tribunal has adopted as the measure of indemnity to be applied on account of damage in respect of cleared land used for crops, the measure of damage which the American courts apply in cases of nuisance or trespass of the type here involved, *viz.*, the amount of reduction in the value of use or rental value of the land caused by the fumigations. In the case of farm land, such reduction in the value of the use is, in general, the amount of the reduction of the crop yield arising from injury to crops, less cost of marketing the same, the latter factor being under the circumstances of this case of negligible importance. (See Ralston v. United Verde Copper Co., 37 Federal Reporter 2d, 180, and 46 Federal Reporter 2d, 1.) Failure of farmers to increase their seeded land in proportion to such increase in other localities, may also be taken into consideration.

The difference between probable yield in the absence of any fumigation and actual crop yield, varying as it does from year to year and from place to place, is necessarily a somewhat uncertain amount, incapable of absolute proof; and the Tribunal has been obliged to base its estimate of damage largely on the fumigation records, meteorological data, statistical data as to crop yields inside and outside the area of probable damage, and other Census records.

As regards the problems arising out of abandonment of properties by their owners, it is to be noted that practically all of such properties, listed in the questionnaire sent out by the former Agent for the United States, Mr. Metzger, appear to have been abandoned prior to the year 1932. However, in order to deal both with this problem and with the problem arising out of failure of farmers to increase their seeded land, the Tribunal, not having to adjudicate on individual claims, estimated, on the basis of the statistical data available, the average acreage on which it is reasonable to say that crops would have been seeded and harvested during the period under consideration but for the fumigations.

As regards the special category of cleared lands used for orchards, the Tribunal is of opinion that no damage to orchards by sulphur dioxide fumigation within the damaged area during the years in question has been proved.

In addition to indemnity which may be awarded for damage through reduction in the value of the use of cleared land measured by decrease in crop yield, it may be contended that special damage has occurred for which indemnity should be awarded by reason of impairment of the soil contents through increased acidity caused by sulphur dioxide fumigations acting directly on the soil or indirectly through increased sulphur content of the streams and other waters. Evidence has been given in support of this contention. The Tribunal is of opinion that such injury to the soil up to this date, due to increased acidity and affecting harmfully the production of crops or otherwise, has not been proved – with one exception, as follows: There is a small area of farming property adjacent to the boundary, west of the river, that was injured by serious increase of acidity of soil due to fumigations. Such injury, though caused, in part, prior to January 1, 1932, may have produced a continuing condition which cannot be considered as a loss for a limited time – in other words, in this respect the nuisance may be considered to have a more permanent effect, in which case, under American law *(Sedgewick on Damages*, 9th Ed. (1920) Sections 932, 947), the measure of damage was not the mere reduction in the value of the use of the land but the reduction in the value of the land itself. The Tribunal is of opinion that such injury to the soil itself can be cured by artificial means, and it has awarded indemnity with this fact in view on the basis of the data available.

* * *

(2) With respect to damage to cleared land not used for crops and to all uncleared (other than uncleared land used for timber), the Tribunal has adopted as the measure of indemnity, the measure of damages applied by American courts, *viz.*, the amount of reduction in the value of the use or rental value of the land. The Tribunal is of opinion that the basis of estimate of damages contended for by the United States, *viz.*, applying to the value of uncleared land a ratio of loss measured by the reduced crop yield on cleared land, has no sanction in any decisions of American courts.

(A) As regards these lands in their use as pasture lands, the Tribunal is of opinion that there is no evidence of any marked susceptibility of wild grasses to fumigations, and very little evidence to prove the respective amounts of uncleared land devoted to wild grazing grass and barren or shrub land, or to prove the value thereof, which would be necessary in order to estimate the value of the reduction of the use of such land. The Tribunal, however, has awarded a small indemnity for damage to about 200 acres of such lands in the immediate neighborhood of the boundary.

It has been contended that the death of trees and shrubs due to fumigation has had an injurious effect on the water storage capacity of the soil and has even created some soil erosion. The Tribunal is of opinion that while there may have been some erosion of soil and impairment of water storage capacity in a limited area near the boundary, it is impossible to determine whether such damage has been due to fires or to mortality of trees and shrubs caused by fumigation.

(B) As regards uncleared land in its use as timberland, the Tribunal has found that damage due to fumigation has occurred to trees during the years 1932 to 1937 inclusive, in varying degrees, over areas varying not only from year to year but also from species to species.

[The Tribunal here described the property in question]

It is uncontroverted that heavy fumigations from the Trail Smelter which destroyed and injured trees occurred in 1930 and 1931; and there were also serious fumigations in earlier years.

[The Tribunal here described technical details of forest conditions]

(a) The Tribunal has adopted as the measure of indemnity, to be applied on account of damage in respect of uncleared land used for merchantable timber, the measure of damages applied by American courts, *viz.*, that since the destruction of merchantable timber will generally impair the value of the land itself, the measure of damage should be the reduction in the value of the land itself due to such destruction of timber; but under the leading American decisions, however, the value of the merchantable timber destroyed is, in general, deemed to be substantially the equivalent of the reduction in the value of the land (see *Sedgwick on Damages*, 9th Ed. 1920, Section 937a). The Tribunal is unable to accept the method contended for by the United States of estimating damage to uncleared timberland by applying to the value of such land as stated by the farmers (after deducting value of the timber) a ratio of loss measured by the reduced crop yield on cleared land. The Tribunal is of opinion, here as elsewhere in this decision, that, in accordance with American law, it is not restricted to the method proposed by the United States in the determination of amount of damages, so long as its findings remain within the amount of the claim presented to it.

[Timber census data omitted]

(b) With regard to damage due to destruction and impairment of growing timber (not of merchantable size), the Tribunal has adopted the measure of damages applied by American courts, *viz.*, the reduction in value of the land itself due to such destruction and impairment.... No evidence has been presented by the United States as to the locations or as to the total amounts of such growing timber existing on January 1, 1932, or as to its distribution into types of conifers – yellow pine, Douglas fir, larch or other trees. While some destruction or impairment, deterioration, and retardation of such growing timber has undoubtedly occurred since such date, it is impossible to estimate with any degree of accuracy the amount of damage. The Tribunal has, however, taken such damage into consideration in awarding indemnity for damage to land containing growing timber.

[The Tribunal rejected claims that fumigations prevented tree seed production and germination. The Tribunal awarded the United States $78,000 as indemnity for its timber claims. The Tribunal then rejected claims for damage to livestock, to property in Northport and with respect to business enterprises as unproven, and also rejected claims based on water contamination as outside its jurisdiction under the Convention]

The United States . . . presents two further items of damages claimed by it, as follows: (Item e) which the United States terms "damages in respect of the wrong done the United States in violation of sovereignty"; and (Item f) which the United States terms, "damages in respect of interest on $350,000 eventually accepted in satisfaction of damage to January 1, 1932, but not paid until November 2, 1935."

[The Tribunal here detailed the United States arguments and then concluded that damages for violation of sovereignty did not fall within its jurisdiction under the Convention. The Tribunal similarly rejected the United States' claim for interest as outside the scope of the Convention.]

In conclusion, the Tribunal answers Question 1 in Article III, as follows: Damage caused by the Trail Smelter in the State of Washington has occurred since the first day of January, 1932, and up to October 1, 1937, and the indemnity to be paid therefor is seventy-eight thousand dollars ($78,000), and is to be complete and final indemnity and compensation for all damage which occurred between such dates. Interest at the rate of six per centum per year will be allowed on the above sum of seventy-eight thousand dollars ($78,000) from the date of the filing of this report and decision until date of payment. This decision is not subject to alteration or modification by the Tribunal hereafter.

The fact of existence of damage, if any, occurring after October 1, 1937, and the indemnity to be paid therefor, if any, the Tribunal will determine in its final decision.

PART THREE

[The Tribunal here reserves until its final decision the question of "whether the Trail Smelter should be required to refrain, from causing damage in the State of Washington in the future and, if so, to what extent" and as an interim measure directs the Trail Smelter to "refrain from causing damage in the State of Washington in the future" to the extent set forth in Part Four of this decision]

PART FOUR

As to Question No. 3, in Article III of the Convention, which is as follows:

> In the light of the answer to the preceding question, what measures or regime, if any, should be adopted or maintained by the Trail Smelter?

The Tribunal is unable at the present time, with the information that has been placed before it, to determine upon a permanent regime, for the operation of the Trail Smelter. On the other hand, in view of the conclusions at which the Tribunal has arrived (as stated in an earlier part of this decision) with respect to the nature, the cause, and the course of the fumigations, and in view of the mass of data relative to sulphur emissions at the Trail Smelter, and relative to meteorological conditions and fumigations at various points down the Columbia River Valley, the Tribunal feels that the information now available does enable it to predict, with some degree of assurance, that a permanent regime based on a more adequate and intensive study and knowledge of meteorological conditions in the valley, and an extension and improvement of the methods of operation of the plant and its control in closer relation to such meteorological conditions, will effectively prevent future significant fumigations in the United States, without unreasonably restricting the output of the plant.

To enable it to establish a permanent regime based on the more adequate and intensive study and knowledge above referred to, the Tribunal establishes the following temporary regime.

[The Tribunal here described the temporary regime which included seasonal sulphur emissions limitations, and hourly emissions limitations. The Tribunal also described a payment schedule]

Nothing in the above paragraphs of Part Four of this decision shall relieve the Dominion of Canada from any obligation now existing under the Convention with reference to indemnity or compensation, if any, which the Tribunal *may find to* be due for damage, if any, occurring during the period from October 1, 1937 (the date to which indemnity for damage is now awarded) to October 1, 1940, or to such earlier date at which the Tribunal may render its final decision.

(Signed) JAN HOSTIE
(Signed) CHARLES WARREN
(Signed) IT. A. E. GREENSHIELDS

ANNEX C

Trail Smelter Arbitral Tribunal March 11, 1941, Decision

[Introductory materials identical to 1938 decision omitted. The Tribunal emphasized that both countries had an equal interest that the resolution of this controversy be fair to meritorious claims while rejecting unwarranted ones and noted the likelihood that the parties might find themselves on opposite sides in future controversies. The Tribunal restated the four questions it was charged with answering, described the procedural history of its deliberations and noted that it had previously rendered final and interim decisions on these questions in its April 16, 1938 decision, hereinafter "previous decision." The Tribunal then reproduced verbatim the geographic description of the region from the previous decision]

In 1896, a smelter was started under American auspices near the locality known as Trail, B.C. In 1906, the Consolidated Mining and Smelting Company of Canada, Limited, obtained a charter of incorporation from the Canadian authorities, and that company acquired the smelter plant at Trail as it then existed. Since that time, the Canadian company, without interruption, has operated the Smelter, and from time to time has greatly added to the plant until it has become one of the best and largest equipped smelting plants on the American continent. In 1925 and 1927, two stacks of the plant were erected to 409 feet in height and the Smelter greatly increased its daily smelting of zinc and lead ores. This increased production resulted in more sulphur dioxide fumes and higher concentrations being emitted into the air. In 1916, about 5,000 tons of sulphur per month were emitted; in 1924, about 4,700 tons; in 1926, about 9,000 tons – an amount which rose near to 10,000 tons per month in 1930. In other words, about 300–350 tons of sulphur were being emitted daily in 1930. (It is to be noted that one ton of sulphur is substantially the equivalent of two tons of sulphur dioxide or SO_2.)

From 1925, at least, to 1937, damage occurred in the State of Washington, resulting from the sulphur dioxide emitted from the Trail Smelter as stated in the previous decision.

[The Tribunal described the International Joint Commission's effort to resolve this controversy, including the indemnity awarded and the Commission's recommendation that Canada "take such further or other action as may be necessary, if any, to reduce

the amount and concentration of SO_2 fumes drifting from its said plant into the United States until it has reduced the amount by some means to a point where it will do no damage in the United States."]

The Consolidated Mining and Smelting Company of Canada, Limited, proceeded after 1930 to make certain changes and additions in its plant, with the intention and purpose of lessening the sulphur contents of the fumes, and in an attempt to lessen injurious fumigations, a new system of control over the emission of fumes during the crop growing season came into operation about 1934. To the three sulphuric acid plants in operation since 1932, two others have recently been added. The total capacity is now of 600 tons of sulphuric acid per day, permitting, if these units could run continually at capacity, the fixing of approximately 200 tons of sulphur per day. In addition, from 1936 units for the production of elemental sulphur have been put into operation. There are at present three such units with a total capacity of 140 tons of sulphur per day. The capacity of absorption of sulphur dioxide is now 600 tons of sulphur dioxide per day (300 tons from the zinc plant gases and 300 tons from the lead plant gases). As a result, the maximum possible recovery of sulphur dioxide, with all units in full operation has been brought to a figure which is about equal to the amount of that gas produced by smelting operations at the plant in 1939. However, the normal shutdown of operating units for repairs, the power supply, ammonia available, and the general market situation are factors which influence the amount of sulphur dioxide treated.

[The Tribunal here provided technical details about SO_2 emissions]

PART TWO

[After a lengthy discussion of the Tribunal's previous decision, opinio juris and decisions from the Permanent Court of International Justice, the Tribunal concluded that arbitral awards were subject to narrow review and that res judicata prevented reconsideration of the indemnity awarded in the previous decision.]

II (a)

The Tribunal is requested to say that damage has occurred in the State of Washington since October 1, 1937, as a consequence of the emission of sulphur dioxide by the smelters of the Consolidated Mining and Smelting Company at Trail, B. C., and that an indemnity in the sum of 834,807 should be paid therefor.

[The Tribunal here gave a metes and bounds description of the alleged damage area, and listed the reports it reviewed]

[The Tribunal] has come to the conclusion that the United States has failed to prove that any fumigation between October 1, 1937, and October 1, 1940, has caused injury to crops, trees or otherwise.

II (b)

The Tribunal is finally requested as to Question I to find with respect to expenditures incurred by the United States during the period July 1, 1936, to September 1, 1940, that the United States is entitled to be indemnified in the sum of $38,657.79 with interest at the rate of five per cent per annum from the end of each fiscal year in which the several amounts were expended to the date of the Tribunal's final decision.

[The Tribunal here rejected the argument that its charge permitted it to award the United States its investigation costs as damages.]

When a State espouses a private claim on behalf of one of its nationals, expenses which the latter may have incurred in prosecuting or endeavoring to establish his claim prior to the espousal are sometimes included and, under appropriate conditions, may legitimately be included in the claim. They are costs, incidental to damage, incurred by the national in seeking local remedy or redress, as it is, as a rule, his duty to do, if, on account of injury suffered abroad, he wants to avail himself of the diplomatic protection of his State. The Tribunal, however, has not been informed of any case in which a Government has sought before an international jurisdiction or been allowed by an international award or judgment indemnity for expenses incurred by it in preparing the proof for presenting a national claim or private claims which it had espoused; and counsel for the United States, on being requested to cite any precedent for such an adjudication, have stated that they know of no precedent. Cases cited were instances in which expenses allowed had been incurred by the injured national, and all except one prior to the presentation of the claim by the Government?*[citations omitted]*

In the absence of authority established by settled precedents, the Tribunal is of opinion that, where an arbitral tribunal is requested to award the expenses of a Government incurred in preparing proof to support its claim, particularly a claim for damage to the national territory, the intent to enable the Tribunal to do so should appear, either from the express language of the instrument which sets up the arbitral tribunal or as a necessary implication from its provisions. Neither such express language nor implication is present in this case.

[The Tribunal reiterated its conclusion that no damages had occurred since October 1, 1937 and therefore awarded no indemnity]

PART THREE

The second question under Article III of the Convention is as follows:

> In the event of the answer to the first part of the preceding question being in the affirmative, whether the Trail Smelter should be required to refrain from causing damage in the State of Washington in the future and, if so, to what extent?

Damage has occurred since January 1, 1932, as fully set forth in the previous decision. To that extent, the first part of the preceding question has thus been answered in the affirmative. As has been said above, the report of the International Joint Commission (1 (g)) contained a definition of the word "damage" excluding "occasional damage that may be caused by SO_2 fumes being carried across the international boundary in air pockets or by reason of unusual atmospheric conditions," as far, at least, as the duty of the Smelter to reduce the presence of that gas in the air was concerned.

The correspondence between the two Governments during the interval between that report and the conclusion of the Convention shows that the problem thus raised was what the parties had primarily in mind in drafting Question No. 2. Whilst Canada wished for the adoption of the report, the United States stated that it could not acquiesce in the proposal to limit consideration of damage to damage as defined in the report (letter of the Minister of the United States of America at Ottawa to the Secretary of State for External Affairs of the Dominion of Canada, January 30, 1934). The view was expressed that "so long as fumigations occur in the State of Washington with such frequency, duration and intensity as to cause injury," the conditions afforded "grounds of complaint on the part of the United States, regardless of the remedial works . . . and regardless of the effect of those works" (same letter).

The first problem which arises is whether the question should be answered on the basis of the law followed in the United States or on the basis of international law. The Tribunal, however, finds that this problem need not be solved here as the law followed in the United States in dealing with the quasi-sovereign rights of the States of the Union, in the matter of air pollution, whilst more definite, is in conformity with the general rules of international law.

Particularly in reaching its conclusions as regards *this* question as well as the next, the Tribunal has given consideration to the desire of the high contracting parties "to reach a solution just to all parties concerned."

As Professor Eagleton puts in (*Responsibility of States in International Law*, 1928, p. 80): "A State owes at all times a duty to protect other States against injurious acts by individuals from within its jurisdiction." A great number of such general pronouncements by leading authorities concerning the duty of a State to respect other States and their territory have been presented to the Tribunal. These and many others have been carefully examined. International decisions, in various matters, from the *Alabama* case onward, and also earlier ones, are based on the same general principle, and, indeed, this principle, as such, has not been questioned by Canada. But the real difficulty often arises rather when it comes to determine what, *pro subjecta materie, is* deemed to constitute an injurious act.

A case concerning, as the present one does, territorial relations, decided by the Federal Court of Switzerland between the Cantons of Soleure and Argovia, may serve to illustrate the relativity of the rule. Soleure brought a suit against her sister State to enjoin use of a shooting establishment which endangered her territory. The court, in granting the injunction, said: "This right (sovereignty) excludes . . . not only the usurpation and exercise of sovereign rights (of another State) . . . but also an actual encroachment which might prejudice the natural use of the territory and the free movement of its inhabitants." As a result of the decision, Argovia made plans for the improvement of the existing installations. These, however, were considered as insufficient protection by Soleure. The Canton of Argovia then moved the Federal Court to decree that the shooting be again permitted after completion of the projected improvements. This motion was granted. "The demand of the Government of Soleure," said the court, "that all endangerment be absolutely abolished apparently goes too far." The court found that all risk whatever had not been eliminated, as the region was flat and absolutely safe shooting ranges were only found in mountain valleys; that there was a federal *duty* for the communes to provide facilities for military target practice and that "no more precautions may be demanded for shooting ranges near the boundaries of two Cantons than are required for shooting ranges in the interior of a Canton." (R. o. 26 I, p. 450, 451; R. O.41, I, p. 137; see D. Schindler, "The Administration of Justice in the Swiss Federal Court in Intercantonal Disputes," *American Journal of International Law*, Vol. 15 (1921), pp. 172–174.)

No case of air pollution dealt with by an international tribunal has been brought to the attention of the Tribunal nor does the Tribunal know of any such case. The nearest analogy is that of water pollution. But, here also, no decision of an international tribunal has been cited or has been found.

There are, however, as regards, both air pollution and water pollution, certain decisions of the Supreme Court of the United States which may legitimately be taken as a guide in this field of international law, for it is reasonable to follow by analogy, in international cases, precedents established by that court in dealing with controversies between States of the Union or with other controversies concerning the quasi-sovereign rights of such States, where no contrary rule prevails in international law and no reason for rejecting such precedents can be adduced from the limitations of sovereignty inherent in the Constitution of the United States.

[The Tribunal here discussed Missouri v. Illinois, *200 U.S. 496, 521 (1906);* Kansas v. Colorado, *185 U.S. 125 (1902); and* New York v. New Jersey, *256 U.S. 296, 309 (1918), to support its conclusion that before a Tribunal exercises extraordinary jurisdiction to control the conduct of a State at the behest of another state, the complaining state must meet a higher threshold of proof than is imposed on the ordinary plaintiff.]*

In the matter of air pollution itself, the leading decisions are those of the Supreme Court in *Georgia v. Tennessee Copper Company*. Although dealing with a suit against private companies, the decisions were on questions cognate to those here at issue. Georgia stated that it had in vain sought relief from the State of Tennessee, on whose territory the smelters were located, and the court defined the nature of the suit by saying: "This is a suit by a State for an injury to it in its capacity of quasi-sovereign. In that capacity, the State has an interest independent of and behind the titles of its citizens, in all the earth and air within its domain."

On the question whether an injunction should be granted or not, the court said (206 U.S. 230):

> It (the State) has the last word as to whether its mountains shall be stripped of their forests and its inhabitants shall breathe pure air.... It is not lightly to be presumed to give up quasi-sovereign rights for pay and... if that be its choice, it may insist that an infraction of them shall be stopped. This court has not quite the same freedom to balance the harm that will be done by an injunction against that of which the plaintiff complains, that it would have in deciding between two subjects of a single political power. Without excluding the considerations that equity always takes into account... it is a fair and reasonable demand on the part of a sovereign that the air over its territory should not be polluted on a great scale by sulphurous acid gas, that the forests on its mountains, be they better or worse, and whatever domestic destruction they may have suffered, should not be further destroyed or threatened by the act of persons beyond its control, that the crops and orchards on its hills should not be endangered from the same source.

Whether Georgia, by insisting upon this claim, is doing more harm than good to her own citizens, is for her to determine. The possible disaster to those outside the State must be accepted as a consequence of her standing upon her extreme rights.

Later on, however, when the court actually framed an injunction, in the case of the Ducktown Company (237 U.S. 474, 477) (an agreement on the basis of an annual compensation was reached with the most important of the two smelters, the Tennessee Copper Company), they did not go beyond a decree "adequate to diminish materially the present probability of damage to its (Georgia's) citizens."

Great progress in the control of fumes has been made by science in the last few years and this progress should be taken into account.

The Tribunal, therefore, finds that the above decisions, taken as a whole, constitute an adequate basis for its conclusions, namely, that, under the principles of international law, as well as of the law of the United States, no State has the right to use or permit the use of its territory in such a manner as to cause injury by fumes in or to the territory of another or the properties or persons therein, when the case is of serious consequence and the injury is established by clear and convincing evidence.

The decisions of the Supreme Court of the United States which are the basis of these conclusions are decisions in equity and a solution inspired by them, together with the regime hereinafter prescribed, will, in the opinion of the Tribunal, be "just to all parties concerned," as long, at least, as the present conditions in the Columbia River Valley continue to prevail.

Considering the circumstances of the case, the Tribunal holds that the Dominion of Canada is responsible in international law for the conduct of the Trail Smelter. Apart from the undertakings in the Convention, it is, therefore, the duty of the Government of the Dominion of Canada to see to it that this conduct should be in conformity with the obligation of the Dominion under international law as herein determined.

The Tribunal, therefore, answers Question No. 2 as follows: (2) So long as the present conditions in the Columbia River Valley prevail, the Trail Smelter shall be required to refrain from causing any damage through fumes in the State of Washington; the damage herein referred to and its extent being such as would be recoverable under the decisions of the courts of the United States in suits between private individuals. The indemnity for such damage should be fixed in such manner as the Governments, acting under Article XI of the Convention, should agree upon.

PART FOUR

The third question under Article III of the Convention is as follows: "In the light of the answer to the preceding question, what measures or regime, if any, should be adopted and maintained by the Trail Smelter?"

Answering this question in the light of the preceding one, since the Tribunal has, in its previous decision, found that damage caused by the Trail Smelter has occurred in the State of Washington since January 1, 1932, and since the Tribunal is of the opinion that damage may occur in the future unless the operations of the Smelter shall be subject to some control, in order to avoid damage occurring, the Tribunal now decides that a regime or measure of control shall be applied to the operations of the Smelter and shall remain in full force unless and until modified in accordance with the provisions hereinafter set forth in Section 3, Paragraph VI of the present part of this decision.

[The Tribunal meticulously described the technical details of the nature and pattern of the fumigations emitting from the Trail Smelter]

(i)

The Tribunal is of opinion that the regime should be given an uninterrupted test through at least two growing periods and one non-growing period. It is equally of the opinion that thereafter opportunity should be given for amendment or suspension of the regime, if conditions should warrant or require. Should it appear at any time that the expectations of the Tribunal are not fulfilled, the regime prescribed in Section 3 *(infra) can* be amended according to Paragraph VI thereof. This same paragraph may become operative if scientific advance in the control of fumes should make it possible and desirable to improve upon the methods of control hereinafter prescribed; and should further progress in the reduction of the sulphur content of the fumes make the regime, as now prescribed, appear as unduly burdensome in view of the end defined in the answer to Question No. 2, this same paragraph can be invoked in order to amend the regime accordingly. Further, under this paragraph, the regime may be suspended if the elimination of sulphur dioxide from the fumes should reach a stage where such a step could clearly be taken without undue risks to the United States' interests.

Since the Tribunal has the power to establish a regime, it must equally possess the power to provide for alteration, modification or suspension of such regime. It would clearly not be a "solution just to all parties concerned" if its action in prescribing a regime should be unchangeable and incapable of being made responsive to future conditions.

(j)

The foregoing paragraphs are the result of an extended investigation of meteorological and other conditions which have been found to be of significance in smoke behavior and control in the Trail area. The attempt made to solve the sulphur dioxide problem presented to the Tribunal has finally found expression in a regime which is how prescribed as a measure of control.

The investigations made during the past three years on the application of meteorological observations to the solution of this problem at Trail have built up a fund of significant and important facts. This is probably the most thorough study ever made of any area subject to atmospheric pollution by industrial smoke. Some factors, such as atmospheric turbulence and the movement of the upper air currents have been applied for the first time to the question of smoke control. All factors of possible significance, including wind directions and velocity, atmospheric temperatures, lapse rates, turbulence, geostrophic winds, barometric pressures, sunlight and humidity, along with atmospheric sulphur dioxide concentrations, have been studied. As said above, many observations have been made on the movements and sulphur dioxide concentrations of the air at higher levels by means of pilot and captive balloons and by airplane, by night and by day. Progress has been made in breaking up the long 'winter fumigations and in reducing their intensity. In carrying finally over to the non-growing season with a few minor modifications a regime of demonstrated efficiency for the growing season, there is a sound basis for confidence that the winter fumigations will be kept under control at a level well below the threshold of possible injury to vegetation. Likewise, for the growing season a regime has been formulated which should throttle at the source the expected diurnal fumigations to a point where they will not yield concentrations below the international boundary sufficient to cause injury to plant life. This is the goal which this Tribunal has set out to accomplish.

[The Tribunal here discussed the details of the operational regime it intended to impose on the Trail Smelter. The regime included monitoring and recordkeeping requirements, as well as setting maximum permissible sulphur emissions under a variety of seasonal and weather condition.]

SECTION 4

While the Tribunal refrains from making the following suggestion a part of the regime prescribed, it is strongly of the opinion that it would be to the clear advantage of the Dominion of Canada, if during the interval between the date of filing of this Final Report and December 31, 1942, the Dominion of Canada would continue, at its own expense, the maintenance of experimental and observational work by two scientists similar to that which was established by the Tribunal under its previous decision, and has been in operation dining the trial period since 1938. It seems probable that a continuance of investigations until at least December 31, 1942, would provide additional valuable data both for the purpose of testing the effective operation of the regime now prescribed and for the purpose of obtaining information as to the possibility or necessity of improvements in it.

The value of this trial period has been acknowledged by each Government.

[The Tribunal here cited various acknowledgements of this point by both governments]

PART FIVE

The fourth question under Article III of the Convention is as follows:

What indemnity or compensation, if any, should be paid on account of any decision or decisions rendered by the Tribunal pursuant to the next two preceding Questions?

The Tribunal is of opinion that the prescribed regime will probably remove the causes of the present controversy and, as said before, will probably result in preventing any damage of a material nature occurring in the State of Washington in the future.

But since the desirable and expected result of the regime or measure of control hereby required to be adopted and maintained by the Smelter may not occur, and since in its answer to Question No. 2, the Tribunal has required the Smelter to refrain from causing damage in the State of Washington in the future, as set forth therein, the Tribunal answers Question No. 4 and decides that on account of decisions rendered by the Tribunal in its answers to Question No. 2 and Question No. 3 there shall be paid as follows: (a) if any damage as defined under Question No. 2 shall have occurred since October 1, 1940, or shall occur in the future, whether through failure on the part of the Smelter to comply with the regulations herein prescribed or notwithstanding the maintenance of the regime, an indemnity shall be paid for such damage but only when and if the two Governments shall make arrangements for the disposition of claims for indemnity under the provisions of Article XI of the Convention; (b) if as a consequence of the decision of the Tribunal in its answers to Question No. 2 and Question No. 3, the United States shall find it necessary to maintain in the future an agent or agents in the area in order to ascertain whether damage shall have occurred in spite of the regime prescribed herein, the reasonable cost of such investigations not in excess of $7,500 in any one year shall be paid to the United States as a compensation, but only if and when the two Governments determine under Article XI of the Convention that damage has occurred in the year in question, due to the operation of the Smelter, and "disposition of claims for indemnity for damage" has been made by the two Governments; but in no case shall the aforesaid compensation be payable in excess of the indemnity for damage; and further it is understood that such payment is hereby directed by the Tribunal only as a compensation to be paid on account of the answers of the Tribunal to Question No. 2 and Question No. 3 (as provided for in Question No. 4) and *not* as any part of indemnity for the damage to be ascertained and to be determined upon by the two Governments under Article XI of the Convention.

PART SIX

Since further investigations in the future may be possible under the provisions of Part Four and of Part Five of this decision, the Tribunal finds it necessary to include in its report, the following provision:

Investigators appointed by or on behalf of either Government, whether jointly or severally, and the members of the Commission provided for in Paragraph VI of Section 3 of Part Four of this decision, shall be permitted at all reasonable times to inspect the operations of the Smelter and to enter upon and inspect any of the properties in the State of Washington which may be claimed to be affected by fumes. This provision shall also apply to any localities where instruments are operated under the present regime or under any amended regime. Wherever under the present regime or under any amended regime, instruments have to be maintained and operated by the Smelter on the territory of the United States, the Government of the United States shall undertake to secure for the Government of the Dominion of Canada the facilities reasonably required to that effect.

The Tribunal expresses the strong hope that any investigations which the Governments may undertake in the future, in connection with the matters dealt with in this decision, shall be conducted jointly.

(Signed) JAN HOSTIE
(Signed) CHARLES WARREN
(Signed) R. A. E. GREENSHIELDS

Index

49th parallel, 18, 21
1951 Convention, 261–263
1967 Protocol, 257–258, 261, 263

affirmative regulatory duties, 282
Africa, 284
African Union, 145
Agenda 21, 148
Ago, Special Rapporteur, 89
Air Quality Agreement between Canada and the United States (1991), 207–208
  Article IV of, 208
  protocols on sulphur and nitrogen emissions of, 208
Air Services Agreement (1979), 90
airshed, just allocation of shared, 37
Al Quaeda terrorists, 305
American Smelting and Refining Company, 21
Amoco Cadiz, 209
Andes, 286
Argentina, 284
Asculum, 266
Australia, 135, 200

balancing of interests, 61–64, 82–83, 119, 199, 230–231, 241
Bamaka Convention, 155
Bankovic, 296, 298–300
Barboza, Special Rapporteur, 62
Barcelona Traction (1970), 90
Basel Ban (1995), 190
Basel Convention on the Transboundary Movement of Hazardous Waste (1989), 189–193, 219
  provisions of, 191–192
  and the Trail Smelter arbitration, 191–194
Beesley, Alan, 43
Belize, 291
"beneficial use," 19
Bennett, Prime Minister R. B., 21
bilateral relations between states, 83
Biosafety Protocol, *See* Convention on Biological Diversity
Birnie, Patricia, 130, 179
Bolivia, 284
Boundary Waters Treaty (1909), 38
  Article 9 of, 28
Boyle, Alan, 63, 130, 179
Brazil, 284
British Property in Spanish Morocco (1925), 232
Bundesgerichtshof (BGH, Federal Court of Justice), 275

Cairo Guidelines (1987), 189
Calder v. Jones, 273

CAN. *See* Climate Change Network
Canadian Pacific, 13
Caron, David, 89
Cartagena Protocol on Biosafety. *See* Convention on Biological Diversity
causality
  in drug trafficking, 290–291
  in *Trail Smelter*, 290
CERCLA. *See* Comprehensive Environmental Response, Compensation and Liability Act (CERCLA)
Chambers, W. Bradnee, 179–180
Chernobyl, 135, 196, 201, 202
Chevron (1984), 246
Chile, 284
Citizens Protective Association (CPA), 17, 21–22, 82, 169–170
Clarke, Justice Lorne, 50–51
"clash of sovereignities," 37, 40
CLC. *See* Convention on Civil Liability for Oil Pollution Damage
"clear and convincing" evidence, 39, 126, 195, 214, 260
Climate Change Network (CAN), 172
Climate Change Treaty, 156
Coase, Ronald, 37
cocaine, 284
Code Penal, Article R 645-1 of, 269–270
Coleville Indians, 113
Columbia, 284, 290, 291
Cominco. *See* Consolidated Mining and Smelting Company
Commission on Environmental Law of the World Conservation Union (IUCN), 44
Commission on Sustainable Development (CSD), 148
compensation. *See also* damage, liability, state liability, *Trail Smelter* compensation
  as a response to transboundary pollution damage, 57
Comprehensive Environmental Response, Compensation and Liability Act (CERCLA), 109–121
Consolidated Mining and Smelting Company, 13–14, 170–171
  and emissions standards, 17
  and restrictions on production, 22–23
Convention for the Prevention and Punishment of Terrorism (1937), 234
Convention of the Law of the Sea (1982), 212–213
  Part XII of, 212–213, 217, 218
  Part XV of, 217
  Article 56 of, 216
  Article 56(1)(b)(iii), 216
  Article 192 of, 218
  Articles 192 and 193 of, 43
  Article 194(2), 217
  Article 195 of, 217–218
  Article 196 of, 217
  Article 219 of, 220
  Article 221 of, 220
  Article 235 of, 213
Convention on Biological Diversity (CBD), 155
  Article 3 of, 41
  Cartegena Protocol on Biosafety, 155, 161–162, 163, 164–166
Convention on Civil Liability for Oil Pollution Damage (CLC, 1969, amended 1992), 77
Convention on Environmental Impact Assessment in a Transboundary Context (EIA Convention), 147, 148, 149
  Article 2(3) of, 147
  Article 3(7) of, 147
  Article 4 of, 147

Convention on Non-Navigational Uses of International Watercourses (1997), 44
Convention on the Protection and Use of Transboundary Watercourses and International Lakes, 147–148
  Article 3 of, 147
  Article 5 of, 147
Convention on Transboundary Air Pollution, Article 8 of, 196
Convention Pertaining to the Status of Refugees (1951), 257–258
Corbin, Arthur L., 263
Corfu Channel (United Kingdom v. Albania), 61, 80, 84, 90, 185, 226, 232–233, 239
Corporate Social Responsibility (CSR), 250
Council Regulation (EC) No. 44/2001, 273
CPA. *See* Citizen's Protective Association (CPA)
Crawford, Special Rapporteur James, 86, 89
CSR. *See* Corporate Social Responsibility (CSR)
Cyprus, 297, 303

damage, 38–39, 54–55 *See also* compensation; liability, state liability
  defining the substance of, 49
  as direct, measurable injury, 53, 181
  level considered wrongful, 39–40
  physical, 54
  proof of, and identification of its cause, 81–82
  as a regulatory tool, 154
Danube Dam case (*Gabčikovo-Nagymaros Project*, Hungary/Slovakia), 140–151 *See also Gabčikovo-Nagyramos* (1997)
  history of, 140–143
  protocols in, 141
  and "state of necessity," 142–143
  and sustainable development, 144–146
  "Variant C" of, 141, 143
Dar es Salaam. *See* Tanzania
Department of External Affairs, 21
Desertification Convention, 42
diligentia quam in suis, 232
"do no harm" principle, 157–159
Draft Articles on Prevention of Transboundary Harm from Hazardous Activities (2001), 62–63, 95–98, 106–107, 113, 131, 137–139
  Article 3 of, 131
  Article 12 of, 131
Draft Articles on the Law of the Non-Navigational Uses of International Watercourses, Article 7 of, 131
Draft Articles on the Responsibility of States for Internationally Wrongful Acts, 85–98, 142, 259 *See also* International Law Commission (ILC)
  Article 8 of, 237
  Article 14 of, 91
  Article 30 of, 91–92
  Article 31 of, 92
  Article 36 of, 92–93
drug trafficking
  and causality, 290–291
  consumer states, 283, 285–286
  cost of Latin American, 284–287
  defined, 283
  harm, 293
  and jurisdiction, 287
  Latin American countries involved with, 284
  producer states, 283, 286–287
  and state liability, 283
  and *Trail Smelter*, 288, 289
Drumbl, Mark, 153

due diligence, 115, 132, 226, 227
 in environmental law, 228–231
 and terrorism, 232–236

Eagleton, Clyde, 59–60
ECE Convention for Long-range Transboundary Air Pollution (1979), 207
 Article 2 of, 207
 Article 6 of, 207
EC-Hormones (WTO Panel, U.S.), 164
ecstasy, 284
EEZ. *See* Exclusive Economic Zone (EEZ)
effects doctrine, 117–120
EIA Convention. *See* Convention on Environmental Impact Assessment in a Transboundary Context (EIA Convention)
emission standards. *See* Consolidated Mining and Smelting Company and emission standards
environmental impact assessment, 146–148
Environmental Protection Agency (EPA), 109–121
Europe, 284
European Commission of Human Rights (ECommHR), 296–307
European Convention for the Protection of Human Rights and Fundamental Freedoms (ECHR), 295
European Court of Human Rights (ECourtHR), 266, 296–307
 Article 1 of, 296–300, 302–303, 305, 306–307
 Article 3 of, 305–306
 Article 17 of, 301
 jurisdiction under, 296–307
 and state responsibility, 303–307
European Union, 156
 Article 2 of the Treaty of the, 145
 Article 174 of the Treaty of the, 146
European-Asbestos, 164
Exclusive Economic Zone (EEZ), 215, 216–217
Exxon Valdez, 210, 219

Factory at Chorzow case (1927–28), 90, 92
Fangataufa, 199
"flag-of-convenience" states, 219
Framework Convention on Climate Change, 42, 155
France, 199
Frank, Jerome, 177
Friedmann, Wolfgang, 178
Friendly Relations Declaration (1970), 234

Gabčikovo-Nagymaros Project (Hungary/Slovakia), 86, 90, 204 *See also Danube Dam* case
Geist, Professor Michael, 272, 273
General Assembly Resolution 2995 (1972), 95
genetically modified organisms (GMOs), 154, 159–166
 US government position on, 160
 WTO action challenging European Union's policy on, 165–166
Geneva Conventions on the Law of the Sea (1958), 42
Georgia v. Tennessee Copper Co., 51
Gesetz gegen den unlauteren Wettbewerb (UWG – Law against Unfair Competition), 274
 Section 14 paragraph 2 of, 275, 277
global climate change, 171–180
Global Climate Coalition, 174
Global Climate Information Project, 174
good-neighborliness. *See sic tuo ut alienum non laedas*
Guantanamo Bay, Cuba, 303

Hague Convention on Jurisdiction and Foreign Judgments in Civil and Commercial Matters, 278–279
Handl, Gunther, 86, 185
Hanson v. Denckla, 271, 275
harm principle, 37, 110, 112, 115, 119, 153, 197
  and no-significant-harm obligation, 127–129
  "significant, substantial, or appreciable" 129–139
Harmon, United States Attorney Judson General, 37
Harmon Doctrine, 184
Hartford Insurance, 118
hazardous waste
  cost of disposal of, 187
  defined by volume generated, 187
hazardous waste transport, and international environmental law, 186–194
Henkin, Louis, 178
Heraclea, 266
heroin, 284
Higgins, Rosalyn, 117, 231
Holmes, Justice Oliver Wendell, 50, 51
Howes, Dean, 28
Huber, Max, 232

Ignatio-Pinto, Judge, 135
Imaginarium, 272–273
Imaginarium, 273
imagniariumnet.com, 272
industrial zones, 20
"inequitable use," 130
Inter-American Court of Human Rights, 266
Intergovernmental Panel on Climate Change (IPCC), 1995 report of, 174
International Council of Environmental Law, 44
International Court of Justice (ICJ), 42, 127, 226
  Advisory Opinion in *Legality of the Threat or Use of Nuclear Weapons* (1996), 42, 127, 142
  judgment on the *Gabčikovo-Nagymaros Project*, 42, 127
  Statute of, 202
International Covenant on Environment and Development, 44
International Criminal Tribunal for the Former Yugoslavia, 237
international human rights law, 184
International Joint Commission (IJC), 15, 27, 38, 39, 197, 232
  and compensation, 28
  Febuary 28, 1931 report of, 28
  Settlement by, 22
international law
  and private actors, 71–73
  and terrorism, 231–239
  Westphalian view of, 226
International Law Commission (ILC), 42–44, 56, 62–63, 78, 227–231 *See also* Draft Articles on State Responsibility
  Report of the Working Group on International Liability for Injurious Consequences arising out of Acts not Prohibited by International Law, 218
International Law Commission (ILC) on the Non-Navigational Uses of International Watercourses, 195
International Liability for Injurious Consequences Arising Out of Acts Not Prohibited by International Law, 43
international regimes, 244–253
international terrorism and transboundary environmental pollution, 225
International Tribunal on the Law of the Sea, 221

"internationally wrongful acts"
defined, 85
forms of, 88
internet, jurisdiction over, 271–277
"invisible injury," 25–26, 175
Iraq, 300
Ireland, 136–137
Issa, 300, 302
IUCN. *See* Commission on Environmental Law of the World Conservation Union

Jamaica, 291
Japan-Agricultural Products (WTO Panel, Japan), 163
Jenks, Wilfred, 52
jurisdiction
and drug trafficking, 287
and the European Court of Human Rights (ECourtHR), 296–307
and internet, in Germany, 273–277
and internet, in the US, 271–273
and multinational enterprises, 293
"just solution," 165, 166, 199

Kammergericht (Higher Regional Court Berlin), 275
Kettle Falls, Washington, 18
Khomeini, Ayatollah, 238
King, Prime Minister, Mackenzie, 21
Kirgis, Frederic, 133
Knox, John, 87, 279
Koskenniemi, Martti, 177–178
Kummer, Katharina, 193
Kyoto Protocol, 171, 244

Lac Lanoux (Lac Lanoux Arbitration (Spain v. France, 1957), 185
LaGrand (Germany v. United States of America, 2001), 90
Lake Lanoux, 80, 84
Lake Roosevelt, 113, 115
Landgericht Munchen I (Regional Court of Munich), 275
Landgericht Nürnberg-Fürth (Regional Court of Nürnberg-Fürth), 275
Laswell, Harold, 178
Leaden, Joe, 169
Legality of Use by a State of Nuclear Weapons in Armed Conflicts, 229
liability, 218 *See also* compensation; damage; state liability; Trail Smelter compensation agreements on civil,
in the Convention of the Law of the Sea (1982), 213
corporate social, 240–253
derivative, Canada's, in *Trail Smelter* arbitration, 103–104
derivative, defined, 100–102
direct, Canada's, in *Trail Smelter* arbitration, 105–107
direct, defined, 102–103
international, and prevention, 227
premise of international, 227–228
and responsibility, 227–228
Liability for Injurious Consequences of Acts not Prohibited by International Law, 196
Llewellyn, Karl, 177, 178
local action rule, 35
Logan Act, 170
Loizidou v. Turkey (1995–96), 90
London Convention, 211, 212, 219
loss, recoverable categories of, 204–205
Lugano Convention, 273

MacNab, Francis, 14
maethamphetamine, 284
marijuana, 284
marine environmental law, development of, 210–213

MARPOL 73/78, 210–212, 219–220
Marshall, John, 40
Mathews, Jessica, 171
McCaffrey, Steve, 87
McDougal, Myres, 178
Menchen, H.L., 37
Mexico, 291
Mickelson, Karen, 112
Miller, Dean, 28
Miller, Russell, 153, 243
Missouri v. Illinois, 50
Model Penal Code, 101–102
Montreal Protocol on Substances that Deplete the Ozone Layer, 94
MOX Plant, 136, 221
Mururoa Atoll, 199

National Research Council of Canada, 21
North Atlantic Treaty Organization (NATO), 298
Nedelsky, 19
"neighborhood" law (good neighborliness principle). *See sic tuo ut alienum non laedas*
Nelson, British Columbia (Canada), 28
nemo tenetur ad impossible, 135
New York v. New Jersey, 50–51
New Zealand, 135–136, 200
Nicaragua v. U.S. (1986) *Military and Paramilitary Activities in and against Nicaragua* (Nicaragua v. United States of America, 1986)"], 90, 233–234, 236, 237–238
"no-harm" principle, 181, 183
non liquet, 36, 73
non-government organizations (NGOs). *See also* non-state actors; private actors and global climate change treaties, 171–172
Non-Navigational Uses of International Watercourses (1994), 43
non-refouling state, 258
non-refoulment, 262
non-state actors, 168–174, 226 *See also* non-government organizations (NGOs); private actors
Northport, Washington (United States), 18, 23, 28
Nuclear Civil Liability Conventions, 205
Nuclear Safety Convention
  Article 6 of, 133
  Chapter 22 of Article 21 of, 134
Nuclear Tests Case (New Zealand v. France, 1973), 156, 200
nuclear tests, atmospheric, by the United Kingdom and the United States, 199–202
nuisance, common law of, 40, 53, 57, 69, 73–74, 282

Oberlandesgericht Bremen (Higher Regional Court Bremen), 276
Oberlandesgericht Frankfurt (Higher Regional Court Frankfurt), 276
Oberlandesgericht Hamburg (Higher Regional Court Hamburg), 277
Oberlandesgericht Munchen (Higher Regional Court Munich), 276
Öcalan v. Turkey, 300
OILPOL 54, 210

Parrish, Austen, 106
Permanent Court of International Justice (PCIJ), 35
permanent regime, 256
Peru, 284, 291
Pichler, 276
"polluter pays" principle, 99, 110, 116
pollution, air, and international law precedents, 48

polycausality, 24
POPs Treaty, 155
Power Shift (Mathews), 171
precautionary principle, 154–159, 244
and quantitative risk assessment, 157, 159
See also (semi)precautionary principle
Preiss, Erica, 146
prevention, and international liability, 227
Prevention of Transboundary Harm from Hazardous Activities, 43, 228
Article 3 of, 229
principle of comity, 118
principle of good neighborliness. *See sic tuo ut alienum non laedas*
private actors, 71–73, 233, 238–239, 292 *See also* non-government organizations (NGOs); non-state actors
private responsibility, 248
and human rights law, 296
Protocol on Liability and Compensation (1999), 190
Pyrrhic Victory, 266
Pyrrhic War, 266
Pyrrus, King of Epirus, 266

quantitative risk assessment, 157 *See also* precautionary principle
Quentin-Baxter, Special Rapporteur Robest, 139

Radio Televizije Srbije (RTS), 298
radioactive contamination, 199–202
Rainbow Warrior (1990), 90
Rao, Special Rapporteur Pemmaraju Sreenivasa, 227
Read, John, 36, 38, 52, 72, 74
Red Sky at Morming (Speth), 172
refugees
Burundi, 262
Haitian, 262
impact on host communities, 260–261
law of, 257–258
as "population of concern" in 2004, 258
Rwandan, 262
and state responsibility, 254, 259–261
and Trail Smelter, 258–267
Report of the Brundtland Commission's Expert Group on Environmental Law (1987), 44
Resolution 49/60 (1994), 235
responsibility
corporate social, 240–253
and liability, 227–228
polititical economy of corporate, 249–253
responsiveness, 247
Restatement (Second) of the Foreign Relations Law of the United States (American Law Institute), 46–47
Restatement (Third) of the Foreign Relations Law of the United States (American Law Institute), 118
Rio Declaration on Environment and Development, 229
Principle 1 of, 144
Principle 2 of, 127
Principle 3 of, 144
Principle 4 of, 144
Principle 5 of, 144, 150
Principle 7 of, 145, 150
Principle 8 of, 145, 150
Principle 9 of, 145
Principle 12 of, 145
Principle 15 of, 155
Principle 17 of, 147
Rio de Janiero Earth Summit, 41
Principle 2 of, 95
Principle 12 of, 120
risk, reciprocity of, 83, 112
risk assessment, 244–245 *See also* quantative risk assessment

River Oder, 35
Rossland, 13
Rubin, Arthur P., 17–18, 26

Sand, Peter, 70
Sellafield, 136
(semi) precautionary principle, summarized, 162 *See also* precautionary principle
"serious consequence," 39, 126, 195, 214, 260
Shelton, Diane, 93–94
sic tuo ut alienum non laedas, 40, 62, 80–82, 120, 185, 214, 228, 256
Smelter Smoke in North America (Wirth), 168
smoke disputes, standard legal process for resolving international, 19–20
smoke easements, 282
"Smoke Eaters," 13–14
smoke injury, 24
smoke zone
  between 1927 and 1937, 23, 24
  bio-regional, 19
social responsibility, and refugees, 254
Soering v. United Kingdom, 305
"solution just." *See* "just solution"
Speth, James Gustave, 172
SPS Agreement. *See* WTO Agreement on Sanitary and Phytosanitary Measures (SPS Agreement)
state, defined, 259
state control, defined, 236–238
state liability, 97–98 *See also* compensation, damages, liability, *Trail Smelter* compensation
  for drug trafficking, 283
  and private actors, 292
  for privately caused harms, 282
state responsibility, 47–55, 56, 59–61, 153, 225, 227, 241 *See also Draft Articles on* the Responsibility of States for Internationally Wrongful Acts
  defined, 85
  and the European Court of Human Rights (ECourtHR), 303–307
  and private actors, 238–239
  for refugees, 254, 259–261
  and state liability, 87
  and terrorism, 236–239
  and Trail Smelter, 296
  and the Trail Smelter Tribunal
Statement of Forest Principles, Principle 1(a) of, 41
Stevens County, Washington (United States), 29
Stockholm Declaration on the Human Environment (1972), 112, 144, 195
  Article 21 of, 229
  Principle 2 of, 144
  Principle 8 of, 144, 150
  Principle 21 of, 41–42, 56, 86, 95, 127
Story Parchment Company v. Paterson Parchment Company, 205
Straddling Fish Stocks Agreement (1995), 155
sulphur dioxide, 14, 27, 38, 183
sustainable development, defined, 144–146
Swiss shooting range precedent, 48

Tadic, 237–238
"tangible economic injury," 20
Tanzania, 262
"targeting test," 272–273, 277, 278–279
Teck Cominco America Inc. (TCAI), 114
Teck Cominco Metals, Ltd (TCML), 109–121
Teheran Hostages 233, 238
Ten Drivers of Environmental Deterioration (Speth), 173
territorial principle, 116–120

terrorism
  and international due diligence, 232–236
  international law on, 231–239
  state responsibility and attribution, 236–239
  state-sponsored, 235–239
  and transboundary environmental pollution, 225
The M/V Saiga (1999), 90
The S.S. Wimbledon (1923), 90
Third United Nations Convention on the Law of the Sea, 209
Three-mile Island, 201
Torrey Canyon 209, 219
tortfeasor, importance of residence of, 273–274
Toys "R" Us v. Step Two, 272–273
Trail smelter, 16
Trail Smelter Arbitration
  adaptive decisional structure of, 154–166
  "analogous" US precedents, 49
  arbitration, 111–121, 158–159, 281–283
  background to, 255–256
  and Basel Convention, 191–194
  basic questions confronting, 66–67, 69–75
  and causal relationship between emissions and damage, 205, 290
  and Citizens' Protective Association, 17, 21–22, 82, 169–170
  claims based on nuisance, 16, 27
  and class interests, 21–22
  compensation for damages, 1917–24, 15
  compensation for damages after January 1, 1932, 38
  compensation for damages through January 1, 1932, 28, 38, 69
  compensation in 1938, 16
  compensation to October 1, 1937, 30
  composition of tribunal, 34, 195–208
  and the Convention of the Law of the Sea (1982), 214–221
  conflict with Stevens County (Washington) farmers, 14, 15
  damage after January 1, 1932, 47
  damage from October 1, 1937 to October 1, 1940, 31
  damages and nuisance theory, 256
  and drug-trafficking, 288, 289
  early history of, 13–14, 27, 33
  and environmental monitoring, 151
  "expert and judicial side" of, 44
  extension of US precedents, 50
  and the failure of private international law, 279–280
  flaws in procedures of the, 69–75
  and international environmental law, 244–245
  and the "just solution", 36, 37, 150
  meteorological investigations of, 30–31
  and permanent sovereignty over natural resources, 182–194
  "political side" of, 41–42
  primary finding of, 16
  and private international environmental harm, 67–69
  questions submitted to, 197–198, 241
  and refugees, 258–267
  and regulatory regime of prevention, 206
  and scientific inquiry, 174–176
  second dispute, 109–121
  and "smoke farmers," 17, 23, 24
  smokestack, 15
  and social conflict, 18–19
  and sovereign rights, 50–51
  specific precedents for, 35–36
  and state responsibility, 296
  and Stevens County farmers, 23
  threshold question of, 47–49
  and transboundary air pollution, 202–204

and US interstate commerce precedents, 50–51
and the US Supreme Court, 50, 74
transboundary air pollution
and compensation, 57
damage from, 198
distinctive feature of, 198
and the Trail Smelter arbitration, 195–208
transboundary environmental pollution
and international terrorism, 225
and international terrorism, shared characteristics, 225
treaty-based conflict resolution, 244–253
Treshow, Michael, 25
Turkey, 297, 303

UEJF & LICRA v. Yahoo!, 269–270
UNHCR Executive Committee (ExCom), 258
Uniform Transboundary Pollution Reciprocal Access Act, 76
United Kingdom, 199
United Nations Conference on the Human Environment, 217
Principle 13 of the Principles of Assessment and Control of Marine Pollution of, 217
United Nations Corporate Norms, 156
United Nations Environment Programme (UNEP, 1982), 189
United Nations Framework Convention on Climate Change (UNFCCC), 168, 171–172, 177, 178
United Nations High Commissioner for Refugees (UNHCR), 258, 260, 262, 264, 300
Executive Committee (ExCom) of, 258
United Nations Scientific Committee on the Effects of Atomic Radiation (UNSCEAR), 202
United States, 199, 289, 303
Universal Declaration of Human Rights (1948), 257, 259, 264
United States Supreme Court, 111, 205

Vanuatu, 171
Vienna Convention on the Law of Treaties, 42
Article 26 of, 301
Article 27 of, 301
Article 31 of, 299
Vienna Conventions on Diplomatic and Consular Relations, 42

Wallace, Perry, 173
Washington, D.C. (United States), 28
Weeramantry, Judge, 144, 148–149, 151, 156
Weiss, Edith Brown, 174, 179
Westphalian prerogative, 168
Wirth, John, 168, 170, 172, 198
Worchester v. Georgia (1832), 40
World Charter for Nature (1981), 155
World Court, 35
WTO Agreement on Sanitary and Phytosanitary Measures (SPS Agreement), 161–166
Article 2.2 of, 163
Article 5.1 of, 163, 164
Article 5.2 of, 164
Article 5.7, 163

Yemen, 305
Yugoslavia, Federal Republic of, 298

zero coca tolerance policies, 286
zinc ores, 13
Zippo Manufacturing v. Zippo Dot Com, 271–272, 273
Zivilprozessordnung (ZPO – Code of Civil Procedure), Section 32 of, 274, 275

www.ingramcontent.com/pod-product-compliance
Lightning Source LLC
LaVergne TN
LVHW090305210926
848480LV00001B/138
* 9 7 8 0 5 2 1 1 2 6 4 2 7 *